2 15 25

Engineering Mechanics

STATICS

VOLUME I

Engineering
Mechanics

STATICS

LAWRENCE E. MALVERN
University of Florida

Prentice-Hall, Inc., Englewood Cliffs, New Jersey

Library of Congress Cataloging in Publication Data

MALVERN, LAWRENCE E. (date)
Engineering mechanics.

Includes index.
CONTENTS: v. 1. Statics. v. 2. Dynamics.
1. Mechanics, Applied. I. Title.
TA350.M343   620.1   75-31958
ISBN 0-13-278663-X

10 9 8 7 6 5 4 3 2 1

Printed in the United States of America

PRENTICE-HALL INTERNATIONAL, INC., *London*
PRENTICE-HALL OF AUSTRALIA, PTY. LTD., *Sydney*
PRENTICE-HALL OF CANADA, LTD., *Toronto*
PRENTICE-HALL OF INDIA PRIVATE LIMITED, *New Delhi*
PRENTICE-HALL OF JAPAN, INC., *Tokyo* .
PRENTICE-HALL OF SOUTHEAST ASIA PRIVATE LIMITED. *Singapore*

Contents

PREFACE ix

1 **BASIC PRINCIPLES AND METHODS OF MECHANICS** 1

1.1 Introduction; Idealizations and Fundamental Concepts *1*
1.2 Vector Quantities; Vector Addition; Rectangular Components *4*
1.3 Force, Mass, and Weight; Laws of Mechanics;
     Dimensions and Units *14*
1.4 Problem Solutions; Numerical Accuracy *26*
1.5 Summary *28*

2 **PARTICLE EQUILIBRIUM** 29

2.1 Introduction; Particle Equilibrium Equations *29*
2.2 Problems of Particle Equilibrium in a Plane *30*

2.3   Coulomb Friction   *40*
2.4   Three-Dimensional Problems of Particle Equilibrium   *50*
2.5   Summary   *57*

3   **VECTOR AND SCALAR PRODUCTS WITH APPLICATIONS   59**

3.1   Introduction   *59*
3.2   Scalar Product; Projection; Rotation of Coordinate Axes   *60*
3.3   Vector Product (Cross Product); Vector Moment of a Force
       About a Point; Moment About an Axis   *69*
3.4   Couple Moment   *82*
       Vector Triple Products and Scalar Triple Product   *96*
3.6   Summary   *98*

4   **RIGID-BODY STATICS   101**

4.1   Equipollence (Rigid-Body Equivalence); Transmissibility:
       Reduction to Resultants   *101*
4.2   Distributed Parallel Force Systems; Center of Mass; Centroid;
       Center of Gravity   *110*
4.3   Rigid-Body Equilibrium: Free-Body Diagram;
       Idealization of Support Reactions; Equations of Equilibrium;
       Statically Determinate Coplanar Problems   *119*
4.4   Rigid-Body Equilibrium Problems in Three Dimensions   *136*
4.5   Summary   *150*

5   **STRUCTURES AND MACHINES   153**

5.1   Connected Bodies;
       Statically Determinate Frames and Machines   *153*
5.2   Statically Determinate Plane Trusses;
       Method of Joints and Method of Sections   *171*
5.3   Space Trusses   *180*
5.4   Stress Resultants in Beams;
       Bending Moment and Shear Force Diagrams   *182*
5.5   Flexible Cables   *197*
5.6   Summary   *206*

6 **WORK, ENERGY, AND THE STABILITY OF EQUILIBRIUM 209**

6.1 Work of a Force;
Conservative Force Fields and Potential Energy   *209*
6.2 Potential Energy Conditions for Equilibrium and for Stability   *217*
Part 1.   One Degree of Freedom   *217*
Part 2.   Systems with More than One Degree of Freedom   *222*
6.3 Virtual Displacements and Virtual Work   *230*
6.4 Summary   *241*

7 **PROPERTIES OF GEOMETRIC SHAPES AND MASSES;**
FLUID STATICS   245

7.1 Introduction   *245*
7.2 Center of Mass   *246*
7.3 Centroids of Volume, Area, Arc Length; Composite Bodies   *248*
7.4 Centroid by Single Integrals   *255*
7.5 Multiple-Integral Evaluation   *262*
7.6 Second Moments of Area   *268*
7.7 Rotation of Axes for Second Moments of Area; Mohr's Circle;
Principal Axes   *276*
7.8 Fluid Statics   *279*
7.9 Mass Moments of Inertia and Products of Inertia   *286*
7.10 Summary   *302*

APPENDICES   i

A **SOME USEFUL TABLES   i**

Table A1.   Six Systems of Units   *i*
Table A2.   Some Conversion Factors   *ii*
Table A3.   Properties of Plane Arcs and Areas   *iii*
Table A4.   Moments of Inertia and Center of Mass for Common
Geometric Shapes of Uniform Density   *v*

B **ANSWERS TO EXERCISES   ix**

INDEX   xvii

Preface

Statics is the first volume of a two-volume textbook on engineering mechanics, to be followed by *Dynamics*. The objectives of this textbook are to give the student an understanding of the principles and methods of mechanics and to develop in him the ability to formulate and solve an engineering mechanics problem in a systematic manner. Both for the promotion of understanding and for the development of problem-solving ability, emphasis is placed on the unity of the subject.

This book grew out of seven years' experience at two state universities with a unified course giving an integrated presentation of statics and dynamics. Although the book preserves the traditional separation into Vol. I, *Statics* (Chapters 1–7), and Vol. II, *Dynamics* (Chapters 8–14), the individual sections have been carefully planned to be clear and meaningful whether they are covered in the sequence presented or in a sequence designed for a unified course including parts of Chapters 8 and 9 of *Dynamics* dealing with particle dynamics and rigid-body translation along with Chapters 5 through 7 of *Statics*.

The introductory Chapter 1 includes the basic concepts and methods of particle mechanics. It provides immediate applications of the vector representations introduced in Sec. 1.2 in the composition and resolution of concurrent forces. In Chapter 2 on particle equilibrium the basic free-body diagram, which is emphasized throughout the book both as an analytical tool and as a standard device for communication among engineers, is introduced.

The traditional statics course includes, in addition to the analysis of equilibrium problems, the development of the basic tools of mechanics—the ideas of moment, couples, and reduction of force systems to resultants. All this is included in Chapters 3 and 4, always keeping in mind that these procedures apply to dynamics as well as to statics.

In Chapter 5 applications of statics to structures and machines are presented. In Chapter 6 work and energy and the stability of equilibrium are introduced. In a short course sections 3.5, 5.3–5.5 and 7.4–7.9 may be omitted without interrupting continuity. Section 6.3 on virtual displacements and virtual work is completely independent of the section on potential energy and stability.

Details of calculation of centroids and second moments have been placed in the last chapter of *Statics*. It is not necessary to master these calculation methods before using the tabulated values of Appendix A in the solutions of the mechanics problems of earlier chapters. Chapter 7 also includes a section on fluid statics and one on mass moments and products of inertia. The last is properly a part of dynamics but was included in the first volume because the topic is frequently covered in a statics course as preparation for a following dynamics course.

Vector notation and vector algebra are used throughout for the statement of principles. This facilitates emphasis on the physical quantities and permits concise, easily remembered statements of the physical principles. In problem solutions, however, component formulations are often more advantageous than solution by vector algebra, and these component methods have been illustrated abundantly in sample problem solutions at the ends of sections as well as in examples in the text.

The two volumes contain ample material for two three-credit semesters or two four-credit quarters. They are arranged for maximum flexibility to facilitate omissions in shorter courses, either in statics or in dynamics, or to be used together in a unified course covering both statics and dynamics. A preliminary version was class-tested for two years by the author and several of his colleagues.

I am grateful to Professors M. A. Eisenberg, C. G. Edson, I. K. Ebcioglu, and S. Y. Lu for advice and for criticism of the preliminary version; to the many students who used the book, for their helpful suggestions; and to Mrs. Edna B. Larrick, who patiently and carefully typed the original version and several revisions.

I am also pleased to acknowledge the support and cooperation of Phyllis Springmeyer of the Prentice-Hall editorial production department, and the excellent and speedy work of George Morris and his associates at Scientific Illustrators in preparing the drawings.

Finally, I wish to record my gratitude to my wife, Marjorie, for her constant inspiration and encouragement.

LAWRENCE E. MALVERN
Gainesville, Florida

CHAPTER ONE

Basic Principles and Methods of Mechanics

1.1 INTRODUCTION; IDEALIZATIONS AND FUNDAMENTAL CONCEPTS

Mechanics is the physical science concerned with the motion and the deformation or flow of material bodies under the action of forces. It is the basic engineering science underlying much of engineering practice. The material bodies of the science may be bodies of solid, liquid, or gaseous material.

The *theory of mechanics*, like all physical theories, deals, however, not with actual physical materials but with various idealized *physical models* of real materials, models capable of being represented by mathematical equations that can be solved to make predictions of the motions and deformations of the physical model material. The usefulness of such theories depends on the extent to which the predictions based on the models are verified by experience with real materials. For engineering purposes, we work with the simplest model that will give a reasonably good representation of observed physical behavior. The complexity required varies with the nature of the application and also with our decision as to what is "reasonably good."

1

In engineering mechanics we use three such (idealized) models:

1. A *particle* or *mass point* (or a collection of such discrete particles).
2. A *rigid body*.
3. A *deformable continuous body*, solid or fluid (also called a *continuous medium* or a *continuum*).

The *mass-point particle*, idealized as possessing mass but occupying no volume, turns out to be a surprisingly successful model of even a very large body when all that we seek to determine is the motion of the mass center of the body. For example, in celestial mechanics the planets may be considered as particles in order to determine their motions around the sun. If we are concerned with the possible rotational motion of the body, however, the mass-point model is inadequate.

The *rigid-body* model assumes that no dimension of the body changes when it is loaded. This may be a satisfactory model for determining the support forces required to hold a loaded structure in place or for predicting the motion of a machine part or of a vehicle under the action of forces, for example, a space vehicle under the action of its attitude-controlling thrusters. The rigid body may be idealized as a three-dimensional body occupying volume, but a thin flat plate or a thin curved shell is often modeled as a rigid body of zero thickness, so that it is a two-dimensional model with its mass distributed over a surface instead of a volume, while a slender bar (straight or curved) may be idealized as a one-dimensional rigid body with its mass distributed along a line or curve. Whether the rigid body is modeled as of one, two, or three dimensions, the distance between any two points in the model is assumed not to change during its loading or motion.

The *deformable continuous medium* or *continuum* is the model used to calculate the deflections of a structural element or of a machine part or to predict the flow field of a fluid. A fluid (liquid or gas) is also sometimes modeled as a collection of discrete mass points simulating the molecules of a real fluid. And a crystalline solid is also often modeled as a collection of particles occupying sites in a lattice to simulate the atoms of a real crystal.

The essence of the *continuum model* (the deformable continuous medium) is that it disregards the molecular structure and assumes that the modeled body is without gaps or empty spaces. The *density* (mass per unit volume) is assumed to be a continuous function of position in the body in the sense of the mathematical definition of a continuous function, except at a finite number of boundary surfaces that may separate one material from another. The deformable continuous body, like the rigid body, is often modeled as two-dimensional, a deformable plate or shell, modeled as of zero thickness but nevertheless possessed of stiffness that resists bending or stretching without altogether preventing such deformation. And structural engineers use a beam model (one-dimensional) with stiffness against bending, stretching, and twisting.

This book on the statics and dynamics of particles and rigid bodies uses almost entirely the first two idealized models.

Engineering mechanics is often classified into three parts as follows:

1. Statics.
2. Dynamics.
   (a) Kinematics.
   (b) Kinetics.
3. Deformable body mechanics (static or dynamic).

Statics deals with unaccelerated motion or *equilibrium*, while *dynamics* deals with accelerated motion. Dynamics is further subdivided into kinematics, the geometry of motion in space and time without the concepts of force and mass, and *kinetics* (or dynamics proper), which relates the forces and mass to the kinematic effects. The first seven chapters, contained in Vol. I of this textbook on engineering mechanics, cover the statics of particles and of rigid bodies.

Statics is often called the science of forces. The methods of combining and simplifying force systems to determine their resultants (see Chapter 4) will be used in subsequent courses in dynamics and in deformable body mechanics as well as in the equilibrium problems of statics.

The student will find that there are only a very few principles that make up the theory of statics, and most students will have met these principles before in a physics course. What distinguishes engineering statics as a discipline is the careful isolation of the body whose equilibrium is being considered in a *free-body diagram*. The importance of careful construction of free-body diagrams cannot be overemphasized. The methods for this will be developed in Chapter 2 for a particle and in Chapter 4 for a rigid body. The analysis of connected bodies in Chapter 5 depends significantly on systematic use of the free-body methods. Not only is the free-body diagram a powerful tool for the analysis of engineering problems in both statics and dynamics, but it is also a means of effective engineering communication. For this reason the student should form the habit of making his diagrams neat and should develop a standard style of presentation that will be readily understood by other engineers.

Mechanics is a science dealing with four primitive elements: *space, time, mass*, and *force*. Other quantities used are defined in terms of these four basic elements. Associated with the four primitive elements are four kinds of dimensions: length $[L]$, time $[T]$, mass $[M]$, and force $[F]$. But only three of them are independent. By virtue of the fundamental postulate of Newton's second law of motion (force equals mass times acceleration; see Sec. 1.3), we must have the relationship

$$[F] = \left[\frac{ML}{T^2}\right] \quad \text{or} \quad [M] = \left[\frac{FT^2}{L}\right] \tag{1.1.1}$$

among the dimensions of the primitive quantities, since acceleration has dimensions

$[L/T^2]$. See the discussion of dimensions and units in Sec. 1.3, where the concepts of mass and force will also be discussed.

Force is a vector quantity. Before taking up the laws of mechanics in Sec. 1.3, we shall review vectors and vector addition in Sec. 1.2.

1.2 VECTOR QUANTITIES; VECTOR ADDITION; RECTANGULAR COMPONENTS

In the three-dimensional space of ordinary geometry, *vector quantities* (or briefly *vectors*) are entities possessing both magnitude and direction and obeying certain laws. The most important of these laws is the law of vector addition to be discussed below. To qualify as a vector quantity, it is not sufficient that a physical entity possess magnitude and direction; it must also compound like a vector. Two forces acting at a point produce the same effect as would be produced by a single force equal to the vector sum of the two forces and acting at the same point. If this were not so, forces would not be vector quantities.

A vector is conventionally represented as an arrow that is pointing in the direction associated with the vector and that has a length proportional to the magnitude of the vector. The simplest example is a directed line joining two points. Other familiar examples are force, velocity, and acceleration. A *scalar* quantity, on the other hand, does not have any direction associated with it, although some scalar quantities may have positive and negative values, e.g., Fahrenheit temperature. In printed text, a vector is usually denoted by a boldface letter. The magnitude of the vector is then usually denoted by the same letter in lightface italic type. In typewritten or handwritten text, the wavy underscore used by copyreaders to designate boldface type is often used for a vector, or the vector may be identified by an arrow above the letter or by underlining. The magnitude of vector **a** is denoted by a (the same letter in italic instead of boldface), or by $|\mathbf{a}|$, and is always a positive number or zero. In script $a = |\underset{\sim}{a}|$.

Two *vectors are equal* if they have the same direction and the same magnitude. Thus a vector is unchanged if it is moved parallel to itself. In some applications, however, the point or the line of action of a vector quantity is important. For example, a force is a vector quantity with a definite point of action; it is a *fixed* or *bound vector*. In rigid-body mechanics, only the line of action of the force is important, not the point; force is then a *sliding vector*. When we say that two forces are equal, however, we mean only that they have the same magnitude and direction; this does not imply that they are mechanically equivalent.

Vector Addition

Any two vector quantities of the same kind (e.g., two forces or two velocities) may be represented as two vectors **a** and **b** so placed that the initial point of **b** coincides with the terminal point of **a**, as in Fig. 1-1. The ***sum*** of **a** and **b** is then defined as the vector **c** extending from the initial point of **a** to the terminal point of **b**:

$$c = a + b. \tag{1.2.1}$$

The actual addition of two vectors is most conveniently performed by using their rectangular components in some coordinate system as in Eq. (1.2.8), but the definition and properties of vector addition do not depend on the introduction of a coordinate system.

From the definition it follows that

$$b + a = a + b,$$

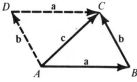

Fig. 1-1 Addition of Vectors

as is indicated by the dashed arrows in Fig. 1-1. Thus vector addition obeys the ***commutative law***. It also follows from the definition that the ***associative law***

$$(a + b) + c = a + (b + c)$$

is satisfied. See Fig. 1-2. The common value of the two expressions is written as the sum **a** + **b** + **c** of the three vectors.

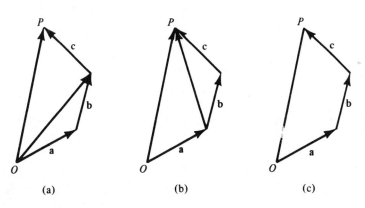

(a)   (b)   (c)

Fig. 1-2 Sum of Three Vectors; Polygon Rule

Figure 1-2(c) illustrates the ***polygon rule*** for adding more than two vectors. The vectors to be added are laid off head to tail; the ***sum vector*** is then the vector from the initial point *O* of the first vector to the terminal point *P* of the last vector. In the example of Fig. 1-2 all three vectors are coplanar (lie in one plane), but this is not necessary, as will be illustrated in Fig. 1-4. When there are just two vectors to add, the polygon rule becomes the ***triangle rule***, illustrated by the triangle *ABC* of Fig.

1-1. The addition of two vectors can alternatively be represented by the ***parallelogram law.*** The two vectors are drawn from the same point (*AB* and *AD* in Fig. 1-1), and the sum is then the diagonal *AC* of the parallelogram formed with the two given vectors as sides.

If a vector is reversed in direction with no change in magnitude, the resulting vector is called the ***negative*** of the original vector. For example, $\overrightarrow{BA} = -\overrightarrow{AB}$ in Fig. 1-1. To subtract **b** from **a**, add the negative of **b** to **a**:

$$\mathbf{a} - \mathbf{b} = \mathbf{a} + (-\mathbf{b}).\qquad(1.2.2)$$

A vector **a** may be multiplied by a scalar c to yield a new vector, $c\mathbf{a}$ or $\mathbf{a}c$, of magnitude $|ca|$. If c is positive, $c\mathbf{a}$ has the same direction as **a**; if c is negative, $c\mathbf{a}$ has the direction of $-\mathbf{a}$. See Fig. 1-3. If c is zero, the product is the ***zero vector*** with zero magnitude and undefined direction. Thus $(0)\mathbf{a} = \mathbf{0}$.

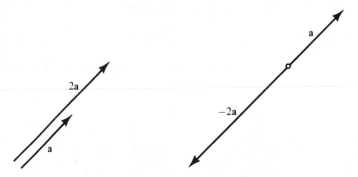

Fig. 1-3 Illustrating Multiplication of Vector **a** by a Scalar

A vector of unit length is called a ***unit vector***. Any vector may be written as the product of the scalar magnitude of the given vector by a dimensionless unit vector in the given direction. Thus $\mathbf{a} = a\hat{\mathbf{e}}$, where $\hat{\mathbf{e}}$ is a dimensionless unit vector in the direction of **a**. A unit vector may be signified by the caret (ˆ) placed over its letter symbol. The caret is often omitted when the context makes it clear that the vector is a unit vector.

Vector Components; Rectangular Components

Two or more vectors whose sum is a given vector are called ***vector components*** of the given vector. It is often convenient to resolve a given vector into three components parallel to three noncoplanar ***base vectors***; the resolution is unique for a given vector and given base vectors. The set of base vectors is called a ***basis***. If the basis consists of mutually orthogonal unit vectors, it is called an ***orthonormal basis***. In this book, the base vectors are usually specified as unit vectors in the positive directions of a set of rectangular Cartesian coordinate axes. The components are then called the **rectangular vector components** of the given vector in that coordinate system. The name rectangular components is also used for the numerical coefficients of the base vectors, that is, for the **measure numbers** a_x, a_y and a_z in Eq. (1.2.4) below.

In Fig. 1-4, for example, $\overline{PA}$, $\overline{AD}$, and $\overline{DE}$ are the rectangular vector components of the vector $\overrightarrow{PE}$, in the directions of the unit base vectors **i**, **j**, **k**. We denote the vector by **a**, its magnitude by a, and the angles it makes with the positive coordinate

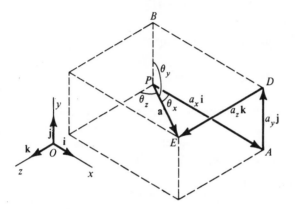

Fig. 1-4   Rectangular Components and Unit Vectors

directions by θ_x, θ_y, θ_z. Then the three **rectangular components** of the vector, that is, the **measure numbers** or numerical coefficients of the base vectors, are

$$a_x = a \cos \theta_x, \qquad a_y = a \cos \theta_y, \qquad a_z = a \cos \theta_z, \qquad (1.2.3)$$

and the rectangular vector components are $a_x\mathbf{i}$, $a_y\mathbf{j}$, $a_z\mathbf{k}$ so that

$$\mathbf{a} = a_x\mathbf{i} + a_y\mathbf{j} + a_z\mathbf{k} \qquad (1.2.4)$$

(vector addition by the polygon rule in Fig. 1-4). Note that $a_x\mathbf{i}$ is a vector, i.e., a directed physical quantity parallel to **i**. The rectangular components are the orthogonal projections of the given vector onto the three coordinate axes. Equation (1.2.4) will be referred to as the **standard form** for a vector in terms of its components.

The word **component** is also often used as a synonym for "projection on an arbitrary direction." For example, a unit vector **n̂** may be given (not necessarily parallel to a coordinate axis); "the component a_n of **a** in the direction of **n̂**" then simply means the orthogonal projection of **a** onto a line having the direction of **n̂**, namely,

$$a_n = a \cos \theta, \qquad (1.2.5)$$

where θ is the angle between the directions of **a** and **n̂**. This equation applies whether or not **a** and **n̂** are drawn from the same point, but the angle θ is more easily visualized if one of the vectors is moved parallel to itself to the same initial point as the other.

The three cosines, $\cos \theta_x$, $\cos \theta_y$, $\cos \theta_z$, are the ***direction cosines*** of the vector. By repeated application of the theorem of Pythagoras we see that the magnitude of the vector is given in terms of the rectangular components by

$$a^2 = a_x^2 + a_y^2 + a_z^2. \tag{1.2.6}$$

Substitution of Eqs. (1.2.3) into this relationship yields

$$\cos^2 \theta_x + \cos^2 \theta_y + \cos^2 \theta_z = 1, \tag{1.2.7}$$

which expresses the fact that the sum of the squares of the direction cosines of any direction is unity.

The addition of two or more vectors is most simply performed in terms of their rectangular components. For example, if

$$\mathbf{c} = \mathbf{a} + \mathbf{b} \quad \text{and} \quad \mathbf{d} = \mathbf{a} - \mathbf{b},$$

then the components of **c** and **d** are given by

$$c_x = a_x + b_x, \qquad d_x = a_x - b_x,$$
$$c_y = a_y + b_y, \qquad d_y = a_y - b_y, \tag{1.2.8}$$
$$c_z = a_z + b_z, \qquad d_z = a_z - b_z.$$

Equations (1.2.8) are special cases of the relationship

$$c_n = a_n + b_n, \tag{1.2.9}$$

which is valid for the components in an arbitrary direction **n̂** (projections), as is evident from Fig. 1-5, where $\mathbf{c} = \mathbf{a} + \mathbf{b}$ and where $a_n = OA$, $b_n = AB$, and $c_n = OB$.

Finite rigid-body rotations are not vectors even though any one rigid-body rotation about an axis can be represented by an arrow lying along the axis, whose length represents (to some scale) the angle of rotation about the axis and whose sense indicates the sense of rotation by the right-hand rule. The reason that finite rotations are not vectors is that they do not compound by vector addition. This is illustrated by a counterexample in Fig. 1-6, where the rigid body is taken to be the rectangle initially lying in the xy-plane with diagonal OA_0. The rectangle is given three

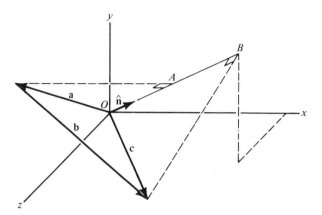

Fig. 1-5   Projection *OB* of Sum **c** Equals Sum of Projections

successive rotations of $\pi/2$ radians (90°), each about one of the coordinate axes. If finite rotations were vector quantities, then the single rotation about a fixed axis, defined by the vector sum $(\pi/2)\mathbf{i} + (\pi/2)\mathbf{j} + (\pi/2)\mathbf{k}$ of the three individual rotations, would take the rectangle from the initial position OA_0 to the final position OA_3. But this is not the case, since evidently the single rotation which takes OA_0 to OA_3 is a 90° rotation about the y-axis, represented as $(\pi/2)\mathbf{j}$.

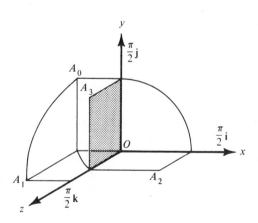

Fig. 1-6   Example Showing that Finite Rotations Are Not Vectors

Although finite rigid-body rotations are not vectors, the angular velocity of a rigid body is a vector (see Chapter 10 in Vol. II); hence the infinitesimal rotation in an infinitesimal time dt can be considered to be a vector, equal to the product of the angular velocity vector by the scalar dt.

Additional concepts of vector algebra will be reviewed in Chapter 3, where the scalar and vector products of two vectors are introduced and applied.

SAMPLE PROBLEM 1.2.1

Determine the resultant $\mathbf{f}^R$ (vector sum) of the two forces shown acting at A in a vertical plane in Fig. 1-7(a). Force magnitudes are in newtons (N). (One pound force equals 4.45 N; see Section. 1.3.)

SOLUTION.   The graphical solution sketched in Fig. 1-7(b) gives

$$f^R = 347\ N \qquad \text{at } \theta_x = 84°.$$

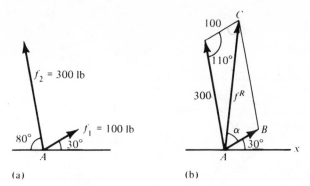

Fig. 1-7 Sample Problem 1.2.1

Trigonometric solution by the law of cosines gives

$$(f^R)^2 = 100^2 + 300^2 - 2(100)(300)\ cos\ 110°$$

$$= 10,000 + 90,000 - 60,000(-0.342) = 120,500.$$

Hence $f^R = 347$ N.
   The law of sines gives

$$\frac{\sin \alpha}{300} = \frac{\sin 110°}{f^R}$$

or, since $\sin 110° = \sin 70°$ and $f^R = 347$ N,

$$\frac{\sin \alpha}{300} = \frac{\sin 70°}{347}.$$

The law of sines can be solved conveniently on a slide rule. Set 70 on the S-scale opposite 347 on the D-scale. Then opposite 300 on the D-scale read on the S-scale

$$\alpha = 54.3°.$$

Hence

$$f^R = 347\ N \qquad at\ \theta_x = 84.3° \quad \textbf{\textit{Answer.}}$$

An alternative trigonometric solution is shown in Fig. 1-8, resolving $\mathbf{f}^R$ into components AD along $\mathbf{f}_1$ and DC perpendicular to $\mathbf{f}_1$. Then

$$BD = f_2 \cos 70° = 102.6 \qquad DC = f_2 \sin 70° = 282$$

$$AD = AB + BD = 100 + 102.6 = 202.7.$$

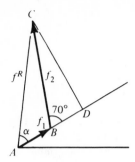

Then α can be determined from

$$\tan \alpha = \frac{DC}{AD} = \frac{282}{202.7} = 1.3912 \qquad \alpha = 54.3°$$

and f^R from

$$DC = f^R \sin \alpha \qquad f^R = 347 \text{ N}.$$

Fig. 1-8 Alternative Solution to Sample Problem 1.2.1

SAMPLE PROBLEM 1.2.2

Write expressions for the sum of the *x*-components and the sum of the *y*-components of the three coplanar vectors in terms of the magnitudes A, B, C (Fig. 1-9).

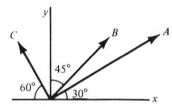

SOLUTION

$$\sum f_x = A \cos 30° + B \cos 45° - C \cos 60°$$
$$= 0.866A + 0.707B - 0.5C.$$

$$\sum f_y = A \sin 30° + B \sin 45° + C \sin 60°$$
$$= 0.5A + 0.707B + 0.866C.$$

Fig. 1-9 Sample Problem 1.2.2

SAMPLE PROBLEM 1.2.3

A force $\mathbf{f}$ of unknown magnitude acts along the line from point $P(6, 8, -6)$ to point $Q(10, 2, -2)$. Express $\mathbf{f}$ in terms of its rectangular components. Coordinates are given in meters.

SOLUTION

$$\mathbf{f} = f\hat{\mathbf{e}}_{PQ} \qquad \text{where the unit vector } \hat{\mathbf{e}}_{PQ} = \frac{\overrightarrow{PQ}}{|\overrightarrow{PQ}|}.$$

$$\overrightarrow{PQ} = \Delta x\, \mathbf{i} + \Delta y\, \mathbf{j} + \Delta z\, \mathbf{k}.$$

Δx is the change in *x from P to Q*, namely, $x_Q - x_P$ (be sure to get these in the correct order), etc.

$$\overrightarrow{PQ} = 4\mathbf{i} - 6\mathbf{j} + 4\mathbf{k} \text{ m} \qquad |PQ| = \sqrt{68} = 8.25 \text{ m}.$$

$$\hat{\mathbf{e}}_{PQ} = 0.486\mathbf{i} - 0.729\mathbf{j} + 0.486\mathbf{k}.$$

Two forces $\mathbf{f}_1$ and $\mathbf{f}_2$ act at A along the lines AB and AC, respectively, as shown in Fig. 1-10 (dimensions are in feet). (a) Express each force in the standard form of Eq. (1.2.4) in terms of unknown magnitudes f_1 and f_2. (b) If $f_1 = 100$ lb and $f_2 = 200$ lb, determine the magnitude and direction cosines of the vector sum $\mathbf{f}^R$.

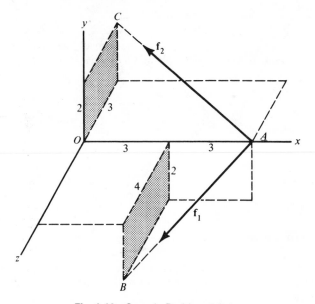

Fig. 1-10   Sample Problem 1.2.4

SOLUTION.   (a) We write the forces as

$$\mathbf{f}_1 = f_1\hat{\mathbf{e}}_1 \qquad \mathbf{f}_2 = f_2\hat{\mathbf{e}}_2,$$

where the unit vectors are

$$\hat{\mathbf{e}}_1 = \frac{\overrightarrow{AB}}{|\overrightarrow{AB}|} = \frac{-3\mathbf{i} - 2\mathbf{j} + 4\mathbf{k}}{\sqrt{29}}$$

$$\hat{\mathbf{e}}_2 = \frac{\overrightarrow{AC}}{|\overrightarrow{AC}|} = \frac{-6\mathbf{i} + 2\mathbf{j} - 3\mathbf{k}}{7}.$$

Thus

$$\hat{e}_1 = -0.567\mathbf{i} - 0.372\mathbf{j} + 0.744\mathbf{k}$$
$$\hat{e}_2 = -0.856\mathbf{i} + 0.286\mathbf{j} - 0.429\mathbf{k},$$

whence

$$\mathbf{f}_1 = -0.567 f_1 \mathbf{i} - 0.372 f_1 \mathbf{j} + 0.744 f_1 \mathbf{k}$$
$$\mathbf{f}_2 = -0.856 f_2 \mathbf{i} + 0.286 f_2 \mathbf{j} - 0.429 f_2 \mathbf{k}   \quad \textbf{Answer.}$$

(b) For $f_1 = 100$ lb and $f_2 = 200$ lb,

$$\mathbf{f}_1 = -56.7\mathbf{i} - 37.2\mathbf{j} + 74.4\mathbf{k}$$
$$\mathbf{f}_2 = -171.2\mathbf{i} + 57.2\mathbf{j} - 85.8\mathbf{k}$$
$$\mathbf{f}^R = \mathbf{f}_1 + \mathbf{f}_2 = -227.9\mathbf{i} + 20.0\mathbf{j} - 11.4\mathbf{k}$$

$$f^R = (51{,}940 + 400 + 130)^{1/2} = \sqrt{52{,}470} = 229.1 \text{ lb}   \quad \textbf{Answer.}$$

$$\cos \theta_x = -\frac{227.9}{229.1} = -0.9948 \qquad \cos \theta_y = \frac{20}{229.1} = 0.0873$$

$$\cos \theta_z = -\frac{11.4}{229.1} = -0.0498   \quad \textbf{Answer.}$$

EXERCISES

1. Two forces act in the *xy*-plane, a force of 300 N at 10° to the *x*-axis and a force of 200 N at 110° to the *x*-axis. Find the magnitude and direction of the sum of the two forces.

2. A force of 100 lb acts in the *xy*-plane at 30° to the *x*-axis. Resolve this force into the sum of two vector components, one at +60° and one at −15° to the *x*-axis in the plane. Illustrate by a sketch.

3. What is the sum of the two forces shown in the figure acting in the *xy*-plane?

4. What must be the two angles α and β in order that the total force on P is a horizontal force of 70 lb? Given forces are coplanar with PQ. (See figure on following page.)

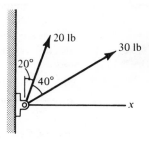

Exercise 3

5. Four coplanar vectors of magnitudes A, B, C, and D make counterclockwise angles of 20°, 100°, 240°, and 345°, respectively, with the

50 lb

Exercise 4

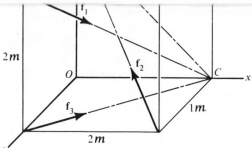

Exercises 8 and 9

x-axis. Express in terms of the four magnitudes A, B, C, and D (a) the sum of the x-components of the four vectors and (b) the sum of the y-components of the four vectors.

6. A force of 50 lb makes a counterclockwise angle of θ with the x-axis in the xy-plane, while a second 100-lb force makes an angle of $\theta + 190°$. If f_x denotes the x-component of the sum of the two forces, determine the value of θ for which f_x is maximum and the value for which it is minimum if the second force in all cases has a magnitude of 100 lb and also lies in the xy-plane.

7. The orthogonal projection of a vector onto an arbitrary line is frequently called "the component of the vector along the line." If a force of 1000 N acts in the xy-plane at 165° to the x-axis, determine the component of the force along a line (a) At 185° to the x-axis in the xy-plane and (b) At 15° to the x-axis in the xy-plane. Illustrate by a sketch.

8. (a) Force $\mathbf{f}_1$ of unknown magnitude f_1 acts along the body diagonal DC of the $2 \times 2 \times 1$ m box shown. Express $\mathbf{f}_1$ in the standard form of Eq. (1.2.4). (b) Write the result for $f_1 = 300$ N.

9. Repeat Ex. 8(a) for (a) $\mathbf{f}_2$, (b) $\mathbf{f}_3$, and (c) $\mathbf{f}_4$.

10. If $f_1 = 300$ N, $f_2 = 150$ N, $f_3 = 100$ N, and $f_4 = 200$ N, (a) what is the vector sum $\mathbf{f}^R$ for all four forces shown in the figure? (b) Show how to determine a unit vector in the direction of $\mathbf{f}^R$. (c) What are the direction cosines of $\mathbf{f}^R$?

11. What is the unit vector in the direction of $\mathbf{a} = 8\mathbf{i} - 10\mathbf{j} + 20\mathbf{k}$?

12. What is the unit vector in the direction opposite to $\mathbf{b} = 12\mathbf{i} + 9\mathbf{j} + 8\mathbf{k}$?

13. A vector of magnitude 100 N is directed along the line from point $A(10, -5, 0)$ to point $B(9, 0, 24)$. Express the vector in terms of the unit vectors $\mathbf{i}$, $\mathbf{j}$, $\mathbf{k}$.

14. A force vector of magnitude 1000 N has direction cosines $\cos \theta_x = 0.7$, $\cos \theta_y = -0.2$ referred to an xyz-system. Express the vector in terms of the orthogonal unit base vectors. (Choose $\cos \theta_z$ positive.)

1.3 FORCE, MASS, AND WEIGHT; LAWS OF MECHANICS; DIMENSIONS AND UNITS

Force, Mass, and Weight

Force is a *push or pull exerted in a given direction*. In Newton's formulation, force is the "action of one body on another body." It is often useful, however, to think of the force as being exerted by a *force field* on a body, without inquiring too closely into what other bodies are responsible for the presence of the force field.

In the rigid-body model we often suppose the force to be a *concentrated force*, that is, a force all acting at one point. Alternatively we may model it as a *distributed force* spread over an area of the surface, or over a volume of the interior if it is an "action-at-a-distance" force (for example, gravitational force). In a deformable body, we must usually consider the force as distributed, since otherwise it could not be supported, although in certain simplified theories of beam bending, for example, we can still model the loads as concentrated forces. Distributed forces will be considered further in Sec. 4.2.

Force is a vector quantity. A concentrated force requires for its complete description the following three characteristics:

Characteristics of a Concentrated Force

1. Magnitude.
2. Direction.
3. Point of application (or line of application in rigid-body mechanics).

In rigid-body mechanics, the specific point of application on the line of action of the force does not matter, since the force will have the same effect on the rigid body no matter where it is applied along its line of action. This is sometimes called the *principle of transmissibility* of a force in rigid-body mechanics (see Sec. 4.1).

The *effects of a force* acting on a body are of two kinds: *external effects* and *internal effects*. Only external effects occur in a rigid body, while a deformable body experiences both external and internal effects. The external effects are of two kinds: (1) acceleration of the body and (2) the bringing into action of other forces, namely, the support reactions, if the body is constrained. The internal effects, which can occur only in a deformable body, are the deformations of the body, that is, the changes in shape or changes of size caused by the acting forces.

Effects of a Force

1. *External effects* (rigid-body effects):
   (a) Acceleration of the body if its constraints do not prevent motion.
   (b) The bringing into action of *support reactions* if the body is constrained so as to prevent or guide the motion.
2. *Internal effects*: deformations (changes in shape or size).

---

make the two analyses separately, even when the deformations are small.

The two kinds of effects of a force furnish two methods by which force magnitudes may be measured: (1) by stretching a calibrated spring (internal effect) or (2) by observing the acceleration imparted to a standard mass. A third method is to balance the force by requiring it to suspend a known weight (using levers or pulleys). The first method is accomplished in engineering practice by means of a load cell or a proving ring. The load cell might consist of a calibrated steel cylinder whose axial shortening under load is measured, while the proving ring is a calibrated steel ring loaded along a diameter to bend it out of its circular shape.

In discussing the characteristics of a force we said that force is a vector quantity. This implies a basic postulate of mechanics, known usually as the **paral-lelogram law of forces,** which states that the effect produced by two forces acting simultaneously at a point is the same as would be produced if the two forces were replaced by a single force equal to the vector sum of the two given forces and acting at the same point of application. See Fig. 1-11. By successive application of the parallelogram law we can conclude that the result produced by any number N of forces acting simultaneously at a point is the same as would be produced by the vector sum of the N forces acting at that point.

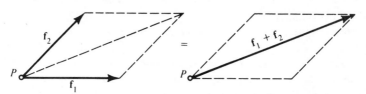

Fig. 1-11   Parallelogram Law of Forces: $f_1 + f_2$ at P Produces Same Effect as f_1 and f_2 at Same Point P

Mass is a nonnegative scalar quantity measuring the inertia of a body. Inertia is the tendency of a body to resist being accelerated, to stay at rest if it is at rest, or to continue on a straight path with constant velocity if it is in motion. We all know from experience that it requires a stronger push to get a car moving than it does to keep it moving. Indeed, if it were not for friction, no push would be required to keep it moving at constant speed on a level road. The property of a body that a force is required to accelerate the body is called the property of *inertia*, and mass is the measure of inertia.

Newton's law of universal gravitation [see Eq. (1.3.2)] states that the attraction between any two particles is proportional to the product of their masses. Thus mass is not only a measure of the inertia of a body but also enters into the determination of the magnitude of the gravitational attraction between the body and any other body. This seems to be a quite different thing from the property of inertia, and one might be tempted to conclude that there were really two kinds of mass—inertial mass and gravitational mass—if experience did not always verify that when two bodies have the same amount of inertial mass they also have the same amount of gravitational mass.

We can measure the mass of a solid body by comparing it to some arbitrary standard mass, for example, the so-called standard kilogram, a platinum-iridium bar carefully preserved near Paris, or an exact copy of the standard kilogram. If the comparison is to be based on the inertial property, we shall have to see to it that the two bodies are each subjected to a force of the same magnitude. Then, if the body we are testing is accelerated just half as much as the standard kilogram, we would say that it possesses twice as much mass, namely, 2 kg, etc. *But in actual practice we determine the mass of a body by placing it on one side of a scale balance and placing enough known masses on the other side to balance the unknown mass*. We thus make use of the basic assumption that inertial mass is equal to gravitational mass. This fact, that in practice we determine the mass of a body by weighing it, leads the unwary to confuse the concepts of weight and mass.

Weight and mass are not the same thing. The weight of a body on or near the earth is the magnitude of the gravitational force exerted on the body by the earth, which is inversely proportional to the square of the distance to the body from the center of the earth. A body has more weight in a valley than the same body has on a mountaintop, but its mass is the same in both positions. An accurate spring balance would show a smaller reading when it supports a mass on a high mountain than when it supports the same mass in a valley. On or near the surface of the moon, the astronaut's "weight" means of course the magnitude of the attraction toward the moon, not toward the earth.

Mass-point or particle model. The inertial properties of a body are frequently treated as though the body were a *particle* with all its mass concentrated at one point—at the *center of mass* of the body.

The total mass m of a finite body is obtained by summing the masses dm of its volume elements, that is, by integration; see Sec. 4.2. The total mass of two bodies or of a body consisting of two parts is the sum of the masses of the separate parts. Definitions and methods for finding the center of mass of a finite body are given in Sec. 4.2. For a symmetrical body of uniform density the center of mass is located at the geometric center of the body.

Laws of Particle Mechanics

The theory of mechanics based on the mass-point particle model is deduced from four postulates abstracted from experience with finite bodies. The first two post-

Four Postulates of Particle Dynamics

1. Parallelogram law of forces.
2. Newton's second law.
3. Newton's third law.
4. Newton's law of universal gravitation.

The four postulates (in modern language) are stated below.

1. The *parallelogram law of forces* (see Fig. 1-11) states that the effect of two forces acting simultaneously at a point is the same as would be produced if the two forces were replaced by a single force equal to their vector sum and acting at the given point.

2. *Newton's second law* for a particle of constant mass m states that in an inertial reference frame the acceleration $\mathbf{a}$ of the particle is proportional to the vector sum of all the forces acting on the particle:

$$\sum \mathbf{f} = km\mathbf{a}. \qquad (1.3.1a)$$

In a consistent set of units [see Eq. (1.3.10) and the following discussion] $k = 1$, and Newton's second law becomes

$$\sum \mathbf{f} = m\mathbf{a}. \qquad (1.3.1b)$$

For all engineering problems to date the co-called fixed stars may be considered an inertial reference frame. *Any other nonrotating reference frame moving at constant velocity with respect to an inertial reference frame will also be inertial.*

Indeed for most ordinary engineering purposes the earth may be considered an inertial reference frame, even though it is rotating and is accelerated relative to the fixed stars. And we shall often speak of "fixed axes," meaning fixed to the earth. But in this era of space travel we frequently need to be more careful.

Newton's first law* is merely a special case of the second law for the case where the velocity is constant in both magnitude and direction. Since acceleration is the

---

* Sir Isaac Newton (1643–1727) published his three laws of motion and the law of gravitation in Book I of his *Principia* (1687) along with many results derived from them, most notably Kepler's laws of planetary motion. Newton's three laws of motion had been anticipated to some degree, notably by Galileo (1564–1642), but never before had they been presented so clearly and their implications carried so far. The implications were worked out by the new calculus, developed by Newton, but they were presented in a geometrical language that could be understood by his contemporaries.

rate of change of velocity, constant velocity means zero acceleration, which implies by Eq. (1.3.1) that the vector sum of the forces is zero. The first law states that a particle under action of no forces would continue to travel in a straight line with constant velocity. This special case is the case of equilibrium of a particle. Note that equilibrium does not necessarily imply that the particle is at rest, although it includes that case when the constant velocity is zero. The first law is important historically as a departure from the long-lived idea of Aristotle that a continuing motion requires a continuing force.

3. **Newton's third law postulates that if one body exerts a force on a second body, then the second body exerts on the first body a force equal in magnitude and opposite in direction to the first force and collinear with it.** This is the meaning concealed by the usual statement "action equals reaction."

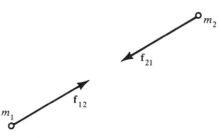

In Fig. 1-12, illustrating Newton's third law, $\mathbf{f}_{12}$ is the force on m_1 exerted by m_2, while $\mathbf{f}_{21}$ is the force on m_2 exerted by m_1. The law implies that $\mathbf{f}_{21} = -\mathbf{f}_{12}$.

Note carefully that the **third law has nothing at all to do with equilibrium. Both bodies may be in accelerated motion.** When one body is in equilibrium under the action of only two forces, the forces must be equal, opposite, and collinear, but despite the similarity of language, this has nothing to do with the third law. The third law deals with interaction forces between **two bodies**, while the equilibrium situation just described is for **one body** under the action of two forces, as illustrated in Fig. 1-13, where $\mathbf{f}_1 = -\mathbf{f}_2$.

Fig. 1-12   Newton's Third Law, $\mathbf{f}_{21} = -\mathbf{f}_{12}$

4. **Newton's law of universal gravitation** postulates that any two particles are attracted toward each other with a force whose magnitude f is directly proportional to the product of their masses and inversely proportional to the square of the distance r between them:

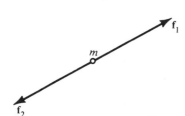

$$f = G\frac{m_1 m_2}{r^2}, \qquad (1.3.2)$$

Fig. 1-13   One Particle in Equilibrium under Two Forces: $\mathbf{f}_2 = -\mathbf{f}_1$

where G is the universal gravitation constant. (By the third law, each body experiences a force equal and opposite to that experienced by the other.)

Center of gravity. An attracting body in general does not possess a center of gravity, where all its gravitational force can be considered to originate. The term *center of gravity* is often applied to the center of mass of a body of engineering interest that is being modeled as a rigid body, since all the earth's gravitational force may be

gravitational force field are unchanged when we replace the distr~~
concentrated force acting at the center of mass. But, of course, the *internal* ~~~
(deformations) will be different. To calculate the accelerations or support reactions
of a rigid body, we may consider all its weight to be concentrated at the center of
mass, but to calculate the deflections of a beam under its own weight, we must
consider the weight as a distributed force.

   The "absolute" free-fall acceleration g of a body under no force but its own
weight (earth's gravity) may be expressed in terms of the universal constant
G. Newton's second law $\mathbf{f} = m\mathbf{a}$ provides the following equation between the
magnitudes W of the force and g of the acceleration:

$$W = mg. \tag{1.3.3}$$

If M denotes the mass of the earth, then Eq. (1.3.2) also gives

$$W = G\frac{Mm}{r^2}. \tag{1.3.4}$$

Equating these two expressions for W and canceling m gives

$$g = \frac{GM}{r^2}. \tag{1.3.5}$$

Thus, g is inversely proportional to the square of the distance from the earth's
center. On the surface of the moon or of another planet besides the earth, Eq.
(1.3.5) gives the free-fall acceleration if M is the mass of the moon (or planet) and
r is the distance from its center. The constant G is universal and applies
anywhere. Experimental measurements give the following *approximate values of
the universal gravitation constant*: $G = 6.67(10^{-11})$ m^3/(kg-sec^2) in the SI absolute
system or $G = 3.44(10^{-8})$ ft^4/(lb-sec^4) in the British gravitational system of
units. More accurate values of the product GM can be obtained from measured
values of g.

   The value of g given by Eq. (1.3.5) is the "absolute" acceleration of gravity as it
would be observed from a nonrotating frame of reference. Since we actually observe
it from a rotating reference frame (the earth), the *apparent acceleration* is slightly

smaller than the absolute value. The following table gives the absolute and apparent free-fall accelerations in ft/sec^2 on the earth's surface at the latitudes 0° (equator) and 90°:

Latitude	Apparent g (ft/sec^2)	Absolute g (ft/sec^2)
0°	32.09	32.20
90°	32.26	32.26

The international gravity formula gives the apparent values at sea level on the earth's surface at latitude λ by

$$g_{app} = 9.78049(1 + 0.0052884 \sin^2 \lambda$$
$$- 0.0000059 \sin^2 2\lambda) \, \text{m/sec}^2. \tag{1.3.6}$$

This empirical formula includes corrections for the deviation of the earth's shape from a perfect sphere. At sea level on the 45th parallel, the apparent values are

$$g_{app} = \begin{cases} 32.174 \, \text{ft/sec}^2 \\ 9.81146 \, \text{m/sec}^2. \end{cases} \tag{1.3.7}$$

For most engineering problems on the earth's surface we can neglect the difference between g and g_{app} and use the three-digit approximations

$$g = 32.2 \, \text{ft/sec}^2 \quad \text{or} \quad g = 9.81 \, \text{m/sec}^2. \tag{1.3.8}$$

The variation with altitude h is given by the formula

$$g = g_0 \left[\frac{r_e}{r_e + h} \right]^2, \tag{1.3.9}$$

where g_0 is the value on the surface of the earth (at radius r_e). Derivation of this equation is left as an exercise.

Dimensions and Units

Dimensions. In Sec. 1.1, we saw that the four primitive elements of mechanics have associated with them four kinds of dimensions: length $[L]$, time $[T]$, mass $[M]$, and force $[F]$. Every physical equation must be dimensionally homogeneous; that is, it must have the same dimensions in every term. Hence, when Newton's second law is written in the form that it takes with a consistent set of units,

$$\mathbf{f} = m\mathbf{a},$$ (1.3.10)

the relationship

$$[F] = \left[\frac{ML}{T^2}\right] \quad \text{or} \quad [M] = \left[\frac{FT^2}{L}\right]$$ (1.3.11)

is implied. In MLT or "absolute" systems M, L, and T are taken as fundamental dimensions, while in FLT or "gravitational" systems F, L, and T are chosen as fundamental.

In making a dimensional check of a formula involving only mechanical quantities, we express all dimensions in terms of the three chosen fundamental dimensions. For example, to check that the dimensions of work (force times distance) are the same as those of the kinetic energy $\frac{1}{2}mv^2$ of a particle, we ask, since $[v] = [L/T]$, whether

$$[FL] = \left[\frac{ML^2}{T^2}\right].$$ (1.3.12)

By using Eq. (1.3.11) we can eliminate F from Eq. (1.3.12) to get

$$\left[\frac{ML^2}{T^2}\right] = \left[\frac{ML^2}{T^2}\right]$$

or alternatively eliminate M from Eq. (1.3.12) to get

$$[FL] = [FL].$$

By either procedure, we see that Eq. (1.3.12) is dimensionally correct, and we also see that in the MLT system the dimensions of work are $[ML^2/T^2]$, while in the FLT system the dimensions of kinetic energy are $[FL]$.

Dimensions and units are not the same thing. The dimension of length $[L]$, for example, may be measured in units of feet, meters, centimeters, miles, or light-years. Six systems of mechanical units (three British and three metric) are listed in Table A1 of Appendix A at the end of this book. Only two of these will be used much in this book: the metric SI system, and the British FLT system in which the unit of mass is the slug.

The ***British systems*** cause a certain amount of confusion because the word "pound" is used with two different meanings: ***pound-mass*** (lbm) and ***pound-force*** (lbf). The pound-mass avoirdupois is legally defined as a certain fraction of the international prototype kilogram (kg) such that

$$1 \text{ kg} = 2.2046 \text{ lbm}, \tag{1.3.13}$$

while the pound-force is legally defined as equal to the weight of a pound-mass in standard position (where $g = 32.174 \text{ ft/sec}^2$). When the pound-force and the pound-mass are both used in Newton's second law, however (with length in ft and time in sec), the system of units is not consistent, so that $k \neq 1$. See Eq. (1.3.17).

The *consistent system that has been most commonly used* by engineers in the United States and other English-speaking countries originated as the British FLT system, where the fundamental units are the pound of force (lb or lbf), the foot (ft), and the second (sec). The unit of mass was then a derived unit, defined as the amount of mass that is given an acceleration of 1 ft/sec^2 when the total force acting on it is 1 lb. The name *slug* is used for this derived unit:

$$1 \text{ slug} = 1 \text{ lb-sec}^2/\text{ft}. \tag{1.3.14}$$

Metric systems; the SI system. The International System of Units (Système International d'Unités), designated SI in all languages, was defined and officially sanctioned by the 1960 Eleventh General Conference on Weights and Measures. The U.S. National Bureau of Standards announced in 1964 that it would henceforth use the SI system, except when use of these units would impair communication. Several American engineering societies and publications as well as governmental agencies are requiring the SI units in published reports.

The mechanical units of the SI system are the MKS absolute system (meter, kilogram, second). In the SI system the newton (abbreviated N) is defined as the force that gives a mass of 1 kg an acceleration of 1 m/s^2. In Vol. I we shall use Eq. (1.3.15) to find the weight of a mass given in kilograms (to three significant figures).

Newton's second law implies that

$$W = mg, \tag{1.3.15}$$

whence

1 kg weighs 9.81 N,   where $g = 9.81 \text{ m/s}^2$,

and

1 slug weighs 32.2 lb,   where $g = 32.2 \text{ ft/sec}^2$.

In the SI system the official abbreviation of second is s, while in the British systems the abbreviation sec is most often used.

The **nonconsistent system with force in pounds-force and mass in pounds-mass** requires $k = 1/g_c$ in Eq. (1.3.1a), where

$$g_c = 32.174 \frac{\text{lbm}}{\text{lbf}} \frac{\text{ft}}{\text{sec}^2},$$

(1.3.16)

so that Newton's second law is

$$\mathbf{f} = \frac{1}{g_c} W_0 \mathbf{a}$$

$$(\text{lbf}) = \left(\frac{1}{(\text{lbm/lbf})(\text{ft/sec}^2)} \right)(\text{lbm})\left(\frac{\text{ft}}{\text{sec}^2} \right)$$

(1.3.17)

where W_0 denotes mass in pounds-mass (equal numerically to the weight in pounds-force in standard position). Note that the units of g_c are not simply ft/sec²; the factor lbm/lbf is needed to make Newton's second law, Eq. (1.3.17), have the same units on both sides. The factor g_c is therefore not really an acceleration but a conversion factor to make the units come out "right," although g_c is numerically equal to the free-fall acceleration in standard position.

Equation (1.3.17) is often written in the form

$$\mathbf{f} = \frac{W}{g} \mathbf{a}$$

(1.3.18)

with the tacit understanding that W means the weight of the body (lbf) at standard position and g means the free-fall acceleration (ft/sec²) at standard position. But this really amounts to using the first system of units, since then $W/g = m$ in slugs. Thus in practice when we say, for example, that a car weighs 3000 pounds, what we really mean is that its weight would be 3000 lbf in standard position.

Some European engineers have also used a nonconsistent metric system where "kilogram" has two meanings: kilogram-mass (kgm) and kilogram-force (kgf). A force of 1 kgf gives a mass of 1 kgm an acceleration of 9.81146 m/s².

Change of units. To change units from one system to another, or to change to nonstandard units, we make use of conversion ratios. To change weight in tons to weight in newtons, we write, for example, for W = 100 tons,

$$W = (100 \text{ tons}) \frac{2000 \text{ lb}}{1 \text{ ton}} \frac{4.45 \text{ N}}{1 \text{ lb}}.$$

(1.3.19)

We cancel the units that appear in both numerator and denominator to obtain $W = 890,000$ N. Thus

$$100 \text{ tons} = 890 \text{ kN} = 0.89 \text{ MN}, \tag{1.3.20}$$

where kN denotes **_kilonewton_** (1000 N) and MN denotes **_meganewton_** (1 million N).

Some conversion factors between metric and British units of length, mass, and force are listed in Table A2 of Appendix A.

SAMPLE PROBLEM 1.3.1

Use the principle of transmissibility to combine the two forces shown acting on the rigid block in Fig. 1-14(a) into a single force that has the same external effect on the body.

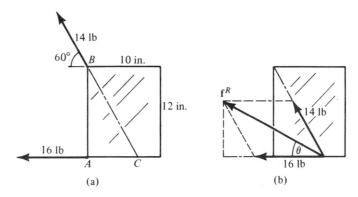

Fig. 1-14 Sample Problem 1.3.1

SOLUTION.   By the principle of transmissibility each force can be moved along its line of action to the intersection at C without changing the external effects. ($AC = 12 \tan 30° = 6.93$ in.). There they can be combined by the parallelogram law as shown in Fig. 1-14(b). The sum $\mathbf{f}^R$ has components

$$16 + 14 \cos 60° = 23 \text{ lb to the left} \quad \text{and} \quad 14 \sin 60° = 12.18 \text{ lb upward}$$

$$\tan \theta = \frac{12.18}{23}, \qquad \theta = 27.8°$$

$$f^R \sin \theta = 12.18, \qquad f^R = 25.1 \text{ lb}.$$

EXERCISES

1. If under their mutual interaction forces an unknown mass m and a known mass m_0 experience accelerations $\mathbf{a}$ and $\mathbf{a}_0$, respectively, determine m.

2. Determine the magnitude and line of action of the single force that has the same external effect on the rigid square block as the three forces shown in the figure.

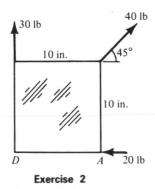

Exercise 2

3. If the free-fall acceleration is 5.32 ft/sec² on the moon, determine for a body that weighed 2000 lb in standard position on the earth the values *on the moon* of the (a) weight in pounds-force, (b) mass in pounds-mass, and (c) mass in slugs.

4. What is the mass of the moon if $g = 1.62$ m/s² at a distance of 1740 km from the center of the moon?

5. (a) Derive Eq. (1.3.9). (b) The earth's radius is 3960 miles. At what elevation above the earth is an astronaut's weight reduced to 81% of his weight on the earth?

6. (a) Compute the value of GM in the SI system if $g = 9.81$ m/s² where the earth's radius is 6.368×10^6 m. (M is the earth's mass, and G is the universal gravitation constant.) (b) Use your result from part (a) to find M if $G = 6.67 \times 10^{-11}$ m³/(kg-s²).

7. Repeat Ex. 6 for the British system if $g = 32.2$ ft/sec² where the earth's radius is 3960 miles and $G = 3.44 \times 10^{-8}$ ft⁴/(lb-sec⁴).

8. What is the ratio of the masses of the earth and moon if the radii are 3960 and 1080 miles and the free-fall accelerations are 32.2 and 5.32 ft/sec², respectively, near the surfaces?

9. What is the force of gravitational attraction between two 10-kg spheres whose centers are 50 cm apart? Give your answer both in newtons and pounds.

10. A cubic foot of water at 70°F weighs 62.4 lb in standard position where $g = 32.174$ ft/sec². At a point where $g = 30.0$ ft/sec², what will be its density in lbm/ft³? In slugs/ft³? What will be its specific weight (i.e., weight per unit volume)?

11. Express the standard sea-level atmospheric pressure of 14.7 lbf/in.² in SI units of kN/m², i.e., kilonewtons per square meter.

12. If the elastic modulus of steel is 30 million lbf/in.², what is its value in SI units of MN/m² (meganewtons per square meter)?

1.4. PROBLEM SOLUTIONS; NUMERICAL ACCURACY

Efficient solution of any engineering problem requires a well-organized procedure. Because of the variety of problems that may be encountered, no single format will be prescribed here for obtaining and presenting the solution. An important part of the engineer's task is to communicate his results to other engineers. If you state your procedures clearly and in sufficient detail that a colleague can understand them, you will find that it helps your own understanding of the problem and leads you to an efficient approach.

Your problem solution should include the following steps:

1. State what is given and what is to be determined, usually illustrated by a neat sketch. Give each unknown a symbol that can be used to represent it both on free-body diagrams and in equations.
2. Draw the appropriate free-body diagrams. See Secs. 2.2 and 4.3 for details.
3. Write equations of equilibrium (or motion) for each free body.
4. Add additional equations to describe the geometry (and/or kinematics and deformation).
5. Solve for the unknowns, using available methods for checking the results.
6. Mark the answers clearly.

This procedural outline may need to be modified for some problems, but it is a good general guide. In writing your solutions, show the formula or principle that is the source of each equation used and make perfectly clear which equations refer to each free body.

Good professional engineering practice calls for clear and neat figures and letters and for uncrowded and systematic arrangement of all computations and diagrams. Squared paper, $8\frac{1}{2}$ by 11 in., and a straightedge for drawing diagrams are recommended. In order that they may be readily checked, all computational steps should be indicated.

Numerical Accuracy

The accuracy required in a practical solution depends on the use to be made of the result. Some calculations of planetary and space vehicle paths require extremely high precision. But most engineering mechanics problems are not so demanding. The accuracy of a result depends on two things: (1) the accuracy of the input data and (2) the computational accuracy. The results can be no more accurate than the input data. In most cases the input figures are not known with an accuracy greater than ±0.2%. Hence there is usually no point in carrying out the computations to any greater accuracy than this. An ordinary slide rule can be read to three significant figures, except at the right-hand end, where the error may be as much as 2 in the third digit (e.g., 997 ± 2), and at the left-hand end, where it can be read to four digits with an error less than 2 in the fourth place (e.g., 1054 ± 2); in either case the error is two-tenths of one percent. Hence slide-rule accuracy is all that can be obtained with input data known only within 0.2%.

Answers should not be given in a way that implies more accuracy than is actually obtained. Even if your computations are performed with a calculator or digital computer with 8- or 10-digit readout, you should round off your final answer. If the input data are accurate to 0.2%, final answers obtained as 99.65157932 and 10.542576 should be shown as 99.7 and 10.54, respectively. This will be understood to mean an accuracy no better than 99.7 ± 0.2 and 10.54 ± 0.02. To include more figures would be a misrepresentation, which has no place in professional engineering practice. For this reason some standard method of rounding off answers like 83.35 and 10.545 should be adopted. Many engineers round such answers to end with an even digit, giving 83.4 and 10.54, when the input data are believed accurate to 0.2%.

1.5 SUMMARY

Following an introduction to the idealizations, concepts, and divisions of mechanics, in Sec. 1.2 we considered in extensive detail the representation of vectors in terms of unit vectors and rectangular components and the addition of vectors. The student should now be able to apply these vector methods in finding the resultant single force equivalent to several applied forces all acting on a body at one point.

In Sec. 1.3 we then considered additional properties of the concepts of force, mass, and weight and the four basic laws of particle mechanics, which will be used over and over and whose content (not the exact wording) should be memorized. The four laws are

1. The parallelogram law of forces (Fig. 1-11).
2. Newton's second law ($\sum \mathbf{f} = m\mathbf{a}$ in consistent units).
3. Newton's third law (Fig. 1-12).
4. Newton's law of universal gravitation [Eq. (1.3.2)].

Newton's third law, requiring the interaction forces between any two bodies to be equal, opposite, and collinear, will be used extensively in both the statics and the dynamics of connected bodies. The second law is the basis of particle dynamics.

In Vol. I, the principal application of the second law will be to determine the relationship between the mass of a body and its weight, by Eq. (1.3.15), which states that in consistent units

$$W = mg, \tag{1.5.1}$$

where g is the magnitude of the free-fall acceleration of the body of mass m under its weight W. On the earth's surface, to three significant figures, this implies that a mass of 1 kg weighs 9.81 N, while a mass of 1 slug weighs 32.2 lb.

Other information on units is summarized in Tables A1 and A2 of Appendix A at the end of this book. Problem solution procedures and numerical accuracy were discussed in Sec. 1.4.

Except for the weight-mass relationship of Eq. (1.5.1), in Vol. I we shall be concerned only with cases where the acceleration is zero, so that Newton's second law reduces to Newton's first law for the equilibrium of a particle, a subject that will be explored in detail in Chapter 2.

CHAPTER TWO

Particle Equilibrium

2.1 INTRODUCTION; PARTICLE EQUILIBRIUM EQUATIONS

The basic principles of particle mechanics were stated in Sec. 1.3. A rigid body may be treated as a particle when all that we seek to determine is the motion of its center of mass, that is, when we are not concerned with rotational effects. Although the title of this chapter is "Particle Equilibrium," many of the problems are for a finite body under the action of *concurrent forces*, that is, forces all passing through one point, so that there is no tendency to cause rotation. Such bodies may be considered as particles for equilibrium analysis.

Particle Equilibrium

A particle is in equilibrium when its acceleration is zero. If the acceleration is kept equal to zero, the particle will travel in a straight line with constant velocity, or if it is initially at rest, it will remain at rest. For zero acceleration $\mathbf{a}$, Newton's second law [Eq. (1.3.1a): $\sum \mathbf{f} = km\mathbf{a}$] reduces to Newton's first law, which is the condition for particle equilibrium:

29

Particle Equilibrium Equation

$$\sum \mathbf{f} = 0. \tag{2.1.1a}$$

A particle is in equilibrium if, and only if, the resultant (vector sum) of all forces **acting on the particle** is equal to zero. This is equivalent to the three rectangular component equations of equilibrium

$$\sum f_x = 0, \qquad \sum f_y = 0, \qquad \sum f_z = 0. \tag{2.1.1b}$$

It is apparent from Eq. (2.1.1b) that at most three unknown quantities can be determined by solving the three equations of equilibrium for a single particle. For coplanar problems with all forces in the xy-plane, the third equation vanishes identically, and only two unknowns can be determined.

A *free-body diagram* should be drawn for each body whose equilibrium is to be analyzed. The diagram should be drawn before the equilibrium equations are written down. This will make it easier to determine exactly which forces (known and unknown) are *acting on the body* and are to be included in the equilibrium equations. This procedure will be illustrated in Sec. 2.2. for coplanar particle equilibrium.

The laws of Coulomb friction will be presented in Sec. 2.3 and applied to some coplanar equilibrium problems. Three-dimensional problems of particle equilibrium are in principle no different from coplanar problems, but because the visualization and representation of the three-dimensional vectors and the geometry involved complicate the solutions, these three-dimensional problems will be postponed to Sec. 2.4.

2.2. PROBLEMS OF PARTICLE EQUILIBRIUM IN A PLANE

Several examples illustrating solution of particle equilibrium problems are given in this section and the two following sections. The student is advised to read through all the examples, since most of the procedures are discussed only in the examples and not in the text. He should also work a good many of the exercises, since experience indicates that that is the only way to master the techniques of engineering mechanics.

Idealizations

In the statics and dynamics of particles and rigid bodies, we use the following idealizations, which experience shows to be quite accurate under the conditions stated. When a segment of a flexible cable or rope is loaded only at the ends and its weight is negligible compared to the end loads, we neglect any sagging and assume that the cable is straight and that the tension force T in the cable acts along the line joining its ends. When cable passes over a pulley whose inertia and bearing friction are negligible in comparison to the applied forces, the tension in the cable is assumed to be the same on both sides of the pulley; this is not true for accelerated massive pulleys or for pulleys with significant bearing friction, as we shall see in the chapters on rotational dynamics. When a supporting bar or link of negligible weight and inertia compared to the bodies supported is connected at only two points in such a way that rotational effects (moments) are not applied at the connections the total force at each of the two connected points is collinear with the line joining the connected points. Such a bar or link is called a ***two-force member***. These idealizations are summarized in Fig. 2-1. Notice that several forces may actually be applied at one of the points A or B of the two-force member, but the vector sum of all the forces at A will be collinear with AB, and the vector sum of all the forces at B will also be collinear with AB. The two end forces must be equal and opposite for equilibrium (or even for acceleration if the mass of the member is neglected). The upper member AB in Fig. 2-1 illustrates compression, while the lower one is in tension. By Newton's third law of action and reaction, a compression member must be pushing back on its supports, while a tension member is pulling on the supports.

Coplanar particle equilibrium equations are

$$\sum f_x = 0, \qquad \sum f_y = 0 \tag{2.2.1}$$

when all forces on the particle in equilibrium act parallel to the xy-plane. An alternative method of solution is to construct the vector sum

$$\sum \mathbf{f} = 0 \tag{2.2.2}$$

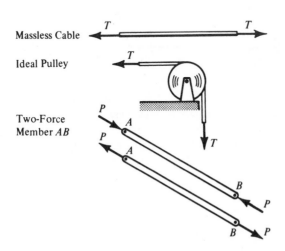

Fig. 2-1 Some Idealizations

graphically by drawing the force polygon; since the sum of the active forces is equal to zero, the polygon must close when the forces are added by the polygon rule (see Sec. 1.2).

A *free-body diagram* should be constructed for each body before its equations of motion or equilibrium are written down. All the known and unknown forces acting on the body should be shown, each one either represented by its components or else represented as a vector. Forces that do not act on the body should not be shown in the free-body diagram. Then when you write the equations of motion or equilibrium for the body, only those forces shown in the free-body diagram for a body should appear in the equations for that body. All forces on the free-body diagram will appear in some equation, although of course a force acting in the *x*-direction will not appear in the *y*- and *z*-component equations.

EXAMPLE 2.2.1

A 500-lb weight is supported by two flexible cables and the applied force P as shown in Fig. 2-2(a). If $P = 600$ lb and $\theta = 15°$, determine the cable tensions T_{AB} and T_{AC}.

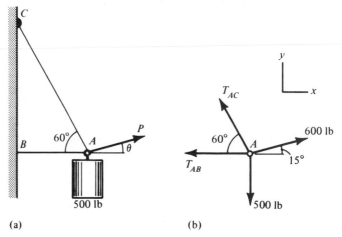

(a)   (b)

Fig. 2-2   (a) Example 2.2.1; (b) Free-Body Diagram

SOLUTION.   The ring A is modeled as a particle in equilibrium under the action of the four forces shown acting on the free-body diagram of Fig. 2-2(b).
Equilibrium equations:

$$[\textstyle\sum f_x = 0]$$

$$600 \cos 15° - T_{AB} - T_{AC} \cos 60° = 0$$

$$[\textstyle\sum f_y = 0]$$ (2.2.3)

$$T_{AC} \sin 60° + 600 \sin 15° - 500 = 0.$$

The second equation gives T_{AC}. Then the first gives T_{AB}.

$$T_{AB} = 380 \text{ lb}, \qquad T_{AC} = 398 \text{ lb} \quad \textbf{\textit{Answer.}}$$

As a check, we construct the force polygon (Fig. 2-3) expressing the fact that $\sum \mathbf{f} = 0$. We first draw to scale the forces that are completely known, and then we close the polygon by drawing two more forces of known direction, so that the four

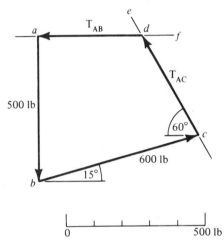

forces add head to tail to zero. That is, we first draw the lines marked *ab* and *bc*. Then we draw *ce* through *c* parallel to T_{AC} and *af* through *a* parallel to T_{AB}. Their intersection locates *d*. Measurement of *cd* and *da* then determines T_{AC} and T_{AB}. In your solutions it is not necessary to include the letters *a, b, . . . ,* etc., which are included here for explanation purposes. If you do a graphical solution, you should indicate the scale in some way. Graphical solutions can furnish quite acceptable answers to most engineering problems if they are performed carefully on a drafting machine or drawing board with the figure drawn sufficiently large. Even if you do the solutions algebraically, a sketch of the polygon will help to determine the reasonableness of your answers. An algebraic sign error or a decimal point error might give a gross error in magnitude or even a wrong direction for the force.

Fig. 2-3 Force Polygon for Example 2.2.1

EXAMPLE 2.2.2 TRIGONOMETRIC SOLUTION OF THREE-FORCE EQUILIBRIUM PROBLEMS

If a particle is in equilibrium under the action of three concurrent forces, these forces must be coplanar, and the force triangle showing that the vector sum of the three forces is zero is especially easy to solve by trigonometry without using components. This furnishes a method to check the component method, and a sketch of the triangle can provide an approximate check on the directions of the forces. A 20-kg mass is suspended as shown in Fig. 2-4. We find the tensions in the strings. Let *A* denote the tension in *AC* and *B* the

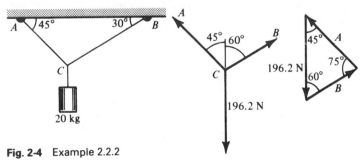

Fig. 2-4 Example 2.2.2

tension in BC. The free-body diagram of joint C is shown with the vector triangle to the right of it. By Eq. (1.3.15) the 20-kg mass weighs $(20)(9.81) = 196.2$ N.

Law of sines:

$$\frac{A}{\sin 60°} = \frac{B}{\sin 45°} = \frac{196.2}{\sin 75°}.$$

(On your slide rule, set 75° on the sine scale opposite 196.2 on the C scale; then read A opposite 60° and B opposite 45°. If you use a calculator instead of a slide rule, calculate $A = 196.2 \sin 60°/\sin 75°$ and $B = 196.2 \sin 60°/\sin 45°$.)

$$A = 175.9 \text{ N}, \qquad B = 143.5 \text{ N} \quad \textbf{Answer.}$$

You can check by components. The component equations are

$$[\textstyle\sum f_x = 0]: \qquad 0.707A + 0.5B - 196.2 = 0$$
$$[\textstyle\sum f_y = 0]: \qquad -0.707A + 0.866B = 0. \tag{2.2.4}$$

EXAMPLE 2.2.3

As an example of connected particles in equilibrium, consider the linkage of three light pin-connected bars, used as a hoisting derrick, as shown in Fig. 2-5. We seek to determine (a) the magnitude of the vertical force P required to hold the system in equilibrium in the position shown and (b) the direction and magnitude of the smallest force (applied at C) that would hold the system in equilibrium.

SOLUTION.   If its weight is negligible, each bar is a two-force member (see Fig. 2-1) and therefore exerts on the pin at either end a force that is in line with the two ends. Figure 2-5(b) shows free-body diagrams of pin B, of pin C, and of the bar BC. The diagram of BC is included only to illustrate that the force f_{BC} on C is equal and opposite to the force f_{BC} on B. This follows from the facts that (1) the two forces acting on bar BC must be equal and opposite for equilibrium and (2) the force exerted on a pin by bar BC must be equal and opposite to the force exerted on the bar by the pin, by Newton's third law. Accordingly, all four of the interaction forces have been labeled by the same symbol f_{BC}, with the different directions indicated by the arrows.

The force triangles for each pin are shown in Fig. 2-5(c). By the law of sines, we have

$$\text{pin } C: \quad \frac{P}{\sin 65°} = \frac{f_{BC}}{\sin 45°} = \frac{f_{CD}}{\sin 70°} \tag{2.2.5}$$

$$\text{pin } B: \quad \frac{f_{BC}}{\sin 30°} = \frac{f_{AB}}{\sin 110°} = \frac{1000 \text{ lb}}{\sin 40°}. \tag{2.2.6}$$

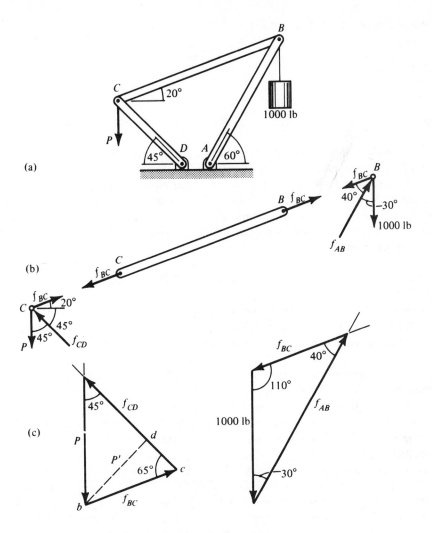

Fig. 2-5 Example 2.2.3: (a) Structure; (b) Free-Body Diagrams; (c) Force Triangles for *C* and *B*

From the equation for pin *B*, we find

$$f_{BC} = \frac{1000 \sin 30^\circ}{\sin 40^\circ} = 778 \text{ lb.}$$

When this is substituted into Eq. (2.2.5), we obtain

$$P = \frac{f_{BC} \sin 65°}{\sin 45°}$$

or

$$P = 997 \text{ lb} \quad \textit{Answer.}$$

You can check the results by component solutions for each pin. The bar forces f_{AB} and f_{CD} can also be determined from Eqs. (2.2.5) and (2.2.6).

To determine the smallest force at C for equilibrium, we examine the force triangle for pin C, Fig. 2-5(c). With f_{BC} known and the direction of f_{CD} known, the third side of the triangle will clearly be shortest if it is perpendicular to the line of action of f_{CD}, as indicated by the dashed line marked P'. Thus the smallest force at C for equilibrium is

$$P' = f_{BC} \sin 65° = 705 \text{ lb} \quad \textit{Answer.}$$

perpendicular to bar CD in the sense of the dotted line from d to b in Fig. 2-5(c). Notice that this choice of direction will also greatly reduce f_{CD}, which will now have magnitude given by the length cd on the vector triangle.

In solving a problem such as Example 2.2.3, the key step is the selection of appropriate free-body diagrams to draw. Since P was the required unknown force, we chose the body that P acts on for the first free-body diagram to draw, namely pin C. But that body has no given force applied to it. We had to choose another body with the known 1000-lb force acting on it in order to get that information into the problem. This technique of dismembering a structure to isolate a number of free-body diagrams will be very important in subsequent chapters. Be sure to construct *separate* isolated free-body diagrams. Do not attempt to draw the diagrams on top of the picture of the structure. Forces f_{AB}, f_{BC}, and f_{CD} should not be drawn on Fig. 2-5(a), for example. It is only when the structure has been taken apart, as in Fig. 2-5(b), that the directions of the interaction forces are meaningful.

The final example of this section is a simple problem chosen to illustrate solution in terms of a variable parameter. Solution in terms of a parameter is important in practice because it leads to a formula that can be used for a whole class of problems.

EXAMPLE 2.2.4

For the system shown in Fig. 2-6(a), we determine (a) the force P required to hold the weight W in equilibrium, as a function of the vertical height h of the supports above pulley B, and (b) the smallest permissible value of h if $d = 6$ m, the mass supported is 500 kg, and the maximum allowable tension force in the cable is 5000 N.

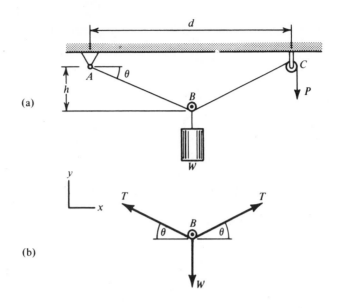

Fig. 2-6 Example 2.2.4: (a) Physical System; (b) Free-Body Diagram of Pulley *B*

SOLUTION.   The free-body diagram of pulley *B* is shown in Fig. 2-6(b). Pulley friction and cable weight have been neglected. Hence the two cable tensions *T* are equal and act along the lines *BA* and *BC* in Fig. 2-6(a). Only one equation of equilibrium is useful since $\sum f_x \equiv 0$:

$$[\sum f_y = 0]: \quad 2T \sin \theta - W = 0. \tag{2.2.7}$$

Since

$$\sin \theta = \frac{h}{[h^2 + (d/2)^2]^{1/2}},$$

we obtain

$$P = T = \frac{1}{2} W \left[1 + \left(\frac{d}{2h} \right)^2 \right]^{1/2} \quad \textbf{Answer.} \tag{2.2.8}$$

We note that as *h* and θ approach zero the required force increases without bound. For the second part of the problem, we substitute $T = 5000$ N and $W = (500)(9.81) = 4905$ N into Eq. (2.2.7), which then gives $\theta = 29.4°$. Then

$$h = \frac{d}{2} \tan \theta = 3 \tan 29.4° = 1.69 \text{ m} \quad \textbf{Answer.}$$

As a check, we note that substitution of the values of $d = 6$ m, $h = 1.69$ m, and $W = 4905$ N into Eq. (2.2.8) gives $T = 4997$ N, which is equal to the assigned value of 5000 N to three significant figures.

Remark on Units

The reader may have noticed that in the examples of this section weights were specified in pounds when British units were employed, while mass was given in kilograms for examples using SI units, requiring use of $W = 9.81$ m to get weight in newtons. We did not consider any examples of $W = 32.2$ m for given mass in slugs, to get weight in pounds, because in the British system it is customary to specify weight.

In Sec. 2.3 we shall present the laws of Coulomb friction and some examples of their use in equilibrium problems.

EXERCISES

1. Determine the tension in cables AC and BC.

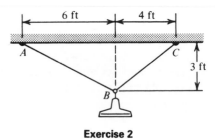

Exercise 2

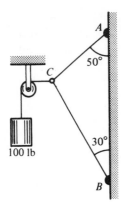

Exercise 1

2. A 30-lb lamp is suspended by two cables AB and BC as shown. Find the tension in each cable.
3. The 1000–lb weight is supported by the horizontal light bar AB, connected to the wall bracket by a smooth pin, and the cable BC as shown. Determine the tension in BC and the force exerted on AB by pin A.

Exercise 3

4. What force must the wall exert on the bolt at *A* to hold it in place when it is acted on by the three forces shown?

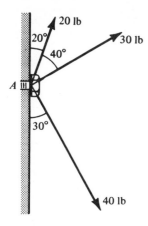

Exercise 4

5. The 600-kg mass is supported by the light bar *AB*, connected to the wall bracket by a smooth pin, and the cable *BC* as shown. Determine the tension in *BC* and the force exerted on *AB* by pin *A*.

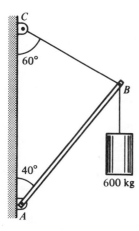

Exercise 5

6. The structural connection is in equilibrium under the action of the four forces shown. If $f_1 = 800$ lb, $f_2 = 400$ lb, and $\theta = 40°$, determine f_3 and f_4.

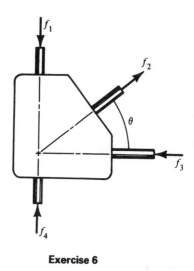

Exercise 6

7. If in Ex. 6 we are given $f_2 = 1000$ lb, $f_4 = 1500$ lb, and $\theta = 30°$, determine f_1 and f_3.

8. The arrangement shown is to be used to pull the pile at *C* out of the ground. (a) Calculate the lifting force exerted by vertical cable *AC* and the tension in *AB* in terms of *P* and θ. (b) Evaluate for $P = 1000$ lb and $\theta = 10°$.

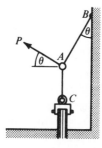

Exercise 8

9. In Fig. 2-2(a), determine the values of *P* and θ required for equilibrium in terms of the cable tensions T_{AB} and T_{AC}.

10. Determine the tension in cable *CE* of the system shown if the ring and cable weights are negligible. Figure is on following page.

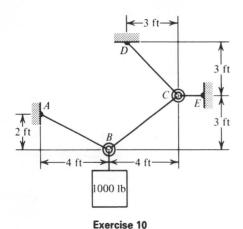

Exercise 10

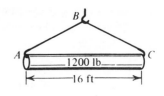

Exercise 11

11. A 16-ft-long pipe weighing 1200 lb is lifted as shown. What is the shortest cable sling *ABC* that can be used if the tension in the cable is not to exceed 1000 lb?

12. The pin-connected equilateral triangle *ABC* shown is formed of light bars and supported by the smooth pin at *A*. Determine (in terms of *W*) the force *P* required for equilibrium and the force in each bar. State whether each bar force is tension or compression.

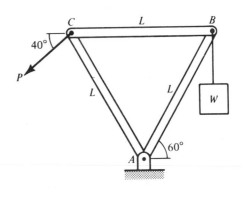

Exercise 12

2.3   COULOMB FRICTION

Friction forces are negligible in many statics problems, but the friction force between two sliding surfaces is never actually zero and is often an important factor.

The direction of the friction force always opposes the relative motion that is occurring or that would occur in the absence of friction. Between lubricated surfaces the *viscous friction* forces tangential to the surface are approximately proportional to the relative speed and to the area of contact but are almost independent of the normal force between the two bodies. When the surfaces are dry, the friction is called *dry friction,* described approximately by the laws of *Coulomb friction.** Dry friction is quite different from viscous friction, since in dry friction the tangential friction force is almost independent of the relative speed and the area of contact, but both the maximum static friction and the sliding friction are proportional to the normal force between the two bodies. And it is usually observed that the maximum

---

* Based on experiments by C. A. Coulomb in 1781.

static friction (no sliding) is greater than the sliding friction force (kinetic friction) between the same two bodies.

Consider the following simple experiment. A block of weight W resting on a horizontal flat surface, as shown in Fig. 2-7(a), is subjected to a gradually increasing horizontal force of magnitude P. The free-body diagram of Fig. 2-7(b) shows the forces acting. Of course the normal force component N and the frictional force component F are really distributed forces over the area of contact. Here we are interested only in external effects and translational motion, so the total of each distributed force is shown as a concentrated force. Although we do not know the point of application of the normal force N, it does not matter because we do not consider rotational effects.

Figure 2-7(d) shows how the friction force F varies with the applied force P. During the static friction range OM the body is in equilibrium; hence $N=W$ and $F=P$. Experience indicates that once the block starts to slide the friction drops to a value slightly less than the maximum static friction F_{max} and then remains approximately constant during the kinetic range $M'K$. In the kinetic range the block accelerates with $P-F=(W/g)a$, where a is the magnitude of the acceleration, while we still have $N=W$, since in this case there is no vertical acceleration.

Figure 2-7(e) shows schematically a highly magnified view of the two sliding surfaces, illustrating that what to the unaided eye appears as a smooth surface actually has a complicated roughness pattern. During static friction some of the protuberances mesh closely and may even weld small areas of the two surfaces together by the cohesive forces. Once these welds are broken and sliding starts, the sliding friction is less than the maximum static friction.

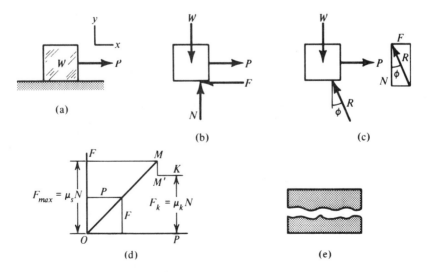

Fig. 2-7 Static and Kinetic Dry Friction

Coefficients of friction are defined as follows. The *coefficient of static friction* μ_s between two surfaces and the *coefficient of kinetic friction* μ_k are defined by

$$F_{max} = \mu_s N \qquad F_k = \mu_k N. \qquad (2.3.1)$$

Note that in general the static friction force F is not equal to $\mu_s N$. Only when the static friction has reached the maximum possible value and *slip is impending* is the friction force F equal to $\mu_s N$. Problems involving *Coulomb friction* fall into three classes:

Friction Problems

    1. Static friction with $F < F_{max}$.
    2. Impending motion $F = \mu_s N$.
    3. Sliding $F = \mu_k N$.

When you do not know which will occur, it is usually best to assume case 1 and solve the statics problem with F as an unknown without using the coefficients of friction.
After you have solved the problem, you must check that your solution gives $F < \mu_s N$. If it gives instead $F > \mu_s N$, you must repeat the solution as a dynamics problem with $F = \mu_k N$.

    Friction problems are often more conveniently analyzed by showing the total reaction force vector of magnitude R at angle φ to the normal as in Fig. 2-7(c), where

$$R = (F^2 + N^2)^{1/2}, \qquad F = N \tan \varphi. \qquad (2.3.2)$$

The critical value of φ at which slip is impending is called the *angle of static friction* φ_s, while the constant value of φ during sliding is called the angle of kinetic friction φ_k. Notice that

$$\mu_s = \tan \varphi_s = \frac{F_{max}}{N} \qquad \mu_k = \tan \varphi_k = \frac{F_k}{N} \qquad (2.3.3)$$

and that *for static friction below the maximum*

$$F < F_{max}, \qquad \varphi < \varphi_s. \qquad (2.3.4)$$

    When the applied forces tend to rotate one body over another, this simple kind of analysis is inadequate; it may be necessary to consider the friction as a distributed force. It should be remembered that Coulomb friction is only an empirical approximation. It gives reasonably good results when the coefficients of friction are accurately known. Since these depend very much on the condition of the surfaces (cleanness and smoothness of finish), it is not possible to quote exact values for given materials. Table 2.3.1 gives fairly typical values for some common materials for smooth dry surfaces; actual values frequently vary by as much as 25% even under apparently comparable conditions and sometimes by almost 100% when the surfaces become wet or greasy. On the other hand, between two surfaces with very strong tendencies to adhere, the coefficient of static friction can be greater than 1.0.

Table 2.3.1 Approximate Dry Friction Coefficients

Materials	μ_s	μ_k
Steel on steel	0.5	0.4
Steel on babbitt	0.4	0.3
Auto tire on pavement	0.9	0.8
Steel on Teflon	0.04	0.04

The reasons for the great variation in friction coefficients under apparently similar conditions becomes apparent when the surfaces are examined under a microscope. The so-called smooth surfaces actually have a random roughness pattern, as was illustrated schematically in Fig. 2-7(e); the roughness pattern may vary radically from one surface to another that appears indentical to the unaided eye. To get accurate results by using friction coefficients you must measure them on the actual bodies for which they are to be used. When this cannot be done, you must allow a considerable margin for error.

EXAMPLE 2.3.1

A 100-lb block is placed on a 30° inclined plane and acted on by a force $P = 80$ lb up the plane as shown in Fig. 2-8. The coefficients of friction between block and plane are known to be $\mu_s = 0.5$, $\mu_k = 0.4$. If this block is released from rest, what will be the friction force? Since we do not know whether the block will slip, we assume that it will not slip and solve the equilibrium problem for the unknown F, assuming a direction for F. If we have guessed wrong on the direction, the solution will give a negative value for F. Figure 2-9 shows the free-body diagram, assuming that F acts down the plane. We choose axes as shown and resolve the weight force into components as shown.

The two equations of equilibrium are

$$[\textstyle\sum f_x = 0] \qquad [\textstyle\sum f_y = 0]$$

$$100 \sin 30° + F - 80 = 0, \qquad N - 100 \cos 30° = 0. \qquad (2.3.5)$$

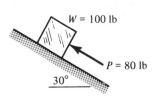

Fig. 2-8 Block on Inclined Plane

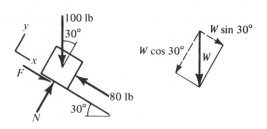

Fig. 2-9 Free-Body Diagram

Hence

$$F = 30 \text{ lb} \qquad N = 86.6 \text{ lb}$$

We check that F is less than the maximum possible $F_{max} = \mu_s N = 43.3$ lb. Hence the assumption is verified, and the direction of F has been correctly guessed, since the solution gave a positive value for F. An alternative graphical solution can be made by drawing the closed vector polygon for $\sum \mathbf{f} = 0$. First lay off the forces that are known both in magnitude and direction. Thus in Fig. 2-10 we first lay off AB and BC to scale, head to tail. Then N (of unknown length) must lie along CD, while F, parallel to CB, must close the polygon to add up to zero. Instead of measuring F from the scale drawing, we can note that

$$80 - F = EB = 100 \sin 30°,$$

whence $F = 30$ lb, as before. An approximate graphical check is useful to detect possible errors in algebraic sign interpretation, errors especially frequent in friction problems.

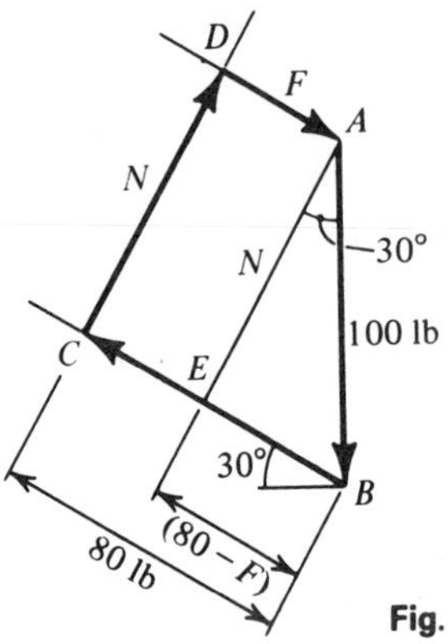

Fig. 2-10 Vector Polygon for Fig. 2-9

EXAMPLE 2.3.2

Figure 2-11 shows a simplified two-dimensional sketch of a 100-lb weight B leaning against a wall, pin-connected by a light link AB to a 200-lb weight on the floor. The inset sketch shows how the connections are actually accomplished by a yoke in the three-dimensional system. We assume symmetry with respect to the vertical plane containing the centers of blocks A and B and analyze it as a two dimensional system. The problem is to determine the largest angle θ for which slip will not occur if the coefficients of static friction at the two surfaces of possible slip are as shown. We neglect friction in the pin connections at A and B and neglect the weight of the link AB, which is then a two-force member (see Fig. 2-1). Free-body diagrams of the three parts of the system are shown in Fig. 2-12. We shall not actually write any equations for the link AB, whose free-body diagram is included only as a reminder of how we know the directions of the forces labeled P in the other two diagrams.

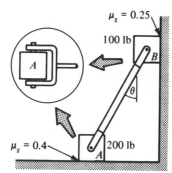

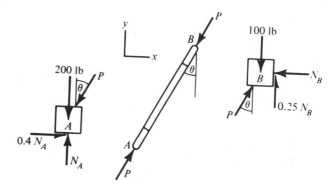

Fig. 2-11 Schematic Illustration of Example 2.3.2

Fig. 2-12 Free-Body Diagrams for Fig. 2-11

Since we are seeking the largest possible angle θ, we analyze the case where slip is impending; hence each friction force is equal to $\mu_s N$, the largest possible static friction, and has been so labeled on the free-body diagrams. Since we assume that slip is not in fact yet occurring, we have an equilibrium problem. We write two equilibrium equations for each body:

A. $[\sum f_x = 0]$: $0.4\,N_A - P \sin \theta = 0$

   $[\sum f_y = 0]$: $N_A - P \cos \theta - 200 = 0$

B. $[\sum f_x = 0]$: $P \sin \theta - N_B = 0$

   $[\sum f_y = 0]$: $P \cos \theta + 0.25\,N_B - 100 = 0,$

$$(2.3.6)$$

a total of four equations for the four unknowns θ, P, N_A, and N_B. We eliminate N_A from the first two equations and N_B from the last two to obtain the two equations below for θ and P.

$$P(2.5 \sin \theta - \cos \theta) = 200$$

$$P(\sin \theta + 4 \cos \theta) = 400,$$

$$(2.3.7)$$

whence

$$\frac{\sin \theta + 4 \cos \theta}{2.5 \sin \theta - \cos \theta} = 2.$$

$$(2.3.8)$$

Clearing of fractions and collecting terms give

$$6 \cos \theta = 4 \sin \theta,$$

whence

$$\tan \theta = 1.5, \qquad \theta = 56.3° \quad \textbf{\textit{Answer.}} \tag{2.3.9}$$

We shall consider now an example of a wedge used to lift a weight. This type of wedge analysis is also used in machine analysis for various kinds of actuators used to position parts of a mechanism.

EXAMPLE 2.3.3   WEDGE

Figure 2-13 shows a light wedge, of wedge angle α, which is used to lift a heavy block of weight W. The coefficient of friction is the same at all sliding surfaces. (The three sliding surfaces have been marked 1, 2, and 3 in the figure.) We wish to determine the horizontal force P required to start the wedge.

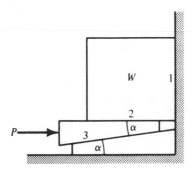

Fig. 2-13   Wedge and Block of Example 2.3.3

SOLUTION.   Figure 2-14 shows free-body diagrams of the block and of the wedge, neglecting the weight of the wedge. The resultant force (R_1, R_2, or R_3) has been shown at each sliding surface, inclined at the friction angle φ_s to the normal, since slip is impending. [Recall that $\mu_s = \tan \varphi_s$ by Eq. (2.3.3)]. The key point in this step is to be sure that the forces are inclined at φ_s in the correct sense. An incorrect assumption here will not automatically lead to a negative P. As shown on the wedge, the tangential friction components ($R_2 \sin \varphi_s$ and $R_3 \sin \varphi_s$) clearly oppose the impending inward motion of the wedge, as they should. Also, on the block, the downward frictional component of R_1 opposes the upward motion of the block, while the horizontal frictional component of R_2 acts to the right on the block to oppose the block's leftward *relative* motion over the wedge. (Actually it is the wedge that moves

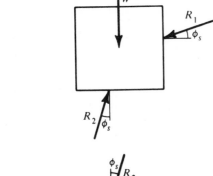

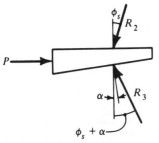

Fig. 2-14 Free-Body Diagrams of Block and Wedge for Example 2.3.3

to the right, and the friction tries to drag the block along with the wedge.) The interaction forces (marked R_2 on both block and wedge) are equal and opposite, as required by Newton's third law.

For the free-body block, the equations of equilibrium are

$$[\textstyle\sum f_x = 0]: \quad R_2 \sin \varphi_s - R_1 \cos \varphi_s = 0 \qquad (2.3.10)$$

$$[\textstyle\sum f_y = 0]: \quad R_2 \cos \varphi_s - R_1 \sin \varphi_s - W = 0. \qquad (2.3.11)$$

The first of these equations gives $R_1 = R_2 \tan \varphi_s$, whence the second becomes

$$R_2 \cos \varphi_s - R_2 \sin \varphi_s \tan \varphi_s = W.$$

Thus

$$R_2 = \frac{W/\cos \varphi_s}{1 - \tan^2 \varphi_s}, \quad R_1 = \frac{W \tan \varphi_s/\cos \varphi_s}{1 - \tan^2 \varphi_s}. \qquad (2.3.12)$$

The equations of equilibrium for the wedge free body are

$$[\textstyle\sum f_x = 0]: \quad P - R_2 \sin \varphi_s - R_3 \sin(\varphi_s + \alpha) = 0 \qquad (2.3.13)$$

$$[\textstyle\sum f_y = 0]: \quad R_3 \cos(\varphi_s + \alpha) - R_2 \cos \varphi_s = 0, \qquad (2.3.14)$$

whence, using the result for R_2, we obtain

$$R_3 = \frac{W/\cos(\varphi_s + \alpha)}{1 - \tan^2 \varphi_s} \qquad (2.3.15)$$

and

$$P = W \left[\frac{\tan \varphi_s + \tan(\varphi_s + \alpha)}{1 - \tan^2 \varphi_s} \right] \quad \textbf{\textit{Answer.}} \qquad (2.3.16)$$

The answer can be given a different form by using Eq. (2.3.1) and a trigonometric identity:

$$\mu_s = \tan \varphi_s \quad \text{and} \quad \tan(\varphi_s + \alpha) = \frac{\tan \varphi_s + \tan \alpha}{1 - \tan \varphi_s \tan \alpha}. \qquad (2.3.17)$$

This leads to

$$P = \frac{[2\mu_s/(1-\mu_s^2)] + \tan\alpha}{1 - \mu_s \tan\alpha} \quad \textbf{\textit{Answer.}} \tag{2.3.18}$$

As a check on the results, consider the overall equilibrium of the free body shown in Fig. 2-15, consisting of the block and wedge together. (The internal forces R_2 do not act on this body and must not be shown.) The equations of equilibrium for this free body are

$$[\Sigma f_x = 0]: \quad P - R_1 \cos\varphi_s - R_3 \sin(\varphi_s + \alpha) = 0 \tag{2.3.19}$$

$$[\Sigma f_y = 0]: \quad R_3 \cos(\varphi_s + \alpha) - R_1 \sin\varphi_s - W = 0. \tag{2.3.20}$$

You can check that substitution of the values found for R_1, R_3, and P, given in Eqs. (2.3.12), (2.3.15), and (2.3.16), reduces each of these equations to an identity.

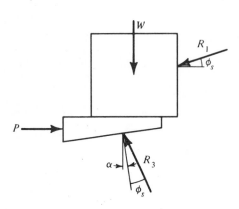

Fig. 2-15 Free Body of Block and Wedge Together

Remark. In Example 2.3.3 we could have represented each sliding surface force by its two components F and N instead of by the resultant R at the known angle φ_s to the normal. This would have given seven unknowns (F_1, N_1, F_2, N_2, F_3, N_3, and P) and only four independent equations of equilibrium for the two free bodies. The required three additional equations would be provided by the friction equation $F = \mu_s N$ for impending motion at each sliding surface.

The analytical solution of a three-dimensional problem in particle statics offers no great difficulty. We have only to write one more component equation of equilibrium. Problems of interconnected bodies become complicated primarily because of the three-dimensional geometry involved. In Sec. 2.4 we consider some equilibrium problem solutions illustrating the geometric considerations.

EXERCISES

1. The block weighs W lb and rests on an inclined plane at angle α to the horizontal. Force P greater than $W \sin\alpha$ acts upward parallel to the plane. (a) If the block is in equilibrium without slipping, what friction force must be acting? (b) If the coefficient of

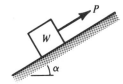

Exercise 1

static friction is μ_s, what is the maximum magnitude P_{max} that P can have without slip occurring?

2. The block of weight W rests on an inclined plane making angle α with the horizontal. Find in terms of W, α, β, and μ_s the force P required to start the block up the plane. The force is applied at angle β to the plane, and μ_s is the coefficient of static friction between block and plane. (Assume a symmetrical block, with P acting in the midplane so that the whole problem is a coplanar problem.)

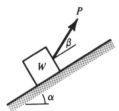

Exercise 2

3. Given: $\mu_s = 0.3$ and $\mu_k = 0.25$. In each case the force acts in the direction shown. (a) What is the force P needed to just start the 200-lb block moving up? (b) What is the forced needed to keep it moving up at a constant speed? (c) What force P is needed to keep it from slipping down? (d) What force P is needed to keep it moving down at constant speed?

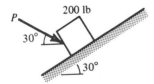

Exercise 3

4. Determine the horizontal force P required to start the 100-lb block moving. The 30-lb upper block is restrained by the rope AB, and the coefficients of static friction are $\frac{1}{2}$ between the two blocks and $\frac{1}{4}$ between the lower block and the floor.

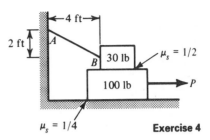

Exercise 4

5. The 100-lb block B rests on the smooth (frictionless) plane inclined at 45°, while the 200-lb block A rests on a horizontal surface where the coefficients of friction with block A are $\mu_s = 0.30$ and $\mu_k = 0.24$. The two are connected by the light rod AB making 30° with the horizontal and connected by smooth pins. Will the system slide if it is released from rest in the position shown? What is the friction force at A?

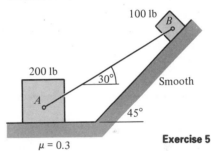

Exercise 5

6. (a) Determine (in terms of α) the smallest force P, acting to the right as shown, that will hold the system in equilibrium (that is, just prevent the 100-lb block from descending) if $\mu_s = 0.35$ at all sliding surfaces. (b) For what value of α is the system self-locking (i.e., with impending downward motion with $P = 0$)?

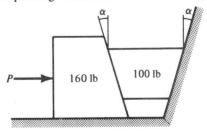

Exercise 6

7. The light wedge of wedge angle α is used to lift the block of weight W. Find the force P required to start the wedge if the coefficient of static friction is μ. The block is constrained by smooth rollers preventing horizontal motion.

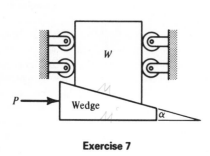

Exercise 7

8. What is the largest angle α for which the wedge of Ex. 7 is self-locking, that is, such that downward motion is impending when $P = 0$?

9. Block A weighs 50 lb. The light bars AB and BC are connected by smooth pins as shown. The coefficient of static friction between the block and the floor is 0.5. Determine the friction force and the two bar forces if a 100-lb vertical load P is applied to pin B.

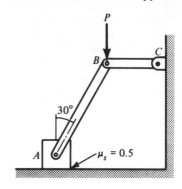

Exercise 9

10. What is the largest load P that the system of Ex. 9 will hold in equilibrium?

2.4 THREE-DIMENSIONAL PROBLEMS OF PARTICLE EQUILIBRIUM

The procedure of solution for particle statics in space is the same as that for plane statics. We draw a free-body diagram, on which each force is represented either by its three components or as a vector, and then write one vector equation or three component equations of equilibrium. Since three-dimensional vector polygons are not easy to construct, graphical solution is not convenient. The biggest difficulty in analytical formulations is in representing the direction of each vector. (See the procedures of Sec. 1.2 for determining a vector through two given points.) Since we have three component equations available for each particle, we can solve for three unknown quantities on each free-body diagram. In equilibrium problems, these unknowns are usually the three components of one unknown force or the three unknown magnitudes of three forces whose directions are known.

It is frequently useful to isolate as a free body a joint or connection point where several members meet instead of isolating the whole structure. This is especially useful when all the members meeting at the joint are *two-force members* (see the discussion of Fig. 2-1). If its weight is negligible, a member connected at both ends by ideal frictionless *ball-and-socket joints* that can exert no rotational moment on the ends and subjected to no other forces except the forces transmitted to it by the two joints will be a two-force member. Many three-dimensional truss structures (see Sec. 5.3) can be analyzed for their primary stresses by treating them as though all joints were ideal ball-and-socket joints even when the actual connections are welded or riveted joints. Example 2.4.1 illustrates the three ideas mentioned in this introduction: (1) representation of a force through two given points, (2) choice of free body, and (3) the idealized two-force member.

EXAMPLE 2.4.1

The 100-lb weight in Fig. 2-16 is supported by a cable from a joint where the three members *AD*, *BD*, and *CD* of a tripod support meet. Neglecting member deformations, we seek to determine each support reaction force at *A*, *B*, and *C*. We assume that the supports *A*, *B*, and *C* and the joint at *D* can be treated as ideal ball-and-socket joints and that the member weights are negligible. We choose the joint at *D* as a free body and draw the free-body diagram in Fig. 2-17, showing the member forces acting along the dashed lines where the members were in Fig. 2-16. We choose the origin *O* on the floor at a point below *D*, with the axes as shown, and we indicate the dimensions that will be needed to get vector representations of the forces.

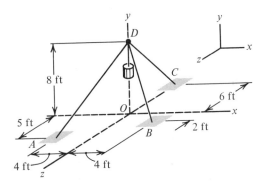

Fig. 2-16 Tripod Support of Example 2.4.1

Each force exerted on *D* by a member is represented in the standard form

$$\mathbf{f} = f\,\hat{\mathbf{e}} = f(e_x\mathbf{i} + e_y\mathbf{j} + e_z\mathbf{k}). \tag{2.4.1}$$

The components e_x, e_y, e_z of the unit vector can be determined by

$$e_x = \frac{d_x}{d}, \qquad e_y = \frac{d_y}{d}, \qquad e_z = \frac{d_z}{d}, \tag{2.4.2}$$

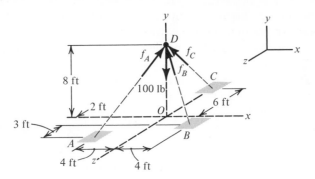

Fig. 2-17 Free-Body Diagram of Joint *D*

where **d** is the relative position vector joining two points on the line of action of the force. Table 2.4.1 summarizes the components for the various forces acting on *D*.

Table 2.4.1 Relative Position and Force Components

f	d	d_x	d_y	d_z	d	f_x	f_y	f_z
f_A	AD	4	8	−5	10.25	$0.3902f_A$	$0.7805f_A$	$-0.4878f_A$
f_B	BD	−4	8	−2	9.165	$-0.4364f_B$	$0.8729f_B$	$-0.2182f_B$
f_C	CD	0	8	6	10	0	$0.8f_C$	$0.6f_C$
$-100\mathbf{j}$	–	–	–	–	–	0	−100 lb	0

The three equations of equilibrium for *D*,

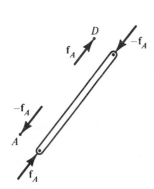

$$[\textstyle\sum f_x = 0]:\quad 0.3902f_A - 0.4364f_B = 0$$
$$[\textstyle\sum f_y = 0]:\quad 0.7805f_A + 0.8729f_B + 0.8f_C - 100 = 0 \qquad (2.4.3)$$
$$[\textstyle\sum f_z = 0]:\quad -0.4878f_A - 0.2182f_B + 0.6f_C = 0,$$

are solved for the three unknown magnitudes f_A, f_B, f_C to obtain

$$f_A = 40.45 \text{ lb},\qquad f_B = 36.17 \text{ lb},\qquad f_C = 46.03 \text{ lb}.$$

Fig. 2-18 Interaction Forces at Ends of Two-Force Member *AD*

Note that the support reaction force vector action *on AD* at *A* is equal to the force vector exerted *by A* at *D*, namely, $\mathbf{f}_A$. This conclusion follows from applying Newton's third law twice: (1) for the interaction forces between *AD* and *D* and (2) for the interaction forces between the support at *A* and member *AD* and recognizing that for the two-force member *AD* to be in equilibrium the forces acting on it must be equal opposite and collinear; see Fig. 2-18. Similar comments apply to the members *BD* and *CD*. Hence the **force vectors exerted by the supports on the members are**

$$\left.\begin{array}{l} \mathbf{f}_A = 15.78\mathbf{i} + 31.6\mathbf{j} - 19.73\mathbf{k} \text{ lb} \\ \mathbf{f}_B = -15.78\mathbf{i} + 31.6\mathbf{j} - 7.89\mathbf{k} \text{ lb} \\ \mathbf{f}_C = 36.8\mathbf{j} + 27.6\mathbf{k} \text{ lb} \end{array}\right\} \quad \textbf{\textit{Answer.}} \qquad (2.4.4)$$

As a check, we note that $\mathbf{f}_A + \mathbf{f}_B + \mathbf{f}_C = 100\mathbf{j}$ lb.

EXAMPLE 2.4.2

The block *P* of weight *W* rests on the inclined plane as shown in Fig. 2-19. The light cord *PQ* passes over a pulley at *Q* and transmits a tensile force *T*. Determine the friction force when the block is released from rest given that

$$W = 12 \text{ lb}, \qquad T = 3 \text{ lb}, \qquad \mu_s = 0.7, \qquad \mu_k = 0.4, \qquad \theta = 30°$$

$$\text{horizontal distance } PM = 5 \text{ ft}, \qquad MN = 10 \text{ ft} \qquad (2.4.5)$$

$$\text{vertical distance } NQ = 10 \text{ ft}.$$

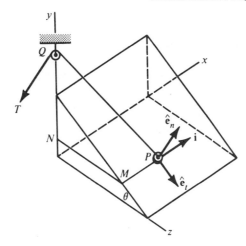

Fig. 2-19 Sketch of Example 2.4.2

We assume first that the block will not slip and determine the friction force by solving the equilibrium problem for the block. Since we do not know the direction of the motion that would occur in the absence of friction, we represent the friction force by two unknown components parallel to the plane: f_x in the *x*-direction and f_t in the downhill tangential direction. The free-body diagram of the block *P* is shown in Fig. 2-20. Because of the difficulty of visualizing the directions involved, the forces have each been represented as a magnitude times a unit vector. We now represent each of the unit vectors in terms of **i**, **j**, **k**, and θ. From Fig. 2-19 we read off

$$\overrightarrow{PQ} = -PM\mathbf{i} + NQ\mathbf{j} - MN\mathbf{k}. \tag{2.4.6}$$

Hence, for the given distances,

$$\hat{\mathbf{e}}_{PQ} = -\tfrac{1}{3}\mathbf{i} + \tfrac{2}{3}\mathbf{j} - \tfrac{2}{3}\mathbf{k}. \tag{2.4.7}$$

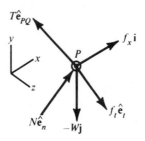

Fig. 2-20   Free-Body Diagram for Example 2.4.2

The normal and the downhill tangent lie in a plane parallel to the yz-plane as shown in Fig. 2-21. Since each of them has unit length, its components will have magnitudes $\cos\theta$ and $\sin\theta$. The algebraic signs are evident from Fig. 2-21. Thus

$$\hat{\mathbf{e}}_n = \cos\theta\mathbf{j} + \sin\theta\mathbf{k} \tag{2.4.8}$$

$$\hat{\mathbf{e}}_t = -\sin\theta\mathbf{j} + \cos\theta\mathbf{k}. \tag{2.4.9}$$

The vector equilibrium equation

$$[\textstyle\sum \mathbf{f} = 0]: \quad (-\tfrac{1}{3}T\mathbf{i} + \tfrac{2}{3}T\mathbf{j} - \tfrac{2}{3}T\mathbf{k}) + (N\cos\theta\mathbf{j} + N\sin\theta\mathbf{k})$$
$$+ (-f_t\sin\theta\mathbf{j} + f_t\cos\theta\mathbf{k}) + f_x\mathbf{i} - W\mathbf{j} = 0 \tag{2.4.10}$$

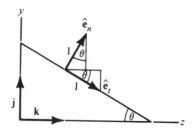

Fig. 2-21   Unit Vector Relationships for Example 2.4.2

then furnishes the three component equations

$$-\tfrac{1}{3}T + f_x = 0$$

$$\tfrac{2}{3}T + N \cos \theta - f_t \sin \theta - W = 0 \tag{2.4.11}$$

$$-\tfrac{2}{3}T + N \sin \theta + f_t \cos \theta = 0.$$

With the given values of $T = 3$ lb and $W = 12$ lb, we obtain $f_x = 1$ lb and

$$N \cos \theta - f_t \sin \theta = 10$$
$$N \sin \theta + f_t \cos \theta = 2, \tag{2.4.12}$$

whence

$$N = 10 \cos \theta + 2 \sin \theta = 9.66 \text{ lb}$$
$$f_t = 2 \cos \theta - 10 \sin \theta = -3.27 \text{ lb} \tag{2.4.13}$$

for $\theta = 30°$ as given. Then

$$f = (f_x^2 + f_t^2)^{1/2} = 3.42 \text{ lb}. \tag{2.4.14}$$

We note that

$$f_{max} = \mu_s N = 6.76 \text{ lb}. \tag{2.4.15}$$

Since the friction force f required for equilibrium is less than f_{max}, our assumption that the block will not slip is verified. The friction force is

$$\left.\begin{array}{l} \mathbf{f} = \mathbf{i} - 3.27\hat{\mathbf{e}}_t \quad \text{or} \\ \mathbf{f} = \mathbf{i} + 1.635\mathbf{j} - 2.83\mathbf{k} \text{ lb}. \end{array}\right\} \quad \textbf{\textit{Answer.}} \tag{2.4.16}$$

EXERCISES

1. The 2000-lb weight W is supported by three light bars AD, BD, and CD connected by smooth socket joints at A, B, C, and D. Determine the supporting forces at A, B, and C and express them in standard form. [A, B, C are in the horizontal xz-plane, while D is 20 ft above the point $E(20, 0, 10)$ ft.]

2. The 3000-lb weight is suspended by three light flexible cables OA, OB, OC tied together at the origin O. Points $B(-4, 12, 3)$ ft and $C(6, 12, 4)$ ft are on the ceiling 12 ft above the level of O, while $A(0, 0, -5)$ ft is on the same level as O. Determine the tensions in the three cables. Figure is on following page.

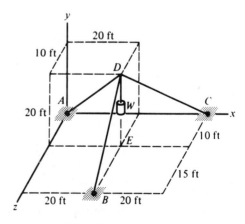

Exercise 1

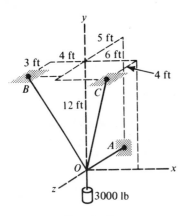

Exercise 2

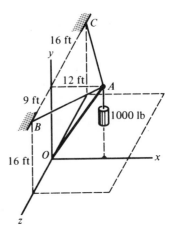

Exercise 4

3. The three light bars are connected by ball-and-socket joints to each other at *D* and to the supporting walls at *A*, *B*, and *C*, so that *BD* and *AD* are parallel to the axes, while *C* is on the *y*-axis, 2 m above *O*. The structure carries two loads at *D*: 9000 N to the left and 12,000 N down. Determine the support reactions.

5. The light bar *OC* lying in the *xz*-plane is supported by a socket joint at *O* and by two light cables *CA* and *CB* at *C* as shown. Determine the cable tensions and the support reaction force at *O*.

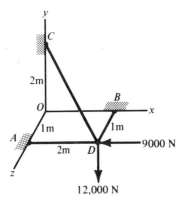

Exercise 3

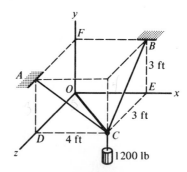

Exercise 5

4. The pin at *A* carries a 1000-lb load and is supported by cables *AB* and *AC* and the bar *OA*. The bar has negligible weight and is attached at *O* by a ball-and-socket joint permitting free rotation. Determine the cable tensions. (The cables are in a horizontal plane 16 ft above the *xz*-plane, while *A* is in the *xy*-plane.)

6. The 200-lb weight is suspended by a light-weight pole *CD* 25 ft long with end *C* placed in the corner of the two walls and the floor as shown. The top of the pole is supported by two horizontal cables attached to the two walls. Find the cable tensions and the force exerted by the pole at *D*.

7. The 150-lb weight is supported by cable *CD* and the two light horizontal bars *AD* and *BD*. The cable is attached at *C*, in the corner formed by two walls and 15 ft above the level

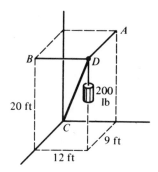

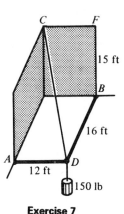

Exercise 6

Exercise 7

of the bars. Determine the cable tension and the support reaction force magnitudes at A and B.

8. Solve Example 2.4.2 (Fig. 2-19) if θ is increased to 45° with the other dimensions unchanged.

9. What tension T is required for impending slip in the system of Example 2.4.2?

10. A jet-propelled sled is headed so that its thrust force T is parallel to the horizontal x-axis shown (perpendicular to the direction of maximum slope on the plane inclined at angle θ to the horizontal plane). If the coefficient of kinetic friction is μ and the sled's weight is W, determine (a) the magnitude of T required to maintain a constant velocity and (b) the angle β between the direction of T and the direction of the straight-line path of the constant-velocity motion.

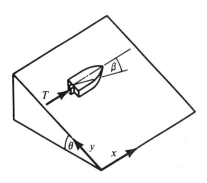

Exercise 10

2.5 SUMMARY

When a rigid body is in equilibrium under the action of concurrent forces its equilibrium can be analyzed by the methods of particle statics. A structure formed by two-force members may be analyzed by drawing a free-body diagram of each connecting pin. In coplanar problems there are two equations of equilibrium for each free-body particle. Hence the methods of particle statics can determine at most two unknowns per free body in coplanar problems, while three-dimensional particle statics can determine three unknowns per free body.

The most important technique which you should have developed from your study of this chapter is the ability to choose the body to isolate for a free-body diagram and represent it properly. In Sec. 2.2 the free-body method was illustrated in coplanar problems. In Sec. 2.3 the laws and methods of Coulomb friction were presented, and finally in Sec. 2.4 the free-body methods were extended to three-dimensional particle statics and use was made of the methods of Sec. 1.2 to represent vectors in space through two given points or at given angles to the coordinate directions.

The free-body method will be further developed in Chapters 4 and 5 for rigid bodies subjected to loads that would cause them to rotate if not constrained. For such problems there will be additional equations of moment equilibrium (related to the tendency to rotate the body) in addition to the components of the equation $\sum \mathbf{f} = 0$, so that more than three unknowns may be determined per free body in three dimensions.

Before taking up rigid-body statics, we shall consider in Chapter 3 two ways to multiply vectors, the scalar product and the vector product, which will be used to define the moment of a force, an important concept for rigid-body statics and dynamics. And in Chapter 4 we shall learn how to reduce systems of forces that are not all concurrent (do not all meet at one point) to simpler systems (resultants) that are equivalent to the given systems for rigid-body analysis purposes.

CHAPTER THREE

Vector and Scalar Products with Applications

3.1 INTRODUCTION

The addition of two vectors and the multiplication of a vector **a** by a scalar c to yield a new vector $c\mathbf{a} = \mathbf{a}c$, of magnitude $|c\mathbf{a}|$ and direction parallel to $+\mathbf{a}$ or to $-\mathbf{a}$ according to whether c is positive or negative, were discussed in Sec. 1.2. We shall now consider multiplication of one vector by another vector. Three different kinds of multiplication of two vectors are useful in mechanics: the scalar product, the vector product, and the tensor product. The names indicate what results from the multiplication. The first two are introduced in this chapter. The *scalar product*, **a·b**, also called the dot product, gives a scalar (a number that may be positive, negative, or zero). The vector product or cross product **a × b** gives a vector. The tensor product will not be needed in this book. Notice that although several ways to multiply by a vector are defined (including the product $c\mathbf{a}$ of a vector times a scalar) there is no definition of division by a vector.

For the scalar product and for the vector product, we shall consider

1. A definition, making no use of components or coordinate axes.

2. A formula for evaluation of the product by using rectangular Cartesian components.
3. Applications.

Although we usually evaluate the product by the component formula, the first definition (making no use of components) is what gives physical significance to the product and leads to useful applications. Some applications of the scalar product and vector product are listed below:

Some Applications of Products of Two Vectors

Scalar product	Work, projection
Vector product	Moment of a force, couple relative velocity in rigid-body rotation, moment of momentum (angular momentum)

The scalar product is introduced in Sec. 3.2 and the vector product in Sec. 3.3. Applications to projection and to moment of a force will be presented in Secs. 3.2 and 3.3. Other applications will appear in later sections of Vol. I and in Vol. II.

3.2 SCALAR PRODUCT; PROJECTION; ROTATION OF COORDINATE AXES

The *scalar product* of any two vectors **a** and **b** is denoted by **a·b** and defined as the number obtained by multiplying the product of the two magnitudes times the cosine of the angle between the two vectors:

$$\mathbf{a \cdot b} = ab \cos \theta. \qquad (3.2.1)$$

The result is a scalar, not a vector; that is why it is called the *scalar product*. It is also called the *dot product* because of the notation. It is not necessary for the two vectors to intersect. For example, in Fig. 3-1 vector **a** is given lying in a horizontal plane M

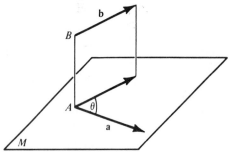

Fig. 3-1 Angle θ between Nonintersecting Vectors **a** and **b**

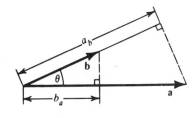

Fig. 3-2 Projections $b_a = b \cos \theta$ and $a_b = a \cos \theta$

through A, while **b** is given parallel to plane M at B. The angle between the two nonintersecting vectors is defined by translating either one of them parallel to itself until they are drawn from the same point, as illustrated in Fig. 3-1.

Since the projection b_a of **b** onto **a** is $b \cos \theta$, while the projection a_b of **a** onto **b** is $a \cos \theta$ (see Fig. 3-2), the product

$$\mathbf{a} \cdot \mathbf{b} = a(b \cos \theta) = b(a \cos \theta)$$
$$= ab_a = ba_b. \tag{3.2.2}$$

Thus we may interpret $\mathbf{a} \cdot \mathbf{b}$ as a times the projection of **b** onto **a** or as b times the projection of **a** onto **b**.

This interpretation of the scalar product by projections can be used to prove that the ***distributive law***

$$\mathbf{a} \cdot (\mathbf{b} + \mathbf{c}) = (\mathbf{a} \cdot \mathbf{b}) + (\mathbf{a} \cdot \mathbf{c}) \tag{3.2.3}$$

holds for any three vectors **a**, **b**, and **c**. Since, by Eq. (1.2.9), the projection of $\mathbf{b} + \mathbf{c}$ onto **a** equals $b_a + c_a$, the sum of the projections of **b** and **c** onto **a**, Eq. (3.2.3) takes the form

$$a(b_a + c_a) = ab_a + ac_a, \tag{3.2.4}$$

which is known to be true by the distributive property of the arithmetic product of numbers. This proves the distributive property of the scalar product. We shall use the distributive property to derive the formula for the scalar product in terms of rectangular components, but first we note two other properties of the scalar product that follow directly from the definition of the scalar product of Eq. (3.2.1).

If m and n are any two numbers, then

$$(m\mathbf{a}) \cdot (n\mathbf{b}) = (ma)(nb) \cos \theta = (mn)(ab \cos \theta) = mn\mathbf{a} \cdot \mathbf{b}.$$

Hence

$$(m\mathbf{a}) \cdot (n\mathbf{b}) = (mn)\mathbf{a} \cdot \mathbf{b} \tag{3.2.5}$$

for any two numbers m, n and any two vectors **a**, **b**. Also, since

$$ab \cos \theta = ba \cos \theta,$$

it follows from the definition of Eq. (3.2.1) that the **scalar product satisfies the commutative law**

$$\mathbf{a \cdot b} = \mathbf{b \cdot a}. \tag{3.2.6}$$

Lest you think that this is too trivial to mention, be forewarned that the vector product $\mathbf{a \times b}$ is not commutative.

For the derivation of the formula, we now need only one more bit of information about scalar products of the mutually orthogonal unit base vectors $\mathbf{i}$, $\mathbf{j}$, $\mathbf{k}$. Since $\cos 0° = 1$, while $\cos 90° = 0$, we have

$$\mathbf{i \cdot i} = 1 \qquad \mathbf{i \cdot j} = 0, \tag{3.2.7}$$

and similarly the scalar product of any other unit vector by itself gives unity, while the scalar product of any two perpendicular unit vectors is zero. Hence, if we write

$$\mathbf{a \cdot b} = (a_x \mathbf{i} + a_y \mathbf{j} + a_z \mathbf{k}) \cdot (b_x \mathbf{i} + b_y \mathbf{j} + b_z \mathbf{k}), \tag{3.2.8}$$

then the distributive property of Eq. (3.2.3) and the property of Eq. (3.2.5) enables us to write $\mathbf{a \cdot b}$ as a sum of nine products of the form of $a_x b_x \mathbf{i \cdot i}$ and $a_x b_y \mathbf{i \cdot j}$, etc. Only three of these nine terms are nonzero because of the orthogonality of the unit vectors. We obtain the following:

Formula for Scalar Product

$$\mathbf{a \cdot b} = a_x b_x + a_y b_y + a_z b_z. \tag{3.2.9}$$

Remember this important formula. It furnishes the method by which we evaluate scalar products. But do not forget Eq. (3.2.1), which gives physical meaning to what we accomplish by using Eq. (3.2.9).

Applications

One of the most common applications of the scalar product is the **projection of a given vector onto a given direction**. By Eq. (3.2.2),

$$a_b = \mathbf{a} \cdot \frac{\mathbf{b}}{b}, \tag{3.2.10}$$

illustrating that the projection of vector **a** onto the direction of **b** is equal to the scalar product of **a** by a *unit vector* **b**/b in the given direction. *Warning*: Do not forget to divide by b if the direction projected onto is specified by a vector **b** that is not a unit vector. If **n̂** is a given unit vector, then the projection of **a** onto a line in the direction of the unit vector **n̂** is

$$a_n = \mathbf{a}\cdot\hat{\mathbf{n}} = a_x n_y + a_y n_y + a_z n_z, \qquad (3.2.11a)$$

where the components n_x, n_y, n_z of the unit vector are the direction cosines of the given direction; see Eq. (1.2.3).

$$n_x = \cos\theta_x \qquad n_y = \cos\theta_y \qquad n_z = \cos\theta_z. \qquad (3.2.11b)$$

The component form of Eq. (3.2.11a) is especially convenient for making the projection when the angle θ between **a** and **n̂** is not known in advance. For example,

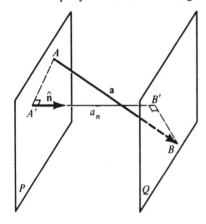

in Fig. 3-3, vector **a** does not even intersect the line whose direction is **n̂**, but Eq. (3.2.11a) enables us to find a_n if we know the components of **a** and of the unit vector **n̂**. Note that geometrically a_n is cut off on the line of **n̂** by two *projecting planes* P and Q through the end points of **a** and making right angles with the line of **n̂**. The projection of a vector **a** onto a given direction is also called the *orthogonal component* of **a** in the given direction or simply the *component* (with orthogonal understood). *Thus a_n is the component of **a** in the direction of **n̂**.*

Fig. 3-3 Projection a_n of a onto Line of Unit Vector **n̂**: $a_n = \mathbf{a}\cdot\mathbf{n}$

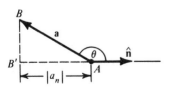

Fig. 3-4 Illustrating Meaning of Negative a_n

The number a_n given by **a·n̂** will be a negative number when $90° < \theta \le 180°$ so that the projection B' of the terminal point B of vector **a** falls on the side of A opposite to the direction of **n̂**, as in Fig. 3-4.

Another geometrical application of the scalar product is the determination of the angle between two vectors given in terms of their components. From the defining Eq. (3.2.1) we have

$$\cos \theta = \frac{\mathbf{a} \cdot \mathbf{b}}{ab}. \tag{3.2.12}$$

The numerator of this formula can be evaluated by Eq. (3.2.9) when the components of each vector are known, and the two magnitudes in the denominator can be calculated by Eq. (1.2.6). For example,

$$b = \sqrt{b_x^2 + b_y^2 + b_z^2}.$$

Since $\cos 90° = 0$, Eq. (3.2.12) furnishes the

Orthogonality Condition

$$\mathbf{a} \cdot \mathbf{b} = 0, \tag{3.2.13}$$

---

a necessary and sufficient condition for two nonzero vectors to be perpendicular. (Note that it is not necessary that the two vectors intersect in order that they be considered perpendicular.)

Rotation of Coordinate Axes

The projection method of Eq. (3.2.11a) furnishes a method to determine the rectangular components of a vector with respect to rotated axes. We consider here only the two-dimensional rotation in the xy-plane. We wish to determine the components v_x' and v_y' of the vector

$$\mathbf{v} = v_x \mathbf{i} + v_y \mathbf{j}$$

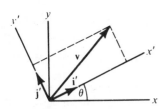

Fig. 3-5 Rotation of Axes in the *xy*-Plane

with respect to the x',y'-axes rotated through counterclockwise angle θ from the x,y-axes. Figure 3-5 shows the old and new axes and the new unit base vectors $\mathbf{i}'$ and $\mathbf{j}'$. We see that

$$\mathbf{i}' = \cos \theta \mathbf{i} + \sin \theta \mathbf{j}$$
$$\mathbf{j}' = -\sin \theta \mathbf{i} + \cos \theta \mathbf{j}. \tag{3.2.14}$$

Hence

$$v'_x = \mathbf{v} \cdot \mathbf{i}' = (v_x \mathbf{i} + v_y \mathbf{j}) \cdot (\cos \theta \mathbf{i} + \sin \theta \mathbf{j})$$

$$v'_y = \mathbf{v} \cdot \mathbf{j}' = (v_x \mathbf{i} + v_y \mathbf{j}) \cdot (-\sin \theta \mathbf{i} + \cos \theta \mathbf{j}).$$

We thus obtain the

Formulas for Rotation of Axes

$$v'_x = v_x \cos \theta + v_y \sin \theta$$

$$v'_y = -v_x \sin \theta + v_y \cos \theta, \qquad\qquad (3.2.15a)$$

---

giving the components v'_x and v'_y with respect to the new axes in terms of the old components v_x and v_y and the angle θ. The inverse transformation giving v_x and v_y in terms of v'_x and v'_y can be derived either by substituting for $\mathbf{i}'$ and $\mathbf{j}'$ according to Eq. (3.2.14) in the expression

$$\mathbf{v} = v'_x \mathbf{i}' + v'_y \mathbf{j}'$$

and collecting terms multiplying $\mathbf{i}$ and $\mathbf{j}$ or alternatively by solving the system of simultaneous equations. Eq. (3.2.15a), for v_x and v_y. The result is

---

Inverse Transformation

$$v_x = v'_x \cos \theta - v'_y \sin \theta$$

$$v_y = v'_x \sin \theta + v'_y \cos \theta. \qquad\qquad (3.2.15b)$$

We conclude this section with three sample problems illustrating applications of the scalar product. In Sec. 3.3 we shall develop the second kind of multiplication of two vectors, the vector product.

SAMPLE PROBLEM 3.2.1

Guy wire *AB* in Fig. 3-6 has a tension force $\mathbf{f}$ of magnitude 140 lb. Determine the component f_{OC} of this force in the direction of line *OC* in the horizontal plane.

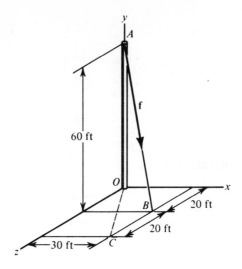

Fig. 3-6   Sample Problem 3.2.1

SOLUTION.   Choose coordinate axes as shown. Then

$$\mathbf{f} = 140\hat{\mathbf{e}}_{AB} \quad \text{and} \quad f_{OC} = \mathbf{f} \cdot \hat{\mathbf{e}}_{OC},$$

where $\hat{\mathbf{e}}_{AB}$ and $\hat{\mathbf{e}}_{OC}$ are unit vectors in the directions of AB and OC:

$$\hat{\mathbf{e}}_{AB} = \frac{\overrightarrow{AB}}{|\overrightarrow{AB}|} = \frac{30\mathbf{i} - 60\mathbf{j} + 20\mathbf{k}}{\sqrt{4900}} = \frac{3}{7}\mathbf{i} - \frac{6}{7}\mathbf{j} + \frac{2}{7}\mathbf{k}$$

$$\hat{\mathbf{e}}_{OC} = \frac{\overrightarrow{OC}}{|\overrightarrow{OC}|} = \frac{30\mathbf{i} + 40\mathbf{k}}{\sqrt{2500}} = \frac{3}{5}\mathbf{i} + \frac{4}{5}\mathbf{k}$$

$$\mathbf{f} = 140\hat{\mathbf{e}}_{AB} = 60\mathbf{i} - 120\mathbf{j} + 40\mathbf{k} \text{ lb}$$

$$f_{OC} = \mathbf{f} \cdot \hat{\mathbf{e}}_{OC} = (60)(\tfrac{3}{5}) + (-120)(0) + (40)(\tfrac{4}{5}) = 68 \text{ lb} \quad \textit{Answer.}$$

SAMPLE PROBLEM 3.2.2

Given points $A(-1, 3, 2)$ and $B(3, 6, -2)$, determine the equation of the plane through A perpendicular to AB.

SOLUTION.   Let $P(x, y, z)$ be a variable point in the plane. Then vector

$$\overrightarrow{AP} = (x + 1)\mathbf{i} + (y - 3)\mathbf{j} + (z - 2)\mathbf{k}$$

lies in the plane and is therefore perpendicular to the vector

$$\overrightarrow{AB} = (3+1)\mathbf{i} + (6-3)\mathbf{j} + (-2-2)\mathbf{k}$$
$$= 4\mathbf{i} + 3\mathbf{j} - 4\mathbf{k}.$$

Hence, by the perpendicularity condition, Eq. (3.2.13), we must have

$$\overrightarrow{AB} \cdot \overrightarrow{AP} = 0,$$

whence

$$4(x+1) + 3(y-3) - 4(z-2) = 0$$

or

$$4x + 3y - 4z + 3 = 0 \quad \textbf{\textit{Answer.}}$$

SAMPLE PROBLEM 3.2.3

Show that the normal to the plane

$$Ax + By + Cz + D = 0$$

has components proportional to (A, B, C).

Solution.   Let (x_1, y_1, z_1) and (x_2, y_2, z_2) be any two points in the plane. Then

$$Ax_2 + By_2 + Cz_2 + D = 0$$

and

$$Ax_1 + By_1 + Cz_1 + D = 0.$$

Subtracting these two equations gives

$$A(x_2 - x_1) + B(y_2 - y_1) + C(z_2 - z_1) = 0,$$

which shows, by the perpendicularity condition, Eq. (3.2.13), that the two vectors

$$A\mathbf{i} + B\mathbf{j} + C\mathbf{k}$$

and

$$(x_2 - x_1)\mathbf{i} + (y_2 - y_1)\mathbf{j} + (z_2 - z_1)\mathbf{k}$$

are perpendicular. Since the last vector joins two arbitrary points in the plane, it follows that $A\mathbf{i}+B\mathbf{j}+C\mathbf{k}$ is perpendicular to every vector lying in the plane. It is therefore perpendicular to the plane. A unit normal $\hat{\mathbf{n}}$ is given by

$$\hat{\mathbf{n}} = \frac{A\mathbf{i}+B\mathbf{j}+C\mathbf{k}}{\sqrt{A^2+B^2+C^2}}.$$

EXERCISES

1. Find $\mathbf{a}\cdot\mathbf{b}$ for $\mathbf{a}=6\mathbf{i}-4\mathbf{j}-6\mathbf{k}$, $\mathbf{b}=4\mathbf{i}-2\mathbf{j}-8\mathbf{k}$.
2. Find $\mathbf{u}\cdot\mathbf{v}$ for $\mathbf{u}=3\mathbf{i}+\mathbf{j}-2\mathbf{k}$, $\mathbf{v}=-\mathbf{i}+2\mathbf{j}-3\mathbf{k}$.
3. Find the projection of $\mathbf{c}=18\mathbf{i}-27\mathbf{j}+81\mathbf{k}$ onto $\mathbf{d}$ if (a) $\mathbf{d}=-\mathbf{i}-2\mathbf{j}+2\mathbf{k}$ and (b) $\mathbf{d}=\mathbf{i}+2\mathbf{j}-2\mathbf{k}$.
4. Find the angle between the two vectors from the origin to the two points $A(4,3,2)$ and $B(-2,4,3)$.
5. A force of magnitude $100\,\text{lb}$ is directed along the line from $A(10,-5,0)$ to $B(9,0,24)$. What is the component of this force in the direction of the vector $\mathbf{a}=-12\mathbf{i}+9\mathbf{j}+8\mathbf{k}$?
6. Evaluate $\mathbf{a}(\mathbf{b}\cdot\mathbf{c})$ for

$$\mathbf{a}=4\mathbf{i}+5\mathbf{j}+3\mathbf{k}$$
$$\mathbf{b}=4\mathbf{i}-5\mathbf{j}+3\mathbf{k}$$
$$\mathbf{c}=3\mathbf{i}+5\mathbf{j}-4\mathbf{k}.$$

7. A 65-lb force $\mathbf{f}$ acts along the diagonal AC as shown. What is the component of $\mathbf{f}$ in the direction of AB?

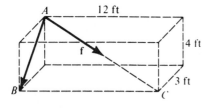

Exercise 7

8. What is the angle between AB and AC in the figure for Ex. 7?

9. Force $\mathbf{f}$ acts along the diagonal of the top of the box shown. What is the component of $\mathbf{f}$ in the direction of AB?

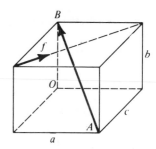

Exercise 9

10. What is the cosine of the angle between $\mathbf{f}$ and AB in Ex. 9?
11. Guy wires AB and AC go from the point A shown in a horizontal plane to the tops of the 12-ft and 4-ft poles. The tension force $\mathbf{f}$ of magnitude $350\,\text{lb}$ acts along AB. Determine the component of $\mathbf{f}$ in the direction of AC.

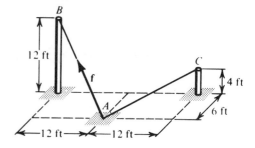

Exercise 11

12. Determine the angle BAC in Ex. 11.
13. Which two vectors in Ex. 6 are perpendicular?
14. Choose x so that the vector $x\mathbf{i} + x\mathbf{j} + 2\mathbf{k}$ will be perpendicular to the vector $18\mathbf{i} - 27\mathbf{j} + 81\mathbf{k}$.
15. If $\mathbf{a} = 10\mathbf{i} + 20\mathbf{j}$, determine the components a'_x and a'_y of $\mathbf{a}$ with respect to the x',y'-axes rotated 30° from the x,y-axes.
16. Derive the inverse transformation equations (3.2.15b).

17. Find the equation of the locus of all points (x, y, z) such that a vector from $A(2, -1, 4)$ to (x, y, z) is perpendicular to the vector from A to $B(3, 3, 2)$.
18. Find the component of the force $\mathbf{f} = 600\mathbf{i} + 1200\mathbf{j} - 900\mathbf{k}$ N in the direction of the normal to the plane $2x - 2y + z = 3$, choosing the positive normal to point away from the origin.

3.3 VECTOR PRODUCT (CROSS PRODUCT); VECTOR MOMENT OF A FORCE ABOUT A POINT; MOMENT ABOUT AN AXIS

The vector product definition requires two parts specifying (1) the direction of the result and (2) the magnitude of the result. The *vector product* or *cross product*

$$\mathbf{c} = \mathbf{a} \times \mathbf{b} \qquad (3.3.1)$$

is defined as the vector $\mathbf{c}$ perpendicular to both $\mathbf{a}$ and $\mathbf{b}$ in the sense that makes $\mathbf{a}, \mathbf{b}, \mathbf{c}$ a right-handed system, with magnitude c given by

$$c = ab \sin \theta, \qquad (3.3.2)$$

where θ is the angle $(0 \le \theta \le 180°)$ between the two vectors. Figure 3-7 shows the sense relationships implied by the "right-handed system" in the definition. The angle θ is measured *from* the first vector listed (namely $\mathbf{a}$ in $\mathbf{a} \times \mathbf{b}$) *to* the second vector. (The angle is measured in the shortest way, so that θ is never greater than 180°. First draw the two vectors from the same point.) If you place your right hand around the common perpendicular with the fingers encircling it in the sense of angle θ, then your thumb will point in the direction of $\mathbf{c}$. (Be sure to use your right hand.) Another way of describing the sense is to say that it is the direction of advance of a screw (with right-hand threads) that is rotated in the sense of angle θ. The direction of the common perpendicular is thus uniquely defined, except when $\theta = 0$ or $\theta = 180°$, but in these two cases the magnitude is zero anyway.

The magnitude $c = ab \sin \theta$ is always nonnegative, since θ satisfies $0 \le \theta \le 180°$. Note that the magnitude is numerically equal to the area $ab \sin \theta$ of the parallelogram formed on $\mathbf{a}$ and $\mathbf{b}$ as sides. The vector product is zero whenever $\mathbf{a}$ and $\mathbf{b}$ are parallel $(\theta = 0$ or $\theta = 180°)$ or when one of the given vectors $\mathbf{a}$ or $\mathbf{b}$ is zero.

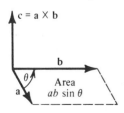

Fig. 3-7 Vector Product

Because of the sense prescription, the *vector product is not commutative*:

$$\mathbf{b} \times \mathbf{a} = -(\mathbf{a} \times \mathbf{b}), \tag{3.3.3}$$

as illustrated in Fig. 3-8. (The angle for $\mathbf{b} \times \mathbf{a}$ would be measured in the opposite sense, *from* the first vector $\mathbf{b}$ *to* the second vector $\mathbf{a}$. The magnitudes are the same, since $ba \sin \theta = ab \sin \theta$, but the two senses are opposite.)

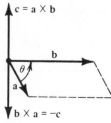

For example, the unit base vectors $\mathbf{i}$, $\mathbf{j}$, $\mathbf{k}$ of a right-handed rectangular coordinate system form the following product, since $(1)(1) \sin 90° = 1$:

$$\mathbf{i} \times \mathbf{i} = 0, \qquad \mathbf{j} \times \mathbf{j} = 0, \qquad \mathbf{k} \times \mathbf{k} = 0$$

$$\mathbf{i} \times \mathbf{j} = \mathbf{k}, \qquad \mathbf{j} \times \mathbf{k} = \mathbf{i}, \qquad \mathbf{k} \times \mathbf{i} = \mathbf{j} \tag{3.3.4}$$

$$\mathbf{j} \times \mathbf{i} = -\mathbf{k}, \qquad \mathbf{k} \times \mathbf{j} = -\mathbf{i}, \qquad \mathbf{i} \times \mathbf{k} = -\mathbf{j}.$$

Fig. 3-8 $\mathbf{b} \times \mathbf{a} = -(\mathbf{a} \times \mathbf{b})$

The algebraic signs for the nonzero products of Eq. (3.3.4) can be conveniently remembered by the scheme shown in Fig. 3-9. Mark the letters i, j, k on a circle, establishing the positive sense on the circle by the order i, j, k. Then, if the two factors multiplied appear in the positive order, the product will be the vector represented by the third letter. If the order is acyclic, that is, reversed (for example, $\mathbf{j} \times \mathbf{i}$), the product will be equal to the negative of the third vector $(-\mathbf{k})$.

We have seen that the vector product is not commutative. It is also *not associative*:

Fig. 3-9 Positive Sense (Cyclic Order)

$$(\mathbf{a} \times \mathbf{b}) \times \mathbf{c} \neq \mathbf{a} \times (\mathbf{b} \times \mathbf{c}). \tag{3.3.5}$$

You can convince yourself of this by putting $\mathbf{a} = \mathbf{b}$; if the vector product were associative, then this would require

$$(\mathbf{a} \times \mathbf{a}) \times \mathbf{c} = \mathbf{a} \times (\mathbf{a} \times \mathbf{c}),$$

but the left-hand side of this equation is always zero because $\mathbf{a} \times \mathbf{a} = 0$, while the right-hand side is in general not zero. It is necessary to keep the parentheses in a triple vector product, since the meaning of $\mathbf{a} \times \mathbf{b} \times \mathbf{c}$ is not clear.

Although the vector product is neither commutative nor associative, the *vector product is distributive*:

$$\mathbf{a} \times (\mathbf{b}_1 + \mathbf{b}_2) = (\mathbf{a} \times \mathbf{b}_1) + (\mathbf{a} \times \mathbf{b}_2). \tag{3.3.6}$$

The proof is omitted,* but we shall use the distributive property to derive a formula for the vector product in terms of the components of the given vectors. For this we need also the result that for any two numbers m and n

$$(m\mathbf{a}) \times (n\mathbf{b}) = (mn)(\mathbf{a} \times \mathbf{b}), \tag{3.3.7}$$

a result immediately apparent from the definition of the vector product.

We write

$$\mathbf{a} \times \mathbf{b} = (a_x\mathbf{i} + a_y\mathbf{j} + a_z\mathbf{k}) \times (b_x\mathbf{i} + b_y\mathbf{j} + b_z\mathbf{k}). \tag{3.3.8}$$

When this is multiplied out, using Eqs. (3.3.6) and (3.3.7), we obtain a nine-term sum containing terms such as

$$a_x b_x \mathbf{i} \times \mathbf{i} = 0 \quad \text{and} \quad a_x b_y \mathbf{i} \times \mathbf{j} = a_x b_y \mathbf{k}$$

by Eqs. (3.3.4). The result is, *for a right-handed coordinate system,*

$$\mathbf{a} \times \mathbf{b} = (a_y b_z - a_z b_y)\mathbf{i} + (a_z b_x - a_x b_z)\mathbf{j} + (a_x b_y - a_y b_x)\mathbf{k}, \tag{3.3.9}$$

which may be remembered conveniently as the expansion of the determinant

$$\mathbf{a} \times \mathbf{b} = \begin{vmatrix} \mathbf{i} & \mathbf{j} & \mathbf{k} \\ a_x & a_y & a_z \\ b_x & b_y & b_z \end{vmatrix} \tag{3.3.10}$$

in terms of the minors of the elements in the first row, namely,

$$\mathbf{a} \times \mathbf{b} = \mathbf{i} \begin{vmatrix} a_y & a_z \\ b_y & b_z \end{vmatrix} - \mathbf{j} \begin{vmatrix} a_x & a_z \\ b_x & b_z \end{vmatrix} + \mathbf{k} \begin{vmatrix} a_x & a_y \\ b_x & b_y \end{vmatrix}. \tag{3.3.11}$$

[If $\mathbf{i}$, $\mathbf{j}$, $\mathbf{k}$ form a left-handed system, then the formulas of Eqs. (3.3.9)–(3.3.11) must be prefixed by a minus sign. But in this book we shall always use a right-handed system.]

For numerical examples, it is usually better to set up the determinant of Eq. (3.3.10) with numerical values and then expand it instead of first writing out the

---

* See, for example, Nathaniel Coburn, *Vector Analysis* (New York: The Macmillan Company, 1955), p. 14.

formula of Eq. (3.3.9). For example, given

$$\mathbf{a} = 6\mathbf{i} - 4\mathbf{j} - 6\mathbf{k} \quad \text{and} \quad \mathbf{b} = 4\mathbf{i} - 2\mathbf{j} - 8\mathbf{k},$$

$$\mathbf{a} \times \mathbf{b} = \begin{vmatrix} \mathbf{i} & \mathbf{j} & \mathbf{k} \\ 6 & -4 & -6 \\ 4 & -2 & -8 \end{vmatrix} \tag{3.3.12}$$

$$= \mathbf{i}(+32 - 12) - \mathbf{j}(-48 + 24) + \mathbf{k}(-12 + 16)$$

$$= 20\mathbf{i} + 24\mathbf{j} + 4\mathbf{k}.$$

When the two given vectors have several zero components it is often easier to apply Eqs. (3.3.4) directly instead of using the formula or the determinant. For example,

$$2\mathbf{i} \times (5\mathbf{i} + 3\mathbf{j} - 4\mathbf{k}) = 10(\mathbf{i} \times \mathbf{i}) + 6(\mathbf{i} \times \mathbf{j}) - 8(\mathbf{i} \times \mathbf{k})$$
$$= 6\mathbf{k} + 8\mathbf{j} = 8\mathbf{j} + 6\mathbf{k}. \tag{3.3.13}$$

Applications

The ***common perpendicular to two vectors*** can be found by the cross product. If $\hat{\mathbf{n}}$ is a unit vector perpendicular to both $\mathbf{a}$ and $\mathbf{b}$, then

$$\hat{\mathbf{n}} = \frac{\mathbf{a} \times \mathbf{b}}{|\mathbf{a} \times \mathbf{b}|}. \tag{3.3.14}$$

Do not forget to divide by the magnitude of the cross product. For example, to the two given vectors $2\mathbf{i}$ and $5\mathbf{i} + 3\mathbf{j} - 4\mathbf{k}$ of Eq. (3.3.13) the common perpendicular unit vector is

$$\hat{\mathbf{n}} = \frac{8\mathbf{j} + 6\mathbf{k}}{10} = \frac{4}{5}\mathbf{j} + \frac{3}{5}\mathbf{k}.$$

Remarks. The two given vectors need not intersect in order to determine a direction perpendicular to both of them by Eq. (3.3.14). So long as $\mathbf{a}$ is not parallel to $\mathbf{b}$ or to $-\mathbf{b}$, Eq. (3.3.14) gives a unique direction for $\hat{\mathbf{n}}$, except that the reverse direction $-\hat{\mathbf{n}}$ is of course also perpendicular to both $\mathbf{a}$ and $\mathbf{b}$. By Eq. (3.3.14), $\mathbf{a}$, $\mathbf{b}$, and $\hat{\mathbf{n}}$ form a right-handed system, while $\mathbf{a}$, $\mathbf{b}$, and $-\hat{\mathbf{n}}$ form a left-handed system.

Moment of a Force (Physical Concepts)

The most important application of the vector product in mechanics is to the calculation of the moment of a force. Before giving the formal definition of the moment of a force by means of the vector product, we shall review the intuitive physical concepts involved. When a force acts on a rigid body that can move, two kinds of external effects may be produced: (1) the body may translate, and (2) the body may rotate. The moment of a force about any point (called the moment center) measures the tendency of the force to cause rotation about some axis through that moment center. Consider first the case of a body supported in bearings, so that the only possible motion is a rotation about the bearing axis. Figure 3-10 illustrates schematically a crank *ABC* rigidly attached to a shaft *AO* that is supported in bearings at *M* and *O*. A force **f** applied at *C* is shown resolved into two vector components $\mathbf{f}_1$ parallel to the shaft, and $\mathbf{f}_2 = \mathbf{f} - \mathbf{f}_1$ in plane *CDE*, a plane perpendicular to the shaft at *E*. Anybody who has ever turned a crank knows that a force like $\mathbf{f}_1$ parallel to the shaft does not contribute to the turning effect. Hence only the component $\mathbf{f}_2$ in the plane perpendicular to the shaft has any turning effect. And at least since the time that Archimedes stated the laws of the lever it has been known that the turning effect of $\mathbf{f}_2$ equals $f_2 d$, the product of the magnitude of the force $\mathbf{f}_2$ times the perpendicular distance d from the force's line of action to the axis; $d = ED$ in Fig. 3-10. Notice that the lever arm d is not equal to the length *AB* of the crank arm unless $\mathbf{f}_2$ is perpendicular to the crank arm *AB*.* And notice also that the turning effect is not fd but $f_2 d$, since the part of **f** represented by $\mathbf{f}_1$ (parallel to the axis) does not contribute to the turning effect.

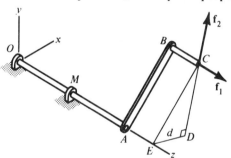

Fig. 3-10   Crank with Shaft *AO*

Figure 3-11 shows an end view of the shaft of the crank of Fig. 3-10. If an arbitrary force acts at *C*, then $\mathbf{f}_1 = f_z \mathbf{k}$ will be parallel to the axis and produce no

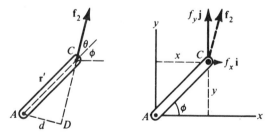

Fig. 3-11   End Views of Crank Shaft (*z*-Axis)

* This was observed by Leonardo da Vinci but was probably known earlier. Archimedes apparently considered only forces perpendicular to the arm.

tendency to turn the crank; accordingly only the components $f_x\mathbf{i}$ and $f_y\mathbf{j}$ have been shown. If x, y, f_x, and f_y are all positive numbers, then $f_y\mathbf{j}$ tends to turn the crank in the positive direction (counterclockwise, increasing φ), while $f_x\mathbf{i}$ tends to turn it in the negative direction. The net positive turning effect is called the **moment about the z-axis**:

$$m_{oz} = xf_y - yf_x$$

$$[M] = [FL],\qquad\qquad (3.3.15a)$$

where $|x|$ and $|y|$ are the perpendicular distances from the z-axis to the lines of action of $f_y\mathbf{j}$ and $f_x\mathbf{i}$, respectively. Notice that the number xf_y will automatically become negative if either (1) $x > 0$ and $f_y < 0$ (downward $f_y\mathbf{j}$) or (2) $x < 0$ and $f_y > 0$; for these two cases $f_y\mathbf{j}$ will tend to turn the crank in the negative sense of decreasing φ; see Fig. 3-12, where it has been assumed that $f_x = 0$. A similar comment can be made about the number yf_x.

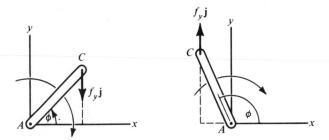

Fig. 3-12 Two Cases of Negative m_{oz}

The **moment about an axis** is thus an algebraic quantity, a number that is positive or negative according to whether it tends to turn the body in the positive direction (of increasing φ, counterclockwise as seen from the positive end of the axis). The formula of Eq. (3.3.15a) automatically provides the correct algebraic sign for any position of crank and any direction of $\mathbf{f}$. Similar expressions can be derived for the moment about an x-axis or a y-axis. For a right-handed coordinate system these formulas can be obtained by cyclic permutation of the subscripts. Thus

$$m_{ox} = yf_z - zf_y \qquad m_{oy} = zf_x - xf_z. \qquad (3.3.15b)$$

In coplanar problems, when all forces act in one plane and the algebraic moment about an axis perpendicular to the plane is calculated, this moment is often called the moment about the point where the axis intersects the plane, given by Eq. (3.3.15a). This should not be confused with the vector moment about a point, to be defined shortly.

Vector Moment About a Point

In the example just discussed the body was mounted in bearings, so that the only possible motion was rotation about the given axis. The turning moment m_{oz} about this axis is an algebraic quantity whose sign indicates the sense of the turning effect (related to the assigned positive direction on the axis by the right-hand rule). To motivate the formal definition of the vector moment about a point, we shall consider briefly the example of Fig. 3-13 where the body is supported by a single ball-and-

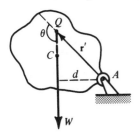

socket joint at A and could rotate freely about an axis in any direction through the point A. What axis it will rotate about depends on the forces acting. If the only force $\mathbf{f}$ acting is the weight W acting at the center of mass C, experience indicates that when released from rest the body will begin to rotate about an axis perpendicular to the vertical plane through AC (the plane determined by A and the line of action of $\mathbf{f}$). Notice that the perpendicular moment arm d of the force $\mathbf{f}$ is equal to

Fig. 3-13   Body Rotating about
Ball-and-Socket Support A

$$d = r' \sin \theta, \tag{3.3.16}$$

where $\mathbf{r}'$ is the relative position vector from A to **any point Q on the line of action of $\mathbf{f}$** and θ is the angle measured **from the direction of $\mathbf{r}'$ to the direction of $\mathbf{W}$**. (The most obvious choice of Q is at C, but any point on the line of action will do. Different points have different r's and different θs, but the product $r' \sin \theta = d$ is the same for all points Q on the line of action.) Thus the moment of $\mathbf{f}$ is in magnitude equal to $Wr' \sin \theta$ and acts to start the body rotating about an axis perpendicular to the plane determined by $\mathbf{f}$ and A. The example of a body supported at one point shows how the force determines the magnitude of the moment about a point and also the direction of the moment vector (parallel to the axis of incipient rotation). The formal definition of the vector moment of a force is not limited to cases of one-point support, however. The body may be completely constrained so that no motion can occur, partially constrained, or completely free to move in any manner.

The **vector moment $\mathbf{m}_A$ of a force $\mathbf{f}$** about a point A (the moment center) is defined as

---

Vector Moment About A

$$\mathbf{m}_A = \mathbf{r}' \times \mathbf{f}$$

$$[m] = [FL], \tag{3.3.17}$$

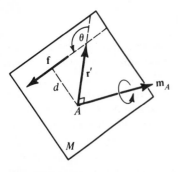

Fig. 3-14   Vector Moment $\mathbf{m}_A$

where $\mathbf{r}'$ is the relative position vector from A to any point on the line of action of $\mathbf{f}$. Figure 3-14 shows how this is consistent with the previous example, since

$$|\mathbf{m}_A| = r'f \sin\theta = fd,$$

where θ is measured from the direction of $\mathbf{r}'$ to the direction of $\mathbf{f}$ and the direction of $\mathbf{m}_A$ is perpendicular to the plane M determined by $\mathbf{f}$ and A, with sense of rotation associated by the right-hand rule with the direction of $\mathbf{m}_A$. It follows from the definition that the moment of a force is unchanged when the force is moved to a different point on the same line of action (keeping the magnitude and direction unchanged). You can imagine that the force is moved by sliding it along the line of action. This does not change the moment about any point.

Rectangular components of $\mathbf{m}_A$ are given by

$$\mathbf{m}_A = \begin{vmatrix} \mathbf{i} & \mathbf{j} & \mathbf{k} \\ x' & y' & z' \\ f_x & f_y & f_z \end{vmatrix}. \tag{3.3.18}$$

Thus

$$m_{Ax} = y'f_z - z'f_y, \qquad m_{Ay} = z'f_x - x'f_z$$
$$m_{Az} = x'f_y - y'f_x, \qquad \mathbf{m}_A = m_{Ax}\mathbf{i} + m_{Ay}\mathbf{j} + m_{Az}\mathbf{k}. \tag{3.3.19}$$

Comparing Eqs. (3.3.19) to Eqs. (3.3.15) we see that in the special case that the moment center is the origin O the rectangular components of the vector moment $\mathbf{m}_O$ are, respectively, equal to the algebraic turning moments m_{OX}, m_{OY}, m_{OZ} about the axes. When the moment center A is not at the origin O, the components of Eqs. (3.3.19) represent algebraic turning moments about axes through A parallel to the x,y,z-axes.

Since

$$m_{Ax} = \mathbf{m}_A \cdot \mathbf{i}, \tag{3.3.20}$$

we see that the algebraic turning moment of $\mathbf{f}$ about an axis through A parallel to the x-axis is equal to the scalar product of $\mathbf{m}_A$ by a unit vector in the direction of the axis. Now the definition of $\mathbf{m}_A$ by Eqs. (3.3.17) made no use of the coordinate axes, which can be chosen as we please. We could have chosen the x-axis parallel to an arbitrary direction specified by a dimensionless unit vector $\hat{\mathbf{e}}_{AB}$. Then we would have, by Eq. (3.3.20), for the **turning moment m_{AB} about any axis through A**

$$m_{AB} = \mathbf{m}_A \cdot \hat{\mathbf{e}}_{AB}.$$

But this equation clearly furnishes the same algebraic value for m_{AB} no matter how the axes are chosen, since it is equal to $|\mathbf{m}_A|$ times the cosine of the angle between $\mathbf{m}_A$ and $\hat{\mathbf{e}}_{AB}$. We thus obtain the following important conclusion:

The algebraic turning moment about any directed axis can be determined by first calculating the vector moment $\mathbf{m}_A$ about any point A **on the axis.** Then m_{AB} is the projection of $\mathbf{m}_A$ onto the axis.

$$m_{AB} = \mathbf{m}_A \cdot \hat{\mathbf{e}}_{AB}. \tag{3.3.21}$$

The algebraic sign of m_{AB} indicates the sense of rotation about AB by means of the right-hand rule applied to the direction $\hat{\mathbf{e}}_{AB}$. For example, in Fig. 3-15, we have $m_{AB} > 0$ and $m_{AC} < 0$, since we have $0 < \alpha_1 < 90°$ and $90° < \alpha_2 < 180°$ relative to the positive directions indicated on bars AB and AC by $\hat{\mathbf{e}}_{AB}$ and $\hat{\mathbf{e}}_{AC}$.

Varignon's theorem states that when two or more forces act at the same point Q then the vector moment about A of the vector sum of the forces is equal to the vector sum of the moments about A of the separate forces. This follows immediately from the distributive character of the vector product, since with $\mathbf{r}' = \overrightarrow{AQ}$,

$$\mathbf{r}' \times (\mathbf{f}_1 + \mathbf{f}_2 + \cdots) = (\mathbf{r}' \times \mathbf{f}_1) + (\mathbf{r}' \times \mathbf{f}_2) + \cdots. \tag{3.3.22}$$

This means, for example, that with a given force $\mathbf{f}$ we can determine the vector moment of $\mathbf{f}$ as the sum of the moments of the vector components of $\mathbf{f}$. Since the projection of a sum equals the sum of the projections (see Fig. 1-5), Varignon's theorem also applies to the algebraic moments about a given axis.

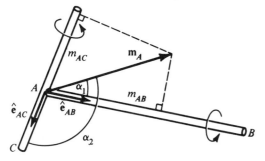

Fig. 3-15 Positive m_{AB} and Negative m_{AC}

SAMPLE PROBLEM 3.3.1

The tension in the guy wire *PO* in Fig. 3-16 is 180 lb. Determine the turning moment about the line *RS* of the force **f** exerted on the pole by the guy wire.

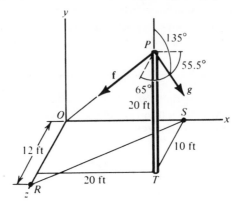

Fig. 3-16   Sample Problems 3.3.1 and 3.3.2

SOLUTION.   Choose axes as shown.   Then

$$\mathbf{f} = 180\hat{\mathbf{e}}_{PO} = 180\frac{-20\mathbf{i}-20\mathbf{j}-10\mathbf{k}}{30}$$

$$= -120\mathbf{i}-120\mathbf{j}-60\mathbf{k} \text{ lb.}$$

We first find $\mathbf{m}_R$ and then project it onto *RS*. We choose $\mathbf{r}' = \overrightarrow{RO}$, since *O* is a point on the line of action of **f**.   Then

$$\mathbf{m}_R = \mathbf{r}' \times \mathbf{f} = -12\mathbf{k} \times (-120\mathbf{i}-120\mathbf{j}-60\mathbf{k})$$

$$= 1440\mathbf{j}-1440\mathbf{i} \text{ lb-ft.}$$

[The multiplication was performed as in the example of Eq. (3.3.13).] By Eq. (3.3.21)

$$m_{RS} = \mathbf{m}_R \cdot \hat{\mathbf{e}}_{RS} = (-1440\mathbf{i}+1440\mathbf{j})\cdot\left(\frac{20\mathbf{i}-12\mathbf{k}}{23.3}\right).$$

$$m_{RS} = -1237 \text{ lb-ft}   \textbf{\textit{Answer.}}$$

SAMPLE PROBLEM 3.3.2

A second force **g** of 100-lb magnitude acts at point *P* in Fig. 3-16 in a direction making angles of 55.5°, 135°, and 65° with the coordinate directions. Determine its turning moments with respect to the coordinate axes through *O*.

Solution

$$\mathbf{g} = 100(\cos 55.5°\mathbf{i} + \cos 135°\mathbf{j} + \cos 65°\mathbf{k})$$

$$= 56.6\mathbf{i} - 70.7\mathbf{j} + 42.3\mathbf{k} \text{ lb.}$$

$$\mathbf{r} = 20\mathbf{i} + 20\mathbf{j} + 10\mathbf{k} \text{ ft.}$$

$$\mathbf{m}_O = \mathbf{r} \times \mathbf{g} = \begin{vmatrix} \mathbf{i} & \mathbf{j} & \mathbf{k} \\ 20 & 20 & 10 \\ 56.6 & -70.7 & 42.3 \end{vmatrix}$$

$$= \mathbf{i}(846 + 707) - \mathbf{j}(846 - 566)$$

$$+ \mathbf{k}(-1414 - 1132)$$

$$= 1553\mathbf{i} - 280\mathbf{j} - 2546\mathbf{k} \text{ lb-ft.}$$

Hence

$$m_{OX} = 1553 \text{ lb-ft}, \qquad m_{OY} = -280 \text{ lb-ft.}$$

$$m_{OZ} = -2546 \text{ lb-ft} \quad \textbf{Answer.}$$

SAMPLE PROBLEM 3.3.3

What should be the line of action of the two-dimensional force $\mathbf{f} = 10\mathbf{i} - 5\mathbf{j}$ acting on block *OABC* in Fig. 3-17 in order that its turning moment about *O* be -15 lb-in.? (This really means about axis *OZ* perpendicular to the plane.)

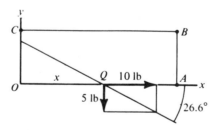

Fig. 3-17 Sample Problem 3.3.3

Solution.   We sketch the line of action inclined at $\theta = \tan^{-1}(-0.5) = -26.6°$ to the horizontal at an undetermined abscissa x in. from *O*. Since we can slide the force anywhere along its line of action without changing its moment, we may as well suppose that it acts at *Q*. Hence, when we resolve it into components, the horizontal component has no moment about *O*, since it passes through *O*. Then $\dot{m}_O = -5x$ lb-in. The negative sign is needed, since for positive x the downward force of 5 lb would tend to cause a clockwise (negative) rotation. Equating this to the given m_O, we obtain

$$-5x = -15 \quad \text{or} \quad x = 3 \text{ in.} \quad \textbf{\textit{Answer.}}$$

The equation of the line is

$$y = -\tfrac{1}{2}(x - 3) \quad \textbf{\textit{Answer.}}$$

SAMPLE PROBLEM 3.3.4

Determine the line of action of force $\mathbf{f} = 3\mathbf{i} - 4\mathbf{j} + 5\mathbf{k}$ N in order that its moment with respect to the origin be $\mathbf{m}_O = 5\mathbf{i} - 10\mathbf{j} - 11\mathbf{k}$ N-m.

SOLUTION. The equations of the line in standard form are

$$\frac{x-a}{3} = \frac{y-b}{-4} = \frac{z-c}{5},$$

where (a, b, c) is some one point on the line and $(3, -4, 5)$ are direction numbers of the line (proportional to its direction cosines).

If $\mathbf{f}$ acts at (x, y, z), any typical point Q on the line, then

$$\mathbf{m}_O = \mathbf{r}_Q \times \mathbf{f} = \begin{vmatrix} \mathbf{i} & \mathbf{j} & \mathbf{k} \\ x & y & z \\ 3 & -4 & 5 \end{vmatrix}$$

$$= (5y + 4z)\mathbf{i} - (5x - 3z)\mathbf{j} - (4x + 3y)\mathbf{k}.$$

Equating this to the given $\mathbf{m}_O$ yields three component equations:

$$5y + 4z = 5$$
$$-5x \quad\quad + 3z = -10$$
$$-4x - 3y \quad\quad = -11.$$

These equations cannot be independent, since they must be satisfied by every point on the line. They represent three planes with one line in common, instead of only one point. (You can check that the determinant of the coefficients equals zero. Eliminating z from the first two equations gives a multiple of the third.) Although there is not a unique solution, we can find one point on the line by arbitrarily assigning one coordinate. If we put $z = 0$, the first two equations give

$y = 1$ m and $x = 2$ m. Hence $(1, 2, 0)$ is a point on the line, which we can substitute for (a, b, c) to obtain the equations of the line in standard form as

$$\frac{x-1}{3} = \frac{y-2}{-4} = \frac{z}{5} \quad \textit{Answer.}$$

EXERCISES

1. Evaluate $\mathbf{a} \times \mathbf{b}$ for $\mathbf{a} = 4\mathbf{i} + 5\mathbf{j} + 3\mathbf{k}$ and $\mathbf{b} = 4\mathbf{i} - 5\mathbf{j} + 3\mathbf{k}$.

2. Evaluate $\mathbf{u} \times \mathbf{v}$ for $\mathbf{u} = -3\mathbf{i} - \mathbf{j} + 4\mathbf{k}$ and $\mathbf{v} = -4\mathbf{i} + \mathbf{j} + 5\mathbf{k}$.

3. Evaluate the following vector products by the direct method of Eq. (3.3.13):
   (a) $2\mathbf{i} \times (3\mathbf{i} + 4\mathbf{j} + 2\mathbf{k})$.
   (b) $3\mathbf{j} \times (2\mathbf{i} + 4\mathbf{j} + 5\mathbf{k})$.
   (c) $2\mathbf{k} \times (3\mathbf{i} + 5\mathbf{j} + \mathbf{k})$.
   (d) $(a\mathbf{i} + b\mathbf{k}) \times f(\frac{3}{5}\mathbf{j} + \frac{4}{5}\mathbf{k})$.

4. Find the area of the parallelogram formed on the sides of the two vectors $\mathbf{a} = 4\mathbf{i} + 3\mathbf{j} + 2\mathbf{k}$ in. and $\mathbf{b} = -2\mathbf{i} + 4\mathbf{j} + 3\mathbf{k}$ in.

5. Find a unit vector $\hat{\mathbf{n}}$ perpendicular to the two vectors of Ex. 4.

6. Find a unit vector $\hat{\mathbf{e}}$ parallel to the line of intersection of the two planes $x - y - z = 0$ and $y - 2z = 0$. (*Hint*: The unit vector must be perpendicular to the normals of the two planes.)

7. Force $\mathbf{f} = 600\mathbf{i} - 900\mathbf{j} + 300\mathbf{k}$ N acts at point $(2, -4, 8)$ m. Determine its moment (a) about point $A(4, 4, 2)$ m and (b) about the line from the origin to A.

8. Force $\mathbf{f} = 30\mathbf{i} + 10\mathbf{j} - 10\mathbf{k}$ lb acts at point $(3, -2, 3)$ ft. Find its moment (a) about the origin O and (b) about the line from O to $B(4, 2, 4)$ ft.

9. The 65-lb force $\mathbf{f}$ acts along the diagonal AB of one end of the $3 \times 4 \times 12$ ft rectanglar box. Determine its moment about the directed line CD.

10. The box shown is $1 \times 2 \times 2$ ft. The 1000-lb force $\mathbf{f}$ acts along AB. Determine its moment about (a) point D and (b) directed axis CD.

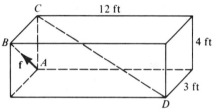

Exercise 9

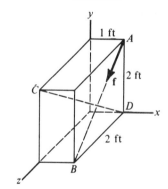

Exercise 10

11. Find the moment of the 100-lb force $\mathbf{f}$ about the directed axis AB. The force acts along CD.

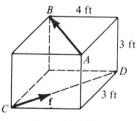

Exercise 11

12. For the data of Ex. 9, Sec. 3.2, determine m_{AB}.

13. The light flexible cable BC exerts a pull of 125 lb as shown, while cable BD exerts a pull of 141.4 lb. The dimensions of the "box" are 4 ft high by 4 ft in the x-direction by 3 ft in the z-direction. Find the total moment of these two forces about the axis OA lying in the xz-plane making a 60° angle with the negative x-axis.

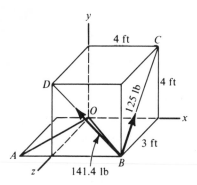

Exercise 13

14. Find the moment of the 100-lb force about point A (meaning about axis AZ perpendicular to the plane) without using the vector product.

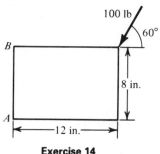

Exercise 14

15. Where should the 100-lb force of Ex. 14 act in order to have a positive (counterclockwise) moment of 320 lb-in. about A?

16. Find where the line of action of the force $\mathbf{f} = 6\mathbf{i} + 2\mathbf{j} - 2\mathbf{k}$ lb intersects the xy-plane if its moment about the origin is $\mathbf{m}_O = -4\mathbf{i} + 48\mathbf{j} + 36\mathbf{k}$ lb-ft.

17. Find where the line of action of the force $\mathbf{f} = -2\mathbf{i} - 3\mathbf{j} + 5\mathbf{k}$ N intersects the xy-plane if its moment about the origin is $\mathbf{m}_O = -10\mathbf{j} - 6\mathbf{k}$ N-m.

3.4 COUPLE MOMENT*

When a force acts on a rigid body, it tends to translate the body and possibly to rotate it. We shall now consider a type of action on a body that consists of a pure torque or pure moment, tending to twist or rotate the body without accelerating its center of mass. In practice such torques are often transmitted by drive shafts, where the moment about the shaft axis is due to distributed tangential forces on the cross section. We shall consider first a simpler case where the moment is produced by a pair of concentrated forces.

Definition: *Two forces* $\mathbf{f}$ *and* $-\mathbf{f}$ *having the same magnitude, parallel lines of action, and opposite senses are said to form a couple.* (See Fig. 3-18.) If the two forces $\mathbf{f}$ and $-\mathbf{f}$ act on the same body, they produce no acceleration of its mass center,

* The theory of couples was introduced in 1803 by L. Poinsot (1777–1859).

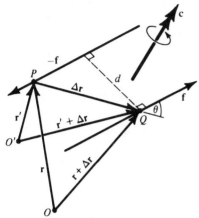

Fig. 3-18   Couple Moment

since their vector sum is zero. They do tend, however, to rotate the body, since the sum of their moments about any moment center is in general not zero. In Fig. 3-18 the total moment of the two forces about moment center O is

$$\mathbf{m}_O = \mathbf{r} \times (-\mathbf{f}) + (\mathbf{r} + \Delta\mathbf{r}) \times \mathbf{f} = \Delta\mathbf{r} \times \mathbf{f},$$

where $\mathbf{r}$ is the position vector of any point P on the line of action of $-\mathbf{f}$ and $\mathbf{r} + \Delta\mathbf{r}$ is the position vector of any point Q on $\mathbf{f}$. For any other moment center O', the total moment is

$$\mathbf{m}'_O = \mathbf{r}' \times (-\mathbf{f}) + (\mathbf{r}' + \Delta\mathbf{r}) \times \mathbf{f} = \Delta\mathbf{r} \times \mathbf{f}$$

also. This shows that the couple moment does not depend on the choice of moment center. ***The couple is a pure torque producing the same turning moment no matter what moment center is considered.*** We denote it by $\mathbf{c}$.

The turning moment of the couple about any directed axis AB is

$$m_{AB} = \mathbf{c} \cdot \hat{\mathbf{e}}_{AB},$$

since if we choose O on AB,

$$[\mathbf{r} \times (-\mathbf{f})] \cdot \hat{\mathbf{e}}_{AB} + [(\mathbf{r} + \Delta\mathbf{r}) \times \mathbf{f}] \cdot \hat{\mathbf{e}}_{AB} = (\Delta\mathbf{r} \times \mathbf{f}) \cdot \hat{\mathbf{e}}_{AB} = \mathbf{c} \cdot \hat{\mathbf{e}}_{AB}.$$

In summary,

Couple Moment

$$\mathbf{c} = \Delta\mathbf{r} \times \mathbf{f} \qquad\qquad (3.4.1a)$$

($\Delta\mathbf{r}$ *from* any point on $-\mathbf{f}$ *to* any point on $\mathbf{f}$).

$$|\mathbf{c}| = fd \qquad [c] = [FL]. \qquad\qquad (3.4.1b)$$

$$m_{AB} = \mathbf{c} \cdot \hat{\mathbf{e}}_{AB}. \qquad\qquad (3.4.1c)$$

The magnitude $|\mathbf{c}|$ equals fd, where $d = |\Delta r| \sin\theta$ is the perpendicular distance between the two forces.

Since the couple moment **c** has the same turning effect about all moment centers, we consider it to be a free vector. It can be drawn anywhere, but it should be perpendicular to the plane of **f** and −**f** and should have a sense indicating the sense of the turning effect produced by the two forces according to the right-hand rule. For this it is convenient to draw it between the two forces. In this book a couple vector will be represented by an arrow with a double point, as in Fig. 3-18. The sense of rotation will usually also be indicated by a curved arrow around the shaft, as in Fig. 3-18. With this representation the couple vector on a free-body diagram can be distinguished from a force. *Couple vectors and force vectors can never be added, since they have different dimensions. On a free-body diagram we would not show the two forces* **f** *and* −**f** *and also the couple vector. We show either the couple vector or the two forces, but not both* as in Fig. 3-18, which is not a free-body diagram but merely illustrates how the couple vector is related to the two forces that it represents.

A given couple vector can be produced by many different possible force pairs. Figure 3-19 illustrates several possible choices for **f**, −**f**, and *d* (all lying in one

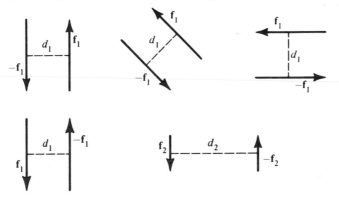

Fig. 3-19 Equivalent Couples in One Plane

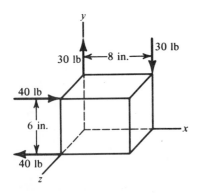

Fig. 3-20 Equivalent Couples in Parallel Planes

plane) that produce the same couple vector **c** (not shown) perpendicular to the plane. Four of the couples illustrated are formed with the same magnitudes $|\mathbf{f}_1|$ and d_1, illustrating that the force pair may be moved about in its plane in any way as long as the product fd and the sense of rotation produced by the two forces is unchanged. The fifth example in Fig. 3-19 has $|\mathbf{f}_2| = \frac{1}{2}|\mathbf{f}_1|$ and $d_2 = 2d_1$, giving the same couple moment. Evidently many more choices are possible, for example, $|\mathbf{f}_3| = m|\mathbf{f}_1|$ and $d_3 = d_1/m$ for any nonzero real number m.

The force pair may also be moved into another parallel plane without changing its couple moment. For example, the two force pairs in Fig. 3-20 are equivalent couples in the sense that they produce the same moment $\mathbf{c} = -240\mathbf{k}$ lb-in. about any moment center whatever. We shall see in Chapter 4 that when two couples have the same moment, they produce the same effect on a rigid body, as we would intuitively expect. The two force pairs in Fig. 3-20 would obviously **not** produce the same effect on a deformable body.

The couple vector produced by a given force pair can be calculated by Eq. (3.4.1a). Alternatively it can be determined by

$$\mathbf{c} = \pm fd\hat{\mathbf{e}}_n, \tag{3.4.2}$$

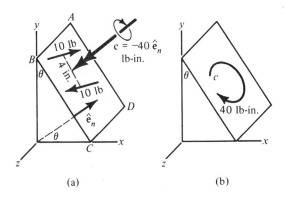

(a)   (b)

Fig. 3-21 Couple in Inclined Plane

where $\hat{\mathbf{e}}_n$ is a unit vector normal to the plane of the force pair, and the algebraic sign is selected to give $\mathbf{c}$ the proper sense according to the right-hand rule. For example, in Fig. 3-21, the two parallel forces of magnitude 10 lb are separated by a distance of 4 in. in plane $ABCD$ parallel to the z-axis. Evidently

$$\hat{\mathbf{e}}_n = \cos\theta\mathbf{i} + \sin\theta\mathbf{j} \tag{3.4.3}$$

and

$$\mathbf{c} = -40\hat{\mathbf{e}}_n$$
$$= (-40\cos\theta)\mathbf{i} - (40\sin\theta)\mathbf{j} \text{ lb-in.}, \tag{3.4.4}$$

where θ is the angle shown between the plane $ABCD$ and the vertical yz-plane. Since many different possible force pairs in a plane all produce the same couple, there is no point in actually showing the force pair. Instead *the couple is often represented by a curved arrow in the plane of the couple as in Fig. 3-21(b).*

Despite the fact that a couple was originally defined by a pair of parallel forces of opposite sense, this is seldom the way we meet a couple in practice. The couple is

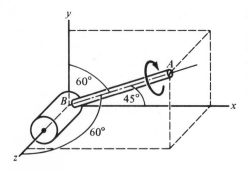

Fig. 3-22 Couple Transmitted by Shaft

usually given as the torque transmitted by a shaft. For example, in Fig. 3-22, we might be given the information that the shaft AB transmits a torque of 100 lb-ft to the cylinder at B, with the sense of the couple **acting on the cylinder** indicated by the curved arrow around the shaft. The vector representation of the couple is then

$$\mathbf{c} = c\hat{\mathbf{e}}_{AB} = -100\hat{\mathbf{e}}_{BA} \text{ lb-ft}$$

$$\hat{\mathbf{e}}_{BA} = \cos 45°\mathbf{i} + \cos 60°\mathbf{j} + \cos 60°\mathbf{k} \qquad (3.4.5)$$

$$\mathbf{c} = -70.7\mathbf{i} - 50\mathbf{j} - 50\mathbf{k} \text{ lb-ft}.$$

The relationship of the couple transmitted by a shaft to the concept of a couple as a pair of forces is illustrated in Fig. 3-23, which shows the cross section of a circular torsion bar. The distributed force acting across the cross section, exerted by the material in front of the pictured plane acting on material behind the plane, causes a couple to be transmitted across the plane but no net force in any direction. The distributed force on any infinitesimal area dA acts in a direction perpendicular to the radius drawn to dA. Figure 3-23(a) shows $d\mathbf{f}$ acting on dA_1 and $-d\mathbf{f}$ acting on the element dA_2 at the same distance r on the other side of O, creating an infinitesimal couple of magnitude $2r|d\mathbf{f}|$ and direction along the axis of the bar.

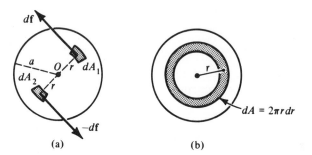

(a)   (b)

Fig. 3-23 (a) Elementary Fore-Pair Contribution to the Couple Transmitted by a Shaft; (b) Area Element for Single Integral

The theory of torsion for a circular bar* shows that the distributed force intensity τ (force per unit area) depends only on r and not on θ. Hence the total couple transmitted by the ring element of area $dA = 2\pi r\, dr$ shown in Fig. 3-23(b) is

$$dc = r\tau(r)\, dA, \qquad (3.4.6)$$

---

*See, for example, E. P. Popov, *Introduction to Mechanics of Solids* (Englewood Cliffs, N.J.: Prentice-Hall, Inc., 1968), Chap. 5.

and the total couple magnitude transmitted is

$$c = 2\pi \int_0^a r^2 \tau(r)\, dr. \tag{3.4.7}$$

The twisting couple c is often called the **torque** transmitted by the shaft. The tangential force intensity τ (force per unit area) is called the **torsional shear stress**. In an **elastic** torsion bar $\tau(r)$ is proportional to r, and the integral of Eq. (3.4.7) is easily evaluated; see Ex. 18 and 19.

Although in most practical problems couples do not present themselves as force pairs, the force-pair representation is a useful concept. It can be used to advantage in the following two operations: (1) moving a force to a new line of action without changing its effect on a rigid body or (2) combining a force with a couple perpendicular to the force.

Moving a force to a parallel line of action. We have already seen that sliding a force to a new position on the same line of action does not change its moment about any moment center. But moving it to a new line of action will change its moment and therefore change its rotational effect on a rigid body. Hence, such a move should never be made without compensating for the changed moment if we wish to avoid changing the effect on a rigid body. This compensation can be made by a couple. For example, in Fig. 3-24, if we wish to move force $\mathbf{f}$ from B to a parallel line of action through A, we can obviously place two equal and opposite forces $\mathbf{f}$ and $-\mathbf{f}$ at A without changing anything. But $-\mathbf{f}$ at A forms with $\mathbf{f}$ at B a couple

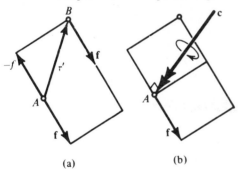

(a)   (b)

Fig. 3-24 Moving Force to Parallel Line

$$\mathbf{c} = \mathbf{r}' \times \mathbf{f}. \tag{3.4.8}$$

Hence the three forces of Fig. 3-24(a) can be represented by the force $\mathbf{f}$ at A and the couple $\mathbf{c}$ (a free vector that can be drawn anywhere) as in Fig. 3-24(b). *Notice that the couple $\mathbf{c}$ that must be added when we move the force to a new line of action through A is just equal to the moment* $\mathbf{m}_A = \mathbf{r}' \times \mathbf{f}$ *that the force produced about A before we moved it.* Since moment $\mathbf{m}_A$ about A is lost when we move $\mathbf{f}$ to A, we must add a couple $\mathbf{c} = \mathbf{m}_A$ to compensate for the lost moment. When we do this, the total moment about any other moment center is also unchanged, and therefore the effect on any rigid body is unchanged. (Of course the effect on a deformable body is changed by moving to a different point of application.)

Combining a force with a perpendicular couple. A couple vector cannot be added to a force vector, since they have different dimensions. But when the couple is represented as a force pair, then another force can be added to one of the forces of the pair. This operation is especially convenient and useful when the couple vector is perpendicular to the force vector. For example, in Fig. 3-25 the given couple vector **c** is perpendicular to the given force vector **f**. To combine the two we first represent the couple by forces **f** and −**f** distance $d = c/f$ apart, as shown in Fig. 3-25(b). The force −**f** is placed at A, so that it cancels the given force **f**, while the added force **f** is placed at B where it will form with −**f** a couple equal to **c**. This means that it is equal and parallel to the given force **f** on a line of action such that its moment about A equals **c**; see Sample Problem 3.4.5. [The perpendicular direction AB could be determined by $\hat{\mathbf{e}}_{AB} = (\mathbf{f} \times \mathbf{c})/|\mathbf{f} \times \mathbf{c}|$, if it is not clear from the sense of **c**, and the perpendicular distance is $d = c/f$. A simpler procedure is available, however, which is illustrated in Sample Problem 3.4.5.] The result of the combination is to replace the given force and couple by a single force equal and parallel to the given force, as shown in Fig. 3-25(c), without changing the effect on a rigid body. The given couple could have been represented by a curved arrow in the plane M perpendicular to **c** as in Fig. 3-21(b) instead of by the couple vector; this is the usual procedure in coplanar problems (see Sample Problem 3.4.2).

Section 3.5, in which we define the scalar triple product and two vector triple products, may be postponed or omitted.

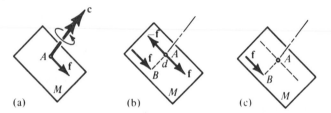

(a)   (b)   (c)

Fig. 3-25   Combining a Force with a Perpendicular Couple

SAMPLE PROBLEM 3.4.1

The block shown in Fig. 3-26 is acted upon by two couples. One couple is given as the pair of 100 N forces shown, acting in the planes of the front and back faces of the block, while

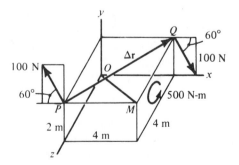

Fig. 3-26   Two Couples of Sample Problem 3.4.1

the second is a 500-N-m couple, represented by the curved arrow in the right-hand face. Determine the total couple vector and the total turning moment about the diagonal *OM*.

SOLUTION.   Let the force at Q be $\mathbf{f}$ and that at P be $-\mathbf{f}$. Then

$$\Delta \mathbf{r} = \overrightarrow{PQ} = 4\mathbf{i} - 4\mathbf{k} \text{ m}$$

$$\mathbf{f} = 50\mathbf{i} - 86.6\mathbf{j} \text{ N}$$

$$\mathbf{c}_1 = \Delta \mathbf{r} \times \mathbf{f} = -346.4\mathbf{i} - 200\mathbf{j} - 346.4\mathbf{k} \text{ N-m}$$

$$\mathbf{c}_2 = 500\mathbf{i} \text{ N-m} \qquad \text{(free vectors)}.$$

$$\text{total } \mathbf{c} = \mathbf{c}_1 + \mathbf{c}_2 = 153.6\mathbf{i} - 200\mathbf{j} - 346.4\mathbf{k} \text{ N-m} \quad \textbf{\textit{Answer.}}$$

$$\hat{\mathbf{e}}_{OM} = \tfrac{2}{3}\mathbf{i} + \tfrac{1}{3}\mathbf{j} + \tfrac{2}{3}\mathbf{k}$$

$$m_{OM} = \mathbf{c} \cdot \hat{\mathbf{e}}_{OM} = -195.2 \text{ N-m} \quad \textbf{\textit{Answer.}}$$

Remark: In Sample Problem 3.4.1 we could just as well have chosen $\mathbf{f}$ to be the force at P and $-\mathbf{f}$ to be the force at Q. Then we would have $\Delta \mathbf{r} = \overrightarrow{QP}$. Since the expressions for $\mathbf{f}$ and $\Delta \mathbf{r}$ would then both have their signs changed, the product $\mathbf{c}_1 = \Delta \mathbf{r} \times \mathbf{f}$ would be unchanged.

SAMPLE PROBLEM 3.4.2

For the coplanar problem shown in Fig. 3-27(a), combine the 200-lb-in. couple with the 50-lb force.

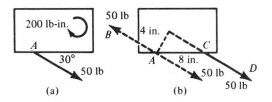

Fig. 3-27 Sample Problem 3.4.2

SOLUTION.   We represent the couple by two 50-lb forces as shown in Fig. 3-27(b) (represented by arrows *AB* and *CD*), separated by a perpendicular distance of 4 in. Then the two forces at *A* cancel, and the result is the single 50-lb force parallel to the given 50-lb force but now on a line intersecting the base at point *C*, 8 in. to the right of *A*.

SAMPLE PROBLEM 3.4.3

Two shafts enter the block shown in Fig. 3-28, each perpendicular to the face through which it enters and transmitting the torques $c_1 = 100$ lb-in. and $c_2 = 200$ lb-in. Find the total turning moment about the directed line OC.

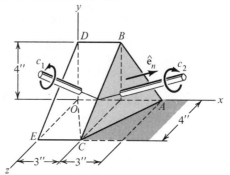

Fig. 3-28   Sample Problem 3.4.3

SOLUTION.   We represent the two couples transmitted as vectors acting on the block. Since the shaft of $\mathbf{c}_1$ makes equal angles with the y- and z-directions,

$$\mathbf{c}_1 = 100(\cos 45°\mathbf{j} + \cos 45°\mathbf{k})$$

$$\mathbf{c}_1 = 70.7\mathbf{j} + 70.7\mathbf{k} \text{ lb-in.}$$

We represent $\mathbf{c}_2$ as $\mathbf{c}_2 = -200\hat{\mathbf{e}}_n$. The normal direction can be determined from

$$\overrightarrow{AB} \times \overrightarrow{AC} = (-3\mathbf{i} + 4\mathbf{j}) \times (-3\mathbf{i} + 4\mathbf{k})$$

$$= 16\mathbf{i} + 12\mathbf{j} + 12\mathbf{k}$$

with magnitude $(256 + 144 + 144)^{1/2} = 23.32$. Thus

$$\hat{\mathbf{e}}_n = \frac{16\mathbf{i} + 12\mathbf{j} + 12\mathbf{k}}{23.32} = 0.6861\mathbf{i} + 0.5146\mathbf{j} + 0.5146\mathbf{k}$$

$$\mathbf{c}_2 = -137.2\mathbf{i} - 102.9\mathbf{j} - 102.9\mathbf{k} \text{ lb-in.}$$

The total couple transmitted is the free vector

$$\mathbf{c} = \mathbf{c}_1 + \mathbf{c}_2 = -137.2\mathbf{i} - 32.2\mathbf{j} - 32.2\mathbf{k} \text{ lb-in.}$$

$$\hat{\mathbf{e}}_{OC} = 0.6\mathbf{i} + 0.8\mathbf{k}$$

$$m_{OC} = \mathbf{c} \cdot \hat{\mathbf{e}}_{OC} = -108 \text{ lb-in.}   \textbf{\textit{Answer.}}$$

SAMPLE PROBLEM 3.4.4

Determine the total moment about the axis AH for the system of three forces and two couples shown acting on the block in Fig. 3-29.

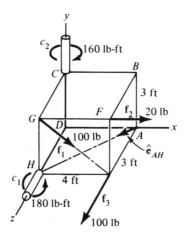

Fig. 3-29   Sample Problem 3.4.4

SOLUTION.   We first find the total vector moment $\mathbf{m}_A$:

$$\mathbf{m}_A = \mathbf{c}_1 + \mathbf{c}_2 + 3\mathbf{k} \times \mathbf{f}_1 + (3\mathbf{j} + 3\mathbf{k}) \times \mathbf{f}_2.$$

(Since $\mathbf{f}_3$ passes through A, it has no moment about A.)

$$\mathbf{f}_1 = 80\mathbf{i} - 60\mathbf{j} \text{ lb}, \qquad \mathbf{f}_2 = 20\mathbf{i} \text{ lb}$$
$$\mathbf{c}_1 = 180\mathbf{k} \text{ lb-ft}, \qquad \mathbf{c}_2 = -160\mathbf{j} \text{ lb-ft}.$$

Hence

$$\mathbf{m}_A = 180\mathbf{k} - 160\mathbf{j} + (240\mathbf{j} + 180\mathbf{i}) + (-60\mathbf{k} + 60\mathbf{j})$$

or

$$\mathbf{m}_A = 180\mathbf{i} + 140\mathbf{j} + 120\mathbf{k} \text{ lb-ft}$$
$$\hat{\mathbf{e}}_{AH} = -0.8\mathbf{i} + 0.6\mathbf{k} \qquad m_{AH} = \mathbf{m}_A \cdot \hat{\mathbf{e}}_{AH}$$
$$m_{AH} = -0.8(180) + 0 + 0.6(120) = -72 \text{ lb-ft} \quad \textbf{\textit{Answer.}}$$

SAMPLE PROBLEM 3.4.5

Combine the force $\mathbf{f} = 8\mathbf{i} + 12\mathbf{k}$ lb acting at $A(3, 0, 0)$ in. with the perpendicular couple $\mathbf{c} = 30\mathbf{i} + 60\mathbf{j} - 20\mathbf{k}$ lb-in. shown in Fig. 3-30.

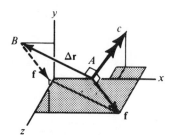

Fig. 3-30   Sample Problem 3.4.5

SOLUTION.   We first verify that $\mathbf{c}$ is perpendicular to $\mathbf{f}$, since

$$\mathbf{c} \cdot \mathbf{f} = 240 + 0 - 240 = 0.$$

Since they are perpendicular, we know that they can be combined into a single force equal to $\mathbf{f}$ on a line of action such that its moment about A equals $\mathbf{c}$. (See the discussion of Fig. 3-25.) Let $B(x, y, 0)$ be the point where the new line of action cuts the xy-plane. Then

$$\mathbf{m}_A = \Delta\mathbf{r} \times \mathbf{f} = [(x-3)\mathbf{i} + y\mathbf{j}] \times \mathbf{f}$$

$$\mathbf{m}_A = \begin{vmatrix} \mathbf{i} & \mathbf{j} & \mathbf{k} \\ x-3 & y & 0 \\ 8 & 0 & 12 \end{vmatrix} = 12y\mathbf{i} - 12(x-3)\mathbf{j} - 8y\mathbf{k}.$$

Equating this to $\mathbf{c}$ gives three component equations

$$12y = 30 \qquad -12(x-3) = 60 \qquad -8y = -20.$$

Hence

$$x = -2 \text{ in.,} \qquad y = 2.5 \text{ in.} \quad \textbf{\textit{Answer.}}$$

(Compare this with the solution to Sample Problem 3.3.4. The difference is that since we are taking moments about A instead of O, we use $\Delta\mathbf{r} = (x-3)\mathbf{i} + y\mathbf{j}$ for the moment arm.)

EXERCISES

1. A couple consists of a force $\mathbf{f} = 10\mathbf{i} - 12\mathbf{j} + 8\mathbf{k}$ lb acting at point $(8, 10, 5)$ ft and force $-\mathbf{f}$ acting at $(10, 7, 6)$ ft. Express the couple as a vector in standard form.

2. Three couples act on a body as follows: $\mathbf{c}_1 = 100\mathbf{i} - 200\mathbf{j} + 150\mathbf{k}$ lb-ft acts at $(2, 3, -1)$ ft, $\mathbf{c}_2 = 40\mathbf{i} + 60\mathbf{j} - 50\mathbf{k}$ lb-ft acts at $(5, 0, 2)$ ft, and $\mathbf{c}_3 = -70\mathbf{i} + 40\mathbf{j} + 25\mathbf{k}$ lb-ft acts at $(-2, -2, -2)$ ft. What single couple $\mathbf{c}$ will have the same turning effect about the point $(2, 2, 1)$ ft as the three given couples acting together?

3. A couple of magnitude 100 N-m acts as shown in a plane whose intercepts on the coordinate axes are each equal to a m. Express the couple vector in standard form. What would be the couple vector if the sense of rotation were reversed?

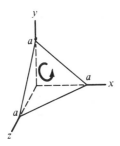

Exercise 3

4. The couple $\mathbf{c}$ of magnitude 250 lb-ft acts as shown in the plane *ABDE*. Express the couple vector in standard form. (Plane *ABDE* is parallel to the z-axis.) What would be the couple vector if the sense of rotation were opposite to that shown but with the same magnitude?

5. Find the turning moment of the couple of Ex. 1 about a directed axis from the origin to the point $(-1, 2, 2)$ ft.

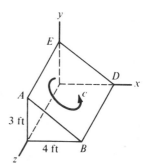

Exercise 4

6. Find the total moment about the origin produced by a force $\mathbf{f} = 70\mathbf{i} - 40\mathbf{k}$ N acting at $(0, -4, 3)$ m and a couple $\mathbf{c} = 200\mathbf{i} - 150\mathbf{j}$ N-m acting at $(3, -4, 0)$ m.

7. Find the total moment about the point $A(2, 2, 1)$ m produced by the force and couple of Ex. 6.

8. Find the total moment about the directed axis *OA* from the origin to the point $A(2, 2, 0)$ m produced by the force and the couple of Ex. 6.

9. Two shafts as shown transmit pure couples to the housing A. Each couple acts around the shaft, transmitting it in the sense shown. What couple must the foundation transmit to the housing so that the total turning effect on the housing should be zero?

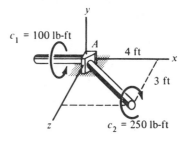

Exercise 9

10. (Coplanar problem.) If the force **f** shown at *B* is moved to *A*, what couple must be added to the given couple *c* so that the rigid-body effects will be unchanged?

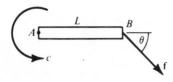

Exercise 10

11. If in the figure shown for Ex. 10, $\theta = 45°$, $f = 200$ lb, $c = 4000$ lb-in., and $L = 72$ in., what single force would have the same rigid-body effect?

12. The bracket shown is held by screws at *A* and *B*. (a) Replace the 100-lb load by a force-couple system at *A*. (b) Represent the couple of part (a) by two horizontal forces at *A* and *B*.

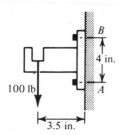

Exercise 12

13. (a) Replace the three forces shown by a single force without changing the rigid-body effect. (b) Where does its line of action intersect *AB*?

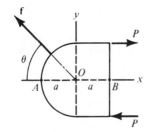

Exercise 13

14. Two couples are transmitted to the cylinder by the shafts *AB* and *OC* as shown. (*AB* is parallel to the *xy*-plane.) Determine the total turning moment about the axis from *O* to *D* of the two couples. (*D* is 2 ft behind the *xy*-plane.)

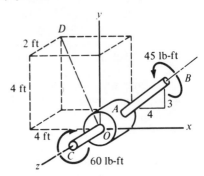

Exercise 14

15. Add the two couples shown, consisting of two 20-lb forces acting along the diagonals of the front and back faces of the box and the two 10-lb forces acting along the left and right face diagonals. The box is 3 ft high by 4 ft by 4 ft.

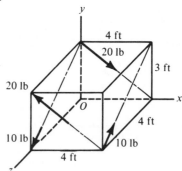

Exercise 15

16. The body shown is acted upon by a system of two forces and one couple. $\mathbf{f}_1 = 10\mathbf{i} + 40\mathbf{j} - 30\mathbf{k}$ lb; $\mathbf{f}_2$ has magnitude of 100 lb and makes angles of 60° with the *x*-direction, 60° with the negative *y*-direction, and 45° with the *z*-direction. (Note axis directions shown.) The couple c_1 of magnitude 100 lb-ft

acts in the inclined front face as shown. Replace the given system by one couple and one force acting at A without changing the rigid-body effects.

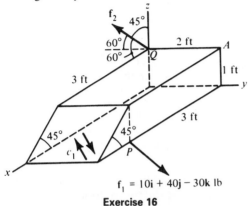

$$f_1 = 10i + 40j - 30k \text{ lb}$$

Exercise 16

17. Force $f_1 = 200i + 100j - 200k$ lb acts at $(2, 0, 4)$ ft. Where must $f_2 = -f_1$ intersect the xy-plane in order to form with f_1 a couple $c = -200i - 200k$ lb-ft?

18. In a uniform elastic circular shaft, the torsional shear stress τ is proportional to the distance r from the center: $\tau = Gr(d\varphi/dL)$, where G is the shear modulus and $d\varphi/dL$ is the angle of twist per unit length. Show that Eq. (3.4.7) then implies the torque-twist relationship $c = J_O G \, d\varphi/dL$, where J_O is the polar moment of inertia of the cross-sectional area defined by $J_O = \int_A r^2 \, dA = \frac{1}{2}\pi a^4$.

19. For a hollow circular shaft with inner radius a and outer radius b, derive formulas for c and J_O comparable to those given in Ex. 18 for a solid shaft, assuming that $\tau = Gr(d\varphi/dL)$ also in the hollow shaft.

20. Derive an expression for the magnitude c of the frictional torque of Coulomb friction on the end of a shaft of diameter D rotating in contact with a flat plate (as at the end of a dry thrust bearing) assuming that the total normal force N is uniformly distributed over the area A, so that $dN = N \, dA/A$ is the normal force on dA.

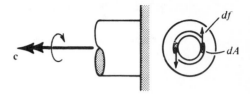

Exercise 20

21. Repeat Ex. 20 for a collar-type thrust bearing where the area of contact is the circular ring $a \leq r \leq b$.

22. As the surfaces of a thrust bearing wear out the friction decreases. It is often assumed that since the wear is proportional to the distance traveled, the normal pressure $p = dN/dA$ is inversely proportional to the distance from the center, $p = k/r$, where the constant k is chosen to make $\int p \, dA = N$. Show that this assumption leads to a predicted frictional torque equal to three-fourths that of Ex. 20.

23. Replace couple $c = 60i + 30k$ N-m and force $f = 2i + 3j - 4k$ N acting at $(5, 7, 0)$ m by a single force without changing the rigid-body effects. Where does the line of action of the resulting single force intersect the xy-plane?

24. Replace the two couples $c_1 = 36i - 20j$ lb-in., $c_2 = 20j - 36k$ lb-in. and the force $f = 3i - 2j + 3k$ lb acting at $(2, 2, 3)$ in. by a single force without changing rigid-body effects. Where does the single force intersect the xy-plane?

25. Force $f_1 = 12i + 18j + 12k$ lb acts at $(a, 0, 0)$ ft, force $-f_1$ acts at $(0, a, 0)$ ft, and force $f_2 = 36i + 36k$ lb acts at $(0, 0, a)$ ft. Replace the three forces by a single force without changing the rigid-body effects. Where does the line of action of the resulting force intersect the xy-plane?

*3.5   VECTOR TRIPLE PRODUCTS AND SCALAR TRIPLE PRODUCT

With three given vectors **a**, **b**, and **c** two different vector triple products can be defined,

$$\mathbf{a} \times (\mathbf{b} \times \mathbf{c}) \neq (\mathbf{a} \times \mathbf{b}) \times \mathbf{c}$$

because the vector product is not associative, as we saw in the example following Eq. (3.3.5). We state here without proof two identities involving the vector triple products:

$$\mathbf{a} \times (\mathbf{b} \times \mathbf{c}) = (\mathbf{a} \cdot \mathbf{c})\mathbf{b} - (\mathbf{a} \cdot \mathbf{b})\mathbf{c} \qquad (3.5.1)$$

$$(\mathbf{a} \times \mathbf{b}) \times \mathbf{c} = (\mathbf{a} \cdot \mathbf{c})\mathbf{b} - (\mathbf{b} \cdot \mathbf{c})\mathbf{a}. \qquad (3.5.2)$$

You can satisfy yourself of the correctness of the two identities by expressing each vector in standard form, e.g., $\mathbf{a} = a_x\mathbf{i} + a_y\mathbf{j} + a_z\mathbf{k}$, carrying out the multiplications and comparing the two sides of the equation.

The **scalar triple product** $\mathbf{a} \times \mathbf{b} \cdot \mathbf{c} = (\mathbf{a} \times \mathbf{b}) \cdot \mathbf{c}$ (often called the **mixed triple product**) can be written without parentheses, since it has meaning only when the cross product is performed first. (You cannot cross-multiply the vector **a** times the scalar **b·c**.) If **a**, **b**, and **c** form a right-handed system, the scalar triple product can be given a geometric interpretation as the volume of the parallelepiped formed on the three vectors as edges. Figure 3-31 shows such a parallelepiped. The area of its base is $A = |\mathbf{a} \times \mathbf{b}| = ab \sin \theta_1$, while its altitude $h = c \cos \theta_2$ is the projection of **c** onto the line of action $\mathbf{a} \times \mathbf{b}$ (positive for a right-handed system). Hence

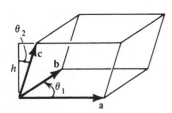

Fig. 3-31 Parallelepiped
Volume = **a** × **b** · **c**

$$(\mathbf{a} \times \mathbf{b}) \cdot \mathbf{c} = |\mathbf{a} \times \mathbf{b}| c \cos \theta_2$$

$$= Ah$$

$$= \text{volume of parallelepiped.}$$

Now the volume of the parallelepiped is the same no matter which of its faces is chosen for base. Thus

$$(\mathbf{a} \times \mathbf{b}) \cdot \mathbf{c} = (\mathbf{b} \times \mathbf{c}) \cdot \mathbf{a} = (\mathbf{c} \times \mathbf{a}) \cdot \mathbf{b} = \text{volume}, \qquad (3.5.3)$$

while if the same three vectors are chosen in an order forming a left-handed system, we get the negative of the volume:

$$(\mathbf{b} \times \mathbf{a}) \cdot \mathbf{c} = (\mathbf{c} \times \mathbf{b}) \cdot \mathbf{a} = (\mathbf{a} \times \mathbf{c}) \cdot \mathbf{b} = -\text{volume}. \qquad (3.5.4)$$

---

* This section may be omitted or postponed.

Because the scalar product is commutative,

$$(\mathbf{b} \times \mathbf{c}) \cdot \mathbf{a} = \mathbf{a} \cdot (\mathbf{b} \times \mathbf{c}),$$

whence the first Eq. (3.5.3) may be written as

$$(\mathbf{a} \times \mathbf{b}) \cdot \mathbf{c} = \mathbf{a} \cdot (\mathbf{b} \times \mathbf{c}),$$

or (omitting the parentheses)

$$\mathbf{a} \times \mathbf{b} \cdot \mathbf{c} = \mathbf{a} \cdot \mathbf{b} \times \mathbf{c}. \tag{3.5.5}$$

This is sometimes stated as *the dot and cross can be interchanged in the scalar triple product.*

By substituting for each of the three vectors in terms of components and carrying out the products, you can verify that

$$\mathbf{a} \times \mathbf{b} \cdot \mathbf{c} = \mathbf{a} \cdot \mathbf{b} \times \mathbf{c} = \begin{vmatrix} a_x & a_y & a_z \\ b_x & b_y & b_z \\ c_x & c_y & c_z \end{vmatrix}. \tag{3.5.6}$$

One possible application of the scalar triple product is the calculation of the algebraic turning moment m_{AB} about the directed axis AB. If A is any point on the axis and $\mathbf{r}'$ is the vector from A to any point on the line of action of $\mathbf{f}$, then

$$\mathbf{m}_A = \mathbf{r}' \times \mathbf{f} \qquad \text{by Eq. (3.3.17)}$$

and

$$m_{AB} = \mathbf{m}_A \cdot \hat{\mathbf{e}}_{AB} \quad \text{by Eq. (3.3.21).}$$

These can be combined into

$$m_{AB} = \mathbf{r}' \times \mathbf{f} \cdot \hat{\mathbf{e}}_{AB} \tag{3.5.7}$$

if desired. This procedure, however, offers no advantage over the two-step calculation by Eqs. (3.3.17) and (3.3.21) in most cases. Indeed the two-step calculation may be preferred because it gives both $\mathbf{m}_A$ and m_{AB}.

In Chapter 4 we shall consider the reduction of equipollent force systems (rigid-body equivalent systems) to a resultant force and/or a resultant couple. The consideration of distributed parallel force systems will lead to the concepts of center of mass, center of gravity, and centroid.

EXERCISES

1. Evaluate $\mathbf{a} \times (\mathbf{b} \times \mathbf{c})$ for $\mathbf{a} = 7\mathbf{i} - 2\mathbf{k}$, $\mathbf{b} = 6\mathbf{i} - \mathbf{j} - 2\mathbf{k}$, $\mathbf{c} = 2\mathbf{i} + 2\mathbf{j}$.
2. Evaluate $(\mathbf{a} \times \mathbf{b}) \times \mathbf{c}$ for the three vectors of Ex. 1.
3. If $\mathbf{a}$ is any vector and $\hat{\mathbf{e}}$ is any unit vector, show that

$$\mathbf{a} = (\mathbf{a} \cdot \hat{\mathbf{e}})\hat{\mathbf{e}} + \hat{\mathbf{e}} \times (\mathbf{a} \times \hat{\mathbf{e}}).$$

This gives a resolution of $\mathbf{a}$ into components parallel and perpendicular to $\hat{\mathbf{e}}$.

4. Evaluate the triple product $\mathbf{a} \times \mathbf{b} \cdot \mathbf{c}$ if (a) $\mathbf{a} = 3\mathbf{i} - 2\mathbf{j} + 5\mathbf{k}$, $\mathbf{b} = \mathbf{i} - \mathbf{j} - \mathbf{k}$, $\mathbf{c} = \mathbf{i} + \mathbf{j} + \mathbf{k}$, (b) $\mathbf{a} = \mathbf{i} - \mathbf{j} + \mathbf{k}$, $\mathbf{b} = -6\mathbf{i} + 6\mathbf{j} - 6\mathbf{k}$, $\mathbf{c} = -2\mathbf{j} + 3\mathbf{k}$.
5. Evaluate the triple product $\mathbf{u} \cdot \mathbf{v} \times \mathbf{w}$ if (a) $\mathbf{u} = \mathbf{i} - \mathbf{j} - \mathbf{k}$, $\mathbf{v} = 3\mathbf{i} - 2\mathbf{j} + 5\mathbf{k}$, $\mathbf{w} = -2\mathbf{i} + 4\mathbf{j} + 4\mathbf{k}$, (b) $\mathbf{u} = \mathbf{i} + 2\mathbf{j} + \mathbf{k}$, $\mathbf{v} = -2\mathbf{i} - 4\mathbf{j} - 2\mathbf{k}$, $\mathbf{w} = \mathbf{i} + \mathbf{j} + \mathbf{k}$.
6. Use the scalar triple product to solve the following exercises from Sec. 3.3: (a) 8(b), (b) 9, (c) 10(b), (d) 11).
7. Verify Eq. (3.5.1) as suggested after Eq. (3.5.2).
8. Verify Eq. (3.5.2) as suggested.

3.6 SUMMARY

The *scalar product* $\mathbf{a} \cdot \mathbf{b}$ was defined in Sec. 3.2 and applied to projection:

$$\mathbf{a} \cdot \mathbf{b} = ab \cos \theta, \tag{3.2.1}$$

where θ is the angle between the directions of $\mathbf{a}$ and $\mathbf{b}$. In terms of components,

$$\mathbf{a} \cdot \mathbf{b} = a_x b_x + a_y b_y + a_z b_z. \tag{3.2.9}$$

The projection a_b of $\mathbf{a}$ onto $\mathbf{b}$ is given by

$$a_b = \mathbf{a} \cdot \frac{\mathbf{b}}{b}. \tag{3.2.10}$$

An additional application calculating the work done by a force will be given in Chapter 6.

The *vector product* $\mathbf{c} = \mathbf{a} \times \mathbf{b}$ has magnitude

$$c = ab \sin \theta \tag{3.3.2}$$

and direction perpendicular to the plane of $\mathbf{a}$ and $\mathbf{b}$ in the sense that makes $\mathbf{a}$, $\mathbf{b}$, $\mathbf{c}$ a right-handed system; see Figs. 3-7 and 3.8. In terms of components,

$$a \times b = \begin{vmatrix} i & j & k \\ a_x & a_y & a_z \\ b_x & b_y & b_z \end{vmatrix}. \tag{3.3.10}$$

The vector moment m_A of a force f about moment center A is defined by

$$m_A = r' \times f, \tag{3.3.17}$$

where r' is the relative position vector *from A to any point on the line of action of* f. The rectangular components of m_A are the algebraic turning moments about axes through A parallel to the x, y, z-axes. The turning moment m_{AB} about any directed axis AB through A is given by the projection

$$m_{AB} = m_A \cdot \hat{e}_{AB}. \tag{3.3.21}$$

A *couple* is a pure turning moment which has no translatory effect. If the couple is formed by two noncollinear forces f and $-f$, then the couple moment vector c is given by

$$c = \Delta r \times f, \tag{3.4.1a}$$

where Δr is a vector from any point on the line of action of $-f$ to any point on the line of action of f. A couple is a free vector.

All the results cited so far in this summary will be used extensively. They should be memorized and their use thoroughly understood. You should also master the technique of adding a couple to maintain rigid-body equivalence when you move a force to a parallel line of action and the technique of combining a force with a perpendicular couple, as discussed at the end of Sec. 3.4. The triple products of Sec. 3.5 will not be used much in Vol. I. They will be used for some of the derivations in Vol. II for rigid-body rotation.

Chapter 4 begins with a section on rigid-body equivalence of force systems and the reduction of a given force system to a simpler equivalent system (resultant), making use of the concepts and methods developed in Secs. 3.3 and 3.4. Resultants for distributed force systems will be considered in Sec. 4.2, with application to locating the center of gravity. Then the methods of analysis for rigid-body equilibrium will be presented, making use of the concepts and techniques of Chapter 3 in addition to the particle statics of Chapter 2.

CHAPTER FOUR

Rigid-Body Statics

4.1 EQUIPOLLENCE (RIGID-BODY EQUIVALENCE); TRANSMISSIBILITY; REDUCTION TO RESULTANTS

The sum of the external forces and the sum of the external moments completely determine the state of accelerated motion or equilibrium of a rigid body, as we shall see in Vol. II. Hence, any two different systems of external forces and couples will have the same effect on a rigid body provided that they are *equipollent*, according to the following definition:

Definition: Two sets of forces and couples are *equipollent* if (1) the vector sum of the forces of the first set is equal to the vector sum of the forces of the second set, and (2) the vector sum of the moments of the forces and couples of the first set about an arbitrary point is equal to the vector sum of the moments of the forces and couples of the second set about the same point.

Equipollent force systems are sometimes called "statically equivalent," but it should be understood that, insofar as the external effects on a rigid body are concerned, equipollent force systems are in fact **dynamically equivalent**, while for internal effects (deformations) in a deformable body, equipollent systems are not even statically equivalent. If you object to the technical term **equipollent**, you would do better to call the systems "rigid-body equivalent" instead of "statically equivalent."

Rigid-body equilibrium will be treated in Secs. 4.3 and 4.4. In this section we shall consider equipollent systems of forces and couples and the reduction to simpler systems (resultants). In Sec. 4.2 we shall apply the methods developed here to distributed force systems.

Transmissibility. The external effects of a force on a rigid body are independent of where the force acts on its given line of action. This statement, often called the **principle of transmissibility**, is an immediate consequence of the fact that sliding a force to a new point on its line of action does not change its contribution to the vector sum of the forces or its moment about any point. The changed force system is equipollent to the system before the sliding occurred. For rigid-body purposes, force may be considered a **sliding vector**. (We may have to imagine a rigid arm attached to the rigid body if we slide the force to a point outside the body itself.)

Change of moment center. In the definition of equipollence the moment sums were required to be equal for the two sets of forces and couples about an **arbitrary point** in space. Of course the moment sums about a different moment center will be different, but the new moment sums for the two equipollent systems must be equal to each other. We shall show that if the two moment sums about one moment center are equal, then the two moment sums about any other center will also be equal, provided that the two force sums are equal.

First, we shall show how the moment sum for one system of forces and couples changes when we change from moment center A to a new moment center B in Fig. 4-1, where the system consists of two concentrated forces and two couples. The same method can be extended to any finite number of forces and couples and to distributed forces as well. Let $\mathbf{d}_{BA}$ be the vector from B to A. Then the moment arm from the new moment center B to P_k is $\mathbf{d}_{BA} + \mathbf{r}'_k$. The couples are free vectors with the same moment about both centers. Thus

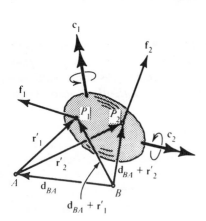

Fig. 4-1   Change of Moment Center

$$\sum \mathbf{m}_A = \mathbf{c}_1 + \mathbf{c}_2 + \mathbf{r}'_1 \times \mathbf{f}_1 + \mathbf{r}'_2 \times \mathbf{f}_2$$

$$\sum \mathbf{m}_B = \mathbf{c}_1 + \mathbf{c}_2 + (\mathbf{d}_{BA} + \mathbf{r}'_1) \times \mathbf{f}_1 + (\mathbf{d}_{BA} + \mathbf{r}'_2) \times \mathbf{f}_2,$$

whence,

Change of Moment Center
$$\sum \mathbf{m}_B = \sum \mathbf{m}_A + \mathbf{d}_{BA} \times (\sum \mathbf{f}). \qquad (4.1.1)$$

It follows from this that if two systems have the same $\sum \mathbf{f}$ and the same $\sum \mathbf{m}_A$, they will also have the same $\sum \mathbf{m}_B$ (different from $\sum \mathbf{m}_A$ in general). ***To prove equipollence it is therefore only necessary to consider one moment center.***

Reduction of a System to One Force and One Couple at an Assigned Point

The replacement of one system of forces and couples by a simpler equipollent system is called ***reduction***. In the general three-dimensional case the most useful simplification is to reduce the given system to a system consisting of one couple and one force acting at an assigned point A, called the ***resultant force and couple at*** A, denoted in this book by $\mathbf{f}^R$ and $\mathbf{c}_A^R$. (Other notations used by some authors are $\mathbf{R}$ for $\mathbf{f}^R$ and $\mathbf{M}$ or $\mathbf{M}_A^R$ for $\mathbf{c}_A^R$.) If this ***resultant set*** is to be equipollent to the given set, it is sufficient that

---

Resultant Force and Couple at A

$$\mathbf{f}^R = \sum \mathbf{f} \qquad (4.1.2)$$

$$\mathbf{c}_A^R = \sum \mathbf{m}_A = \sum \mathbf{c} + \sum (\mathbf{r}' \times \mathbf{f}), \qquad (4.1.3)$$

---

where $\mathbf{r}'$ is the vector from A to any point on the line of action of $\mathbf{f}$. ***The summations are to be understood in the general sense of including integrals when some of the forces are distributed forces.*** In the examples of this section we shall consider only concentrated forces. Some distributed force examples will be given in later sections. Notice that the resultant force $\mathbf{f}^R$ does not depend on the choice of point A but that the resultant couple does depend on the point. By Eq. (4.1.1), the resultant set at a different point would be

$$\mathbf{f}^R = \sum \mathbf{f}$$
$$\mathbf{c}_B^R = \mathbf{c}_A^R + \mathbf{d}_{BA} \times \mathbf{f}^R. \qquad (4.1.4)$$

When the resultant force $\mathbf{f}^R$ and the resultant couple $\mathbf{c}_A^R$ at a point happen to be perpendicular, they can be combined and replaced by a single force at another point, as we saw in connection with Fig. 3-25. (See also Sample Problem 3.4.5). Although this leads to a resultant that might be considered simpler than the resultant force and couple at A, for most purposes it is more useful to prescribe the point A as a physically identifiable point on the body or on the reference frame than to make the additional simplification to a single force. Some exceptions occur in the special cases of coplanar systems and of parallel force systems where it is especially easy to make the reduction to a single force.

Wrench resultant. When the resultant force and resultant couple are parallel they are said to constitute a **wrench**, and no further reduction is possible. If the force and couple vectors have the same sense along their line of action, the wrench is called a **positive wrench**; if they have opposite senses, it is a negative wrench (see Fig. 4-2). The most general set of forces and couples can always be reduced to a wrench, since c_A^R can be resolved into the sum of a vector component perpendicular to f^R and a component parallel to f^R. The perpendicular component can then be combined with f^R to give a single force at another point A', equal to f^R. The parallel component then forms with f^R a wrench that may be considered as a force along a line through A' and a twist about the same line. Although any force system can be reduced to a wrench, we shall usually prefer to stop the reduction with the resultant set f^R, c_A^R at a chosen point A.

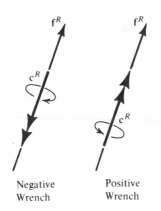

Negative Wrench   Positive Wrench

Fig. 4-2   Wrench Resultant

Special cases. In certain special cases it is easy and useful to give a simpler resultant system than the force-couple pair at an assigned point.

1. **Resultant is a couple.** When $\sum f = 0$, the resultant is simply c^R, a free vector.
2. **Concurrent forces** at A (no couples). The resultant at A is simply $f^R = \sum f$.
3. **Coplanar system.** All forces in the xy-plane and all couples perpendicular to the plane. If $\sum f \neq 0$, the resultant at A is in general a force in the plane and a couple perpendicular to it, which are easily combined into a single force acting at another point. If $\sum f = 0$, the resultant is a couple perpendicular to the plane.
4. **Parallel force system.** If all the forces have the same sense, the resultant at A is in general a force and a couple, but they are easily combined into a single force acting at another point. If some of the forces have opposite senses, we could have $\sum f = 0$ so that the resultant could be a single couple (free vector) as in case 1.

Sample Problem 4.1.1 illustrates the general case and uses the cross product to calculate moments. The other two sample problems are concerned either with a coplanar system or with a parallel system, where (when $\sum f \neq 0$) we reduce the system to a single force. The solutions given for these special cases also illustrate the calculation of moments about axes without using the cross product.

In Sec. 4.2 we shall treat examples of distributed force systems and the concepts of center of mass and center of gravity.

SAMPLE PROBLEM 4.1.1

Reduce the force system consisting of two couples and three forces, shown acting on the block in Fig. 4-3, to an equipollent force-couple resultant at A.

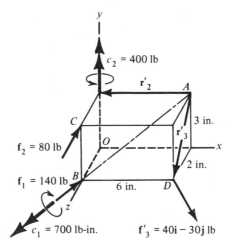

Fig. 4-3   Sample Problem 4.1.1

SOLUTION

$$\hat{\mathbf{e}}_{BA} = \frac{6\mathbf{i} + 3\mathbf{j} - 2\mathbf{k}}{7}$$

$$\mathbf{f}_1 = 140\mathbf{e}_{BA} = 120\mathbf{i} + 60\mathbf{j} - 40\mathbf{k}\ \text{lb}$$

$$\mathbf{f}_2 = \qquad\qquad -80\mathbf{k}\ \text{lb}$$

$$\mathbf{f}_3 = \qquad\quad 40\mathbf{i} - 30\mathbf{j}\ \text{lb}$$

$$\mathbf{f}^R = \sum \mathbf{f} = 160\mathbf{i} + 30\mathbf{j} - 120\mathbf{k}\ \text{lb}\quad \textbf{\textit{Answer.}}$$

$$\mathbf{c}_A^R = \mathbf{c}_1 + \mathbf{c}_2 + \mathbf{r}_2' \times \mathbf{f}_2 + \mathbf{r}_3' \times \mathbf{f}_3.$$

Choose $\mathbf{r}_1' = 0$, $\mathbf{r}_2' = -6\mathbf{i}$ in., and $\mathbf{r}_3' = -3\mathbf{j} + 2\mathbf{k}$ in.

$$\mathbf{c}_1 = -600\mathbf{i} - 300\mathbf{j} + 200\mathbf{k}\ \text{lb-in.}$$

$$\mathbf{c}_2 = \qquad\quad 400\mathbf{j}\ \text{lb-in.}$$

$$\mathbf{r}_2' \times \mathbf{f}_2 = \quad -480\mathbf{j} \text{ lb-in.}$$

$$\mathbf{r}_3' \times \mathbf{f}_3 = \quad 60\mathbf{i} + \ 80\mathbf{j} + 120\mathbf{k} \text{ lb-in.}$$

$$\mathbf{c}_A^R = -540\mathbf{i} - 300\mathbf{j} + 320\mathbf{k}\text{lb-in.} \quad \textbf{\textit{Answer.}}$$

SAMPLE PROBLEM 4.1.2

A block is loaded by the coplanar system of four forces and two couples shown in Fig. 4-4(a). Determine (a) the resultant at A and (b) a single force resultant. Where does the single force resultant intersect AD?

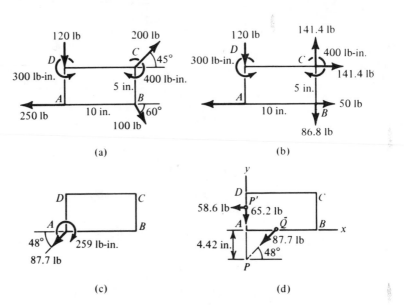

(a)   (b)

(c)   (d)

Fig. 4-4 Sample Problem 4.1.2

SOLUTION.   (a) *Force-couple resultant at A*. We can first resolve the forces at B and C into components as shown in Fig. 4-4(b). Then

$$f_x^R = \Sigma f_x = -250 + 141.4 + 50 = -58.6 \text{ lb}$$

$$f_y^R = \Sigma f_y = -120 + 141.4 - 86.6 = -65.2 \text{ lb.}$$

or

$$\mathbf{f}^R = 87.7 \text{ lb at } 48° \text{ to } BA$$

as shown in Fig. 4-4(c).

$$C_{AZ}^R = \sum C_z + \sum m_{AZ}$$
$$= 300 - 400 + 10(141.4) - 5(141.4) - 10(86.6)$$
$$= -259 \text{ lb-in.}$$

(The moment magnitudes were calculated by $m = fd$, and signs were prefixed to give the correct sense. Alternatively, the formula $m_z = xf_y - yf_x$ could be used.)

(b) *Single-force resultant.* We represent the resultant $\mathbf{f}^R$ by drawing its two components at unknown point P on AD, say y in. above A, as shown at P' in Fig. 4-4(d). There its moment about A will be $58.6y$ lb-in. Hence, for equipollence to the resultant at A,

$$58.6y = -259 \text{ lb-in.}$$
$$y = -4.42 \text{ in.}$$

Hence the point P is actually 4.42 in. below A, and the line of action of the single 87.7-lb force equipollent to the given system is the line QP shown.

SAMPLE PROBLEM 4.1.3

Five forces act parallel to the z-axis at points A, B, C, D, and O as shown in Fig. 4-5. Reduce the system to a single force. (Coordinates are in feet.)

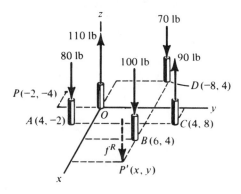

Fig. 4-5 Sample Problem 4.1.3

SOLUTION

Force at	f_z	m_x	m_y
A	-80	160	320
B	-100	-400	600
C	$+90$	$+720$	-360
D	-70	-280	-560
O	$+110$	0	0
Total	-50 lb	$+200$ lb-ft	-100 lb-ft

$$\mathbf{f}^R = \sum \mathbf{f} = -50\mathbf{k} \text{ lb.}$$

Assume that $\mathbf{f}^R$ acts at point $P'(x, y)$ shown for positive x and y by the dashed arrow in Fig. 4-5. To be equipollent to the given system it must have the same moment about O as the given system and therefore the same moments m_x and m_y as the given system. Thus

$$-50y = 200, \qquad 50x = -100,$$

$$y = -4 \text{ ft}, \qquad x = -2 \text{ ft.}$$

The resultant is therefore a 50-lb downward force through $P(-2, -4)$ ft   ***Answer.***

EXERCISES

1. (a) Reduce the given system of three forces and two couples to an equipollent resultant at A. $f_1 = 80$ lb, $f_2 = 100$ lb, $f_3 = 50$ lb, $c_1 = 300$ lb-ft,   and $c_2 = 250$ lb-ft. The couples act in the right face and the front face of the "box" as shown. (b) What is the moment of this resultant about the axis AD?

2. Given the answer of Ex. 1, what is the equipollent resultant at B?

3. The block shown is acted upon by forces of 40 and 50 lb along two edges and a 100-lb force along the diagonal of the top. In addition a

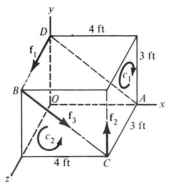

Exercise 1

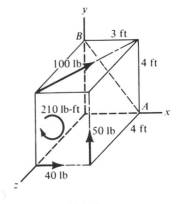

Exercise 3

couple of magnitude 210 lb-ft acts in the front
face of the block with the sense indicated. (a)
Determine a resultant system consisting of a
single force at *A* and a couple. (b) What is the
turning moment of this resultant about the
axis *AB*?

4. The 14-ft beam is acted on by the three
parallel forces and the couple shown.
Reduce this system to a force-couple combination (a) at *A* and (b) at *B*. (c) Replace the
given system by an equipollent single force,
and determine where its line of action intersects *AB*.

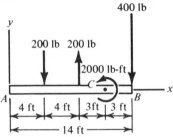

Exercise 4

5. Solve parts (a) and (b) of Ex. 4 with the 2000-
lb-ft couple replaced by an upward force of
400 lb at *C*.

6. (a) Reduce the coplanar system shown to a
force-couple combination at *A*. (b) Determine a single force equipollent to the given
system and find where its line of action crosses
AB.

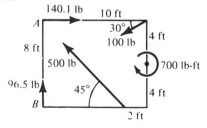

Exercise 6

7. Determine a single force equipollent to the
system of four forces shown parallel to the
y-axis. Each square in the *xz*-plane is 1 m on
a side.

8. Reduce the system of four forces and two
couples to a force-couple system at point *A*.

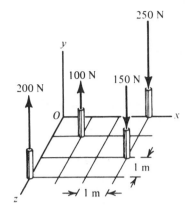

Exercise 7

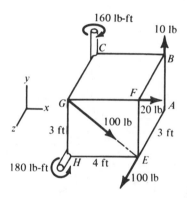

Exercise 8

9. (a) Reduce the system shown (two forces and
two couples) to an equipollent system at the
point *A*, consisting of a single force and a
single couple. (b) Find the total turning
moment about the axis *AB*.

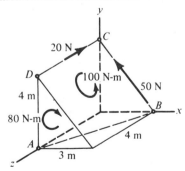

Exercise 9

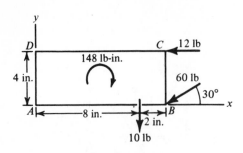

Exercise 11

10. The frame is acted upon by a positive wrench and a negative wrench as shown. (a) Find the resultant at A, and (b) determine the turning moment of this resultant about the directed axis OA.

11. The block $ABCD$ is acted upon by the three forces shown and the clockwise couple of 148 lb-in. Find a single force equivalent to the given system, and locate the point where its line of action intersects AB.

12. What is the resultant in Ex. 7 if the 100-N force is replaced by a 200-N force?

13. Determine the wrench equipollent to a couple $c = 14i + 2j - 6k$ N-m and a force $f = 7i + 4j - k$ N acting at the origin.

14. Determine the single force equipollent to six forces parallel to the vertical z-axis as follows: 5000 lb downward at the origin, 6000 lb downward at $(20, 0, 0)$ ft, 8000 lb upward at $(15, 6, 0)$ ft, 4000 lb downward at $(7, 15, 0)$ ft, 7000 lb downward at $(20, 22, 0)$ ft, and 9000 lb downward at $(0, 22, 0)$ ft.

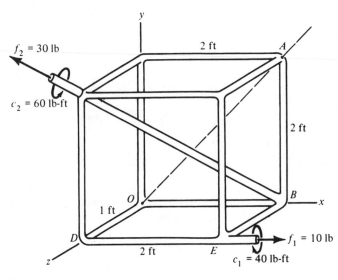

Exercise 10

4.2. DISTRIBUTED PARALLEL FORCE SYSTEMS; CENTER OF MASS; CENTROID; CENTER OF GRAVITY

When a beam is modeled as a one-dimensional body, a distributed load is specified as a *force intensity*, that is, a force per unit length. In the following two examples the beam lies along the x-axis, and the downward force per unit length is denoted by $w(x)$, so that the downward force df on a length dx is given by $w(x)\,dx$. The first example is for a uniform force distribution, $w(x) = \text{const.}$, but in general the force intensity varies with x. Such a distributed parallel force system can be replaced by a single force, equivalent to the given distributed force system for the purposes of rigid-body mechanics, as the following two examples illustrate.

EXAMPLE 4.2.1

The beam AB in Fig. 4-6 carries a uniformly distributed downward load of intensity $w(x) = 100\ \text{lb/ft}$ on its left 12 ft. We replace the distributed load by an equipollent single force as follows: We choose the y-axis downward; then $f_y = f$, and

$$f^R = \Sigma f_y = \int_0^{12} 100\,dx = 1200\ \text{lb}.$$

Since the perpendicular moment arm is x for the elemental downward force $df = 100\,dx$ acting on dx, the moment of this elemental force is $x(df) = 100x\,dx$, and the total clockwise moment about A is

$$+\,\Big)\ \Sigma\,m_A = \int_0^{12} 100x\,dx = 7200\ \text{lb-ft}.$$

Hence f^R must act at x_R ft to right of A, where

$$1200x_R = 7200 \quad \text{or} \quad x_R = 6\ \text{ft} \quad \textbf{\textit{Answer.}}$$

The equipollent force is the 1200-lb downward force shown dashed in the middle of the distributed load diagram (equal to the "area" of the load diagram: 12 ft times 100 lb/ft, as we would intuitively expect).

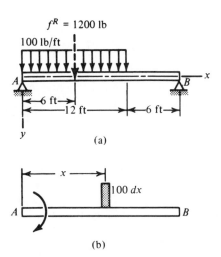

(a)

(b)

Fig. 4-6   Resultant of Uniform Parallel Force Distribution

EXAMPLE 4.2.2

The same 18-ft beam AB is loaded by a triangular load distribution on its left 12 ft, as shown in Fig. 4-7. We replace this by an equipollent single force as follows: The load intensity $w(x)$ varies linearly from zero at $x = 0$ to 300 lb/ft at $x = 12$ ft. By similar triangles,

$$\frac{w(x)}{300} = \frac{x}{12} \quad \text{or} \quad w(x) = 25x$$

$$f^R = \sum f_y = \int_0^{12} 25x \, dx = 1800 \text{ lb}$$

[again equal to the "area" of the load diagram: $\frac{1}{2}(12)(300)$]. For equipollence,

$$1800x_R = + \,\rangle\, \sum m_A = \int_0^{12} xw(x)\, dx = \int_0^{12} 25x^2\, dx = 14{,}400 \text{lb-ft}$$

$$x_R = 8 \text{ ft.}$$

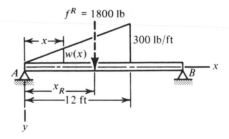

Fig. 4-7   Resultant of Triangular Load Distribution

Warning: Although the distributed load in each of these examples is equipollent to a single concentrated force, the single force will not produce the same deflections of the beam as the given distributed force, because beam deflection is not a rigid-body effect.

Generalization. When a beam is loaded by a distributed parallel load $q_y(x)$, force acting in the y-direction, per unit length in the x-direction, on the interval $a < x < b$, the resultant is a concentrated force $f_y^R \mathbf{j}$ of magnitude equal to the "area" under the load curve, located at x_R, such that

$$f_y^R = \int_a^b q_y(x)\, dx \qquad x_R f_y^R = \int_a^b xq_y(x)\, dx. \qquad (4.2.1)$$

For rigid-body purposes the distributed load may be replaced by its resultant. (The abscissa x_R of the resultant is equal to the coordinate $\bar{x}$ of the centroid of the load diagram, defined by $\bar{x}A = \int_a^b x\,dA$; see Sec. 7.3.) In the two examples above, $q_y(x) = w(x)$, since the y-axis was downward. If we had chosen the y-axis upward, we would have written $q_y(x) = -w(x)$, since $w(x)$ was specified as downward.

Center of Mass

The mass center of a collection of N mass-point particles of masses $m_1, m_2, \ldots, m_N$ and position vectors $\mathbf{r}_1, \mathbf{r}_2, \ldots, \mathbf{r}_N$ is the point with position $(\bar{x}, \bar{y}, \bar{z})$ defined by

---

Center of Mass

$$m\bar{x} = \sum_{k=1}^{N} m_k x_k, \qquad m\bar{y} = \sum_{k=1}^{N} m_k y_k, \qquad m\bar{z} = \sum_{k=1}^{N} m_k z_k, \qquad (4.2.2)$$

where

$$m = m_1 + m_2 + \cdots + m_N \qquad (4.2.3)$$

or

$$m\bar{\mathbf{r}} = \sum_{k=1}^{N} m_k \mathbf{r}_k, \qquad \text{where} \quad \bar{\mathbf{r}} = \bar{x}\mathbf{i} + \bar{y}\mathbf{j} + \bar{z}\mathbf{k}. \qquad (4.2.4)$$

The term $m_k x_k$ in the first sum is the ***first moment*** of the mass m_k with respect to the yz-plane, the mass m_k multiplied by its algebraic moment arm measured from the yz-plane. The abscissa $\bar{x}$ is the abscissa where the total mass m of the collection would have to be located in order to have the same total first moment. Similar interpretations apply to $\bar{y}$ and $\bar{z}$.

In the continuous mass distribution model of a body the mass of each infinitesimal volume element is $dm = \rho\,dV$, where $\rho = \rho(x, y, z, t)$ is the density or mass per unit volume. The **mass** m of a body occupying volume V is given by the volume integral

$$m = \int_V dm = \int_V \rho\,dV. \qquad (4.2.5)$$

The volume integral may be evaluated at any time when ρ is known as a function of position by introducing a coordinate system, for example, a rectangular Cartesian system where the volume element dV is a rectangular block with sides dx, dy, dz,

$$dV = dx\, dy\, dz, \tag{4.2.6}$$

as in Fig. 4-8. Then the volume integral becomes an iterated *triple integral*

$$m = \iiint_V \rho(x, y, z)\, dx\, dy\, dz. \tag{4.2.7}$$

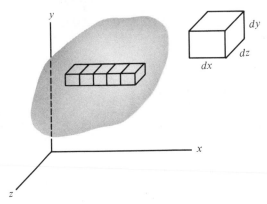

Fig. 4-8 Special Choice of Volume Elements to Form a
Triple Integral $\iiint f\, dx\, dy\, dz$

The details of choosing the variable limits on the inner integrals may be found in a calculus textbook. We shall not actually need to evaluate any such triple integrals for some time. And, in fact, the mass center can frequently be located by a single integral (see Sec. 7.4). All that is needed here is the concept of an integral as a sum. *We can write the volume integral of Eq. (4.2.5) with a single integral sign, representing a single sum over all the elements* when we are not involved in evaluation by successive integrations with respect to the three coordinates.

The total mass of two bodies or of a body consisting of two parts is the sum of the masses of the separate parts:

$$\text{total } m = \int_{V_1+V_2} \rho\, dV = \int_{V_1} \rho\, dV + \int_{V_2} \rho\, dV = m_1 + m_2. \tag{4.2.8}$$

This may be extended to any number N of bodies (or parts) to give the same form of equation as Eq. (4.2.3) for N particles.

The ***center of mass*** * or ***mass center*** of a body can now be defined as the point with coordinates $(\bar{x}, \bar{y}, \bar{z})$ given by

---

Center of Mass $(\bar{x}, \bar{y}, \bar{z})$

$$m\bar{x} = \int_V x \, dm \qquad m\bar{y} = \int_V y \, dm \qquad m\bar{z} = \int_V z \, dm \qquad (4.2.9)$$

or

$$\bar{x} = \frac{\int_V x\rho \, dV}{\int_V \rho \, dV} \qquad \bar{y} = \frac{\int_V y\rho \, dV}{\int_V \rho \, dV} \qquad \bar{z} = \frac{\int_V z\rho \, dV}{\int_V \rho \, dV}. \qquad (4.2.10)$$

The integrals on the right-hand side of Eqs. (4.2.9) define the ***first moments of the mass*** distribution with respect to the yz-plane, the xz-plane, and the xy-plane, respectively. The ***moment arm*** of the element dm, measured from the yz-plane, is x, and $\bar{x}$ is the moment arm of a particle of mass m that has the same first moment with respect to the yz-plane as the distributed mass. Similar comments apply to the other moment arms $\bar{y}$ and $\bar{z}$. Note that the first moments may be positive, negative, or zero. For example, ***if the origin is chosen at the center of mass, all the first moments are zero.***

Centroid

Notice that if the density ρ is uniform throughout the body, ρ can be factored out of all the integrals in Eq. (4.2.10) and canceled from the numerators and denominators. Then the center of mass will be at the ***centroid*** of the volume (the point defined by omitting the density factors). When ρ varies from point to point, however, the center of mass will as a rule not be located at the centroid of the volume. If a body possesses a plane of geometric symmetry, the ***centroid will lie in the plane of symmetry***, since the first moment of the volume with respect to the plane of symmetry will be zero. For example, $\int_V x \, dV = 0$ if the yz-plane is the plane of symmetry. Since for a uniform mass distribution ($\rho = $ const.) the mass center is at the centroid, ***the mass center lies in any plane of symmetry of a body of uniform density.***

   Tables of centroids and centers of mass are given in Tables A3 and A4 of Appendix A for bodies of various simple geometric shapes. Methods of locating the centroid of a composite body formed from two or more simple bodies are presented in Sec. 7.3. Integration methods are presented in Secs. 7.4 and 7.5.

---

* The center of mass concept for a body (as distinguished from the center of gravity) was introduced by L. Euler in work of 1758–1760.

Center of Gravity

We are now in a position to demonstrate that *in a uniform gravitational field* the distributed weight of a body is equipollent to a single force whose line of action passes through the center of mass, as was stated in Sec. 1.3. In Fig. 4-9 the uniform gravitational field is assumed to act in the negative z-direction. We shall show that the line of action of the resultant intersects the xy-plane at $(\bar{x}, \bar{y}, 0)$ and hence passes through the center of mass $C(\bar{x}, \bar{y}, \bar{z})$.

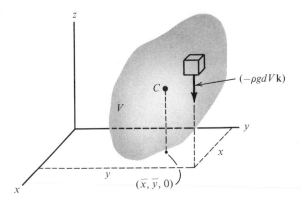

Fig. 4-9 Center of Gravity in Uniform Parallel Gravitational Field

The downward force on the volume element of mass $(\rho\, dV)$ is its weight $(\rho g\, dV)$. The total downward force is

$$W = f^R = g \int_V \rho\, dV = mg, \tag{4.2.11}$$

where m is the total mass. The turning moments about the x- and y-axes are

$$\sum m_x = -\int_V y\rho g\, dV = -mg\bar{y}$$

$$\sum m_y = +\int_V x\rho g\, dV = mg\bar{x} \tag{4.2.12}$$

since, by Eq. (4.2.10)

$$m\bar{x} = \int_V x\rho\, dV \qquad m\bar{y} = \int_V y\rho\, dV. \tag{4.2.13}$$

If the resultant downward force mg acts at (x_R, y_R, z_R), then for equipollence we must have

$$mgx_R = \sum m_x = mg\bar{x}$$
$$-mgy_R = \sum m_y = -mg\bar{y}. \tag{4.2.14}$$

Hence $x_R = \bar{x}$ and $y_R = \bar{y}$, and the line of action of the resultant weight intersects the xy-plane at $(\bar{x}, \bar{y}, 0)$. It therefore passes through the mass center $(\bar{x}, \bar{y}, \bar{z})$.

Methods of locating centers of mass are discussed in Sec. 7.2–7.5. For certain simple arcs, areas, and volumes the centroid (center of mass for a body of uniform density) may be found in Tables A3 and A4 of Appendix A.

In Sec. 4.3 we shall consider the conditions for equilibrium of a rigid body.

SAMPLE PROBLEM 4.2.1

If the downward distributed load on a 10-ft beam varies linearly from 20 lb/ft at $x = 0$ to 40 lb/ft at $x = 10$ ft, determine the total downward force and the position x_R of its line of action. See Fig. 4-10. Two methods of solution will be illustrated.

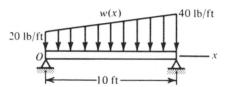

Fig. 4-10 Sample Problem 4.2.1

SOLUTION 1. The downward distributed load $w(x)$ is a linear function

$$w(x) = A + Bx \text{ lb/ft}.$$

Since

$$w(0) = 20 \text{ lb/ft} = A \quad \text{and} \quad w(10) = 40 \text{ lb/ft} = A + 10B,$$

we must have $A = 20$ and $B = 2$, so that

$$w(x) = 20 + 2x \text{ lb/ft}.$$
$$f^R = \int_0^{10} (20 + 2x)\, dx = 300 \text{ lb} \quad \textit{Answer.}$$

$$\Sigma \, m_0 = \int_0^{10} xw(x)\, dx = \int_0^{10} (20x + 2x^2)\, dx$$

$$= [10x^2 + \tfrac{2}{3}x^3]_0^{10} = 1667 \text{ lb-ft.}$$

$$300x_R = 1667 \quad \text{or} \quad x_R = 5.56 \text{ ft} \quad \textbf{\textit{Answer.}}$$

SOLUTION 2.   We make use of the result stated in connection with Eq. (4.2.1) that the resultant is equal to the area of the load diagram and acts through the centroid of the load diagram. To simplify the calculation we divide the trapezoidal load diagram into a triangular part A_1 and a rectangular part A_2, as illustrated in Fig. 4-11. Table A3 of Appendix A locates the centroid of the triangle at two-thirds of the way from the vertex to the base. Thus

$$f_1^R = A_1 = \tfrac{1}{2}(10 \text{ ft})(20 \text{ lb/ft}) = 100 \text{ lb}, \qquad x_1 = 6.67 \text{ ft.}$$

$$f_2^R = A_2 = (10 \text{ ft})(20 \text{ lb/ft}) = 200 \text{ lb}, \qquad x_2 = 5 \text{ ft.}$$

$$f^R = f_1^R + f_2^R = 300 \text{ lb} \quad \textbf{\textit{Answer.}}$$

$$300x_R = x_1 f_1^R + x_2 f_2^R = 667 + 1000 = 1667$$

$$x_R = 5.56 \text{ ft} \quad \textbf{\textit{Answer.}}$$

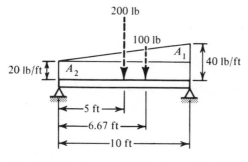

Fig. 4-11 Replacement of Distributed Load by Two Concentrated Loads

---

EXERCISES

(Exercises on centers of mass are given in Sec. 7.3)

1. If a beam has a downward linear load variation from $w = 100$ kN/m at $x = 0$ to $w = 300$ kN/m at $x = 12$ m, determine the equipollent single force.

2. If a beam of length L carries a downward distributed load $w(x) = kx^2/L^2$, determine the equipollent single force.

3. If a beam of length L carries a downward

distributed load $w(x) = w_0[1 - (x^2/L^2)]$, determine the equipollent single force.

4. A uniform 20-ft beam weighs 600 lb and carries a single concentrated downward force $P = 1000$ lb at $x = 5$ ft from the left end. Where

does the single force equipollent to P and the weight act?

5. If the beam of Fig. 4-7 is a uniform 18-ft beam weighing 450 lb, where does the single force equipollent to the weight and the given load act?

4.3 RIGID-BODY EQUILIBRIUM; FREE-BODY DIAGRAM; IDEALIZATION OF SUPPORT REACTIONS; EQUATIONS OF EQUILIBRIUM; STATICALLY DETERMINATE COPLANAR PROBLEMS

The basic tool in the analysis of the motion or equilibrium of a body is the free-body diagram. In this chapter we shall consider equilibrium analysis for a single rigid body, where the choice of what body to isolate in the free-body diagram is usually no problem; we shall consider the whole body in most cases. In the following chapter we shall consider the analysis of structures and machines made of connected rigid bodies. Then the art of choosing an appropriate set of free bodies to isolate becomes more important, and the power of the free-body method is even more apparent.

The *free-body diagram** is a simplified sketch of the body with the known and unknown external force and couple vectors acting on it shown with some indication of their directions. Alternatively, each acting force or couple may be represented by its components. (In coplanar problems a couple may be shown as a curved arrow.) The body should be shown removed from any supports, and wherever a support has been removed an appropriate idealized support reaction force and/or couple should be shown. The diagram should show all *external* forces and couples that *act on the body* but no forces or couples not acting on the body from outside the body. *Forces exerted by the body are not shown, and internal forces between parts of the isolated body are not shown.*

Idealized support reactions are simplified representations (usually as concentrated forces and couples) of the actual support forces, which are usually distributed forces over a small area. The idealization is necessary for two reasons. First, the actual distribution of the support forces over the small area is unknown, and, second, it would be too complicated to use in the problem solution even if we knew it. But, as we saw in Sec. 4.1, the most complicated possible system of forces is equipollent to a system of one force and one couple acting at a chosen point. This means that, *insofar as the external effect on a rigid body is concerned, the most general support reaction can be represented by a force and a couple acting at a point.*

---

* The name "free-body diagram" was apparently first used by I. P. Church, a professor at Cornell University in 1887, although the idea of "isolating the system" goes back at least to the time of d'Alembert, Euler, and the Bernoullis.

In coplanar problems, where all forces act in the xy-plane and all couple vectors are perpendicular to the xy-plane, the most general support reaction can be represented by three reaction components: two force components and the z-component of the couple (which may be indicated by a curved arrow indicating the sense of the couple).

Many support reactions can be further simplified because of the nature of the support. For example, an idealized frictionless ball-and-socket joint is incapable of transmitting any couple. Such a support would be represented only by a single force vector or by its three components. In coplanar problems the counterpart of the ball-and-socket joint is the smooth pin or hinge, where the support reaction is a force vector or its two components. Figure 4-12 summarizes some special kinds of idealized supports for coplanar problems, tabulated according to the number of unknown reaction components that the support is capable of exerting.

In the third group, supports with one reaction component, *the conventional roller symbols shown are often used for supports that can exert either upward or downward force.* Such symbols should usually not be interpreted literally as actual rollers.

Supports with Three Reaction Components

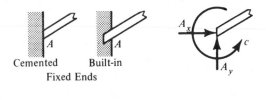

Supports with Two Reaction Components

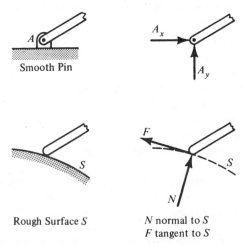

Fig. 4-12 Idealized Coplanar Reactions; Sketch at Right Shows Reaction Components on Free Body

Supports with One Reaction Component

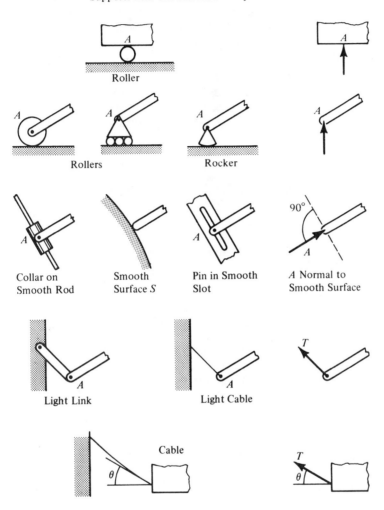

Fig. 4-12   (Continued)

For modeling actual supports we sometimes assume a simpler type of reaction than the most general reaction that the support could exert. This is illustrated by the hinge support at *A* in the two examples of Fig. 4-13 where a rectangular plate in a vertical plane is supported by one hinge at *A* in case (a) and by two hinges at *A* and *B* in case (b). Since we are treating this as a coplanar equilibrium problem, we treat the hinge support in the first case as a fixed end (see Fig. 4-12). But in case (b) the couple reactions have been omitted, and we have moreover assumed that the hinge supports are so spaced that only the hinge at *A* exerts a force along the hinge axis. Solutions for the hinge reactions for these two cases will be given at the end of this section in Sample Problem 4.3.1. The reason that we can omit the couple reactions at A and B in Fig. 4-13(b) is that they will be negligible if the hinges are well aligned. The support reactions are actually developed by the elastic response of the deformed material of the supports. To develop a significant couple reaction at A the hinge must rotate (about an axis through A perpendicular to the plane) far enough to cause enough elastic reaction of the hinge pin to produce a significant couple. See the exaggerated illustrative sketch of the hinge at A for case (a) in Fig. 4-13(c).

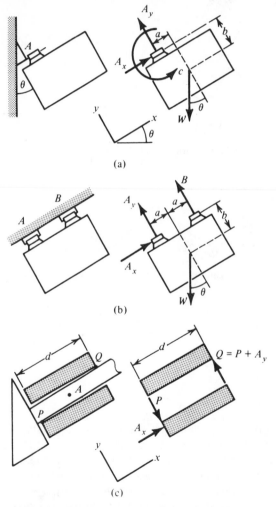

Fig. 4-13 Hinge-Supported Rectangular Plate: (a) One Hinge; (b) Two Hinges; (c) Hinge Couple Formed by Forces at P and Q

In Fig. 4-13(c) the normal forces P and Q acting at points P and Q are equipollent to the force A_y and couple c given by

$$A_y = \tfrac{1}{2}(Q - P) \qquad c = \tfrac{1}{2}(Q + P)d$$

acting at point A, halfway between points P and Q. These, with A_x, are the reactions shown at A in Fig. 4-13(a). Because rotation is inhibited by the support at B in the second case, Fig. 4-13(b), no significant couples are developed if the hinges are well aligned. We have therefore omitted the couple reactions in Fig. 4-13(b).

It is important to make such reasonable simplifications as those of Fig. 4-13(b) whenever possible in order to reduce the number of unknown reaction components. The methods of rigid-body statics and dynamics permit solution for at most six unknowns on each free body, and this number is reduced to at most three in coplanar problems. The reason for this is that the basic equations of rigid-body mechanics are two vector equations, giving only six independent component equations per free body. In coplanar problems three of the equations are satisfied identically, leaving only three independent equations per free body. It is possible to solve a mechanics problem with more unknowns than this, but it is impossible to solve it by methods of rigid-body mechanics alone. Additional equations are furnished by the methods of deformable body analysis.

Rigid-body dynamics theory will be developed in Chapters 11 and 12 of Vol. II. There it will be shown that the system of applied external forces and couples on a rigid body is equipollent to an *effective-force* system consisting of two vectors, an *effective-force* vector and an *effective-couple* vector. The *effective force* is equal to the rate of change, $\dot{\mathbf{p}} = m\bar{\mathbf{a}}$ of the linear momentum $\mathbf{p} = m\bar{\mathbf{v}}$. The body has mass m, while $\bar{\mathbf{a}}$ and $\bar{\mathbf{v}}$ are, respectively, the acceleration and velocity of the center of the mass of the body. The *effective couple* is the rate of change of the angular momentum, which will be zero for motion without rotation.

Definition: A rigid body is said to be in equilibrium if, with respect to an inertial frame of reference, the body is not rotating and its center of mass has a constant velocity. Then each material point of the body is unaccelerated. We shall consider mainly problems where the body is actually at rest. When a rigid body is in equilibrium the system of external forces and couples acting on it is equipollent to zero, as will be indicated by Eqs. (4.3.1).

Equilibrium Equations

When the body is unaccelerated, the effective force and effective couple will both be zero. In this case we draw the free-body diagram, and write the two vector

---

Equilibrium Equations

$$\sum \mathbf{f}_{ext} = 0$$

$$\sum (\mathbf{m}_O)_{ext} = 0,$$

(4.3.1)

---

where the subscript "ext" means "externally applied." Equations (4.3.1) are equivalent to the six component

---

Equations of Equilibrium

$$\sum f_x = 0, \qquad \sum f_y = 0, \qquad \sum f_z = 0$$

$$\sum m_{Ox} = 0, \qquad \sum m_{Oy} = 0, \qquad \sum m_{Oz} = 0,$$

(4.3.2)

---

where the subscript "ext" has been omitted for simplicity. No additional independent equations can be written by taking moments about another moment center B, since by Eq. (4.1.1)

$$\sum \mathbf{m}_B = \sum \mathbf{m}_O + \mathbf{d}_{BO} \times (\sum f).$$

(4.3.3)

Hence however $\sum \mathbf{f} = 0$ and $\sum \mathbf{m}_O = 0$, it follows also that $\sum \mathbf{m}_B = 0$ without placing any more conditions on the acting forces. *Hence in the general three-dimensional case there are for each free body at most six independent equations of equilibrium.* The number of useful independent equations is even smaller in special cases where some of the six are of the form $0 = 0$.

Coplanar problems involve only three independent equations for each free body. If all forces act in the xy-plane, the component equations $\sum f_z = 0$, $\sum m_x = 0$, and $\sum m_y = 0$ each reduce to $0 = 0$, leaving only the three

---

Coplanar Equilibrium Equations

$$\sum f_x = 0, \qquad \sum f_y = 0, \qquad \sum m_A = 0,$$

(4.3.4)

---

where the notation m_A means the moment about an axis through A parallel to the z-axis. *No additional independent equations can be obtained by taking moments*

about another point B. But we can substitute another moment equation for one or more of the force component equations.

Alternative Coplanar Equilibrium Conditions

$$\sum f_x = 0, \qquad \sum m_A = 0, \qquad \sum m_B = 0 \qquad\qquad (4.3.5)$$

(for AB not parallel to the y-axis), or

$$\sum m_A = 0, \qquad \sum m_B = 0, \qquad \sum m_C = 0 \qquad\qquad (4.3.6)$$

(for A, B, C not collinear).

Equations (4.3.5) imply that $\sum f_y = 0$, since the first two equations of Eqs. (4.3.5) imply that the coplanar external force system is equipollent to a single force in the y-direction at A, while $\sum m_B = 0$ shows that this resultant must be zero. Alternatively, the first equation of Eqs. (4.3.6) shows that the coplanar system is equipollent to a single force resultant passing through A. The second equation shows that if this resultant is nonzero, it must pass through B, while the third equation shows that it must also pass through C. Since A, B, and C are not collinear, the resultant force must be zero, when all three equations are satisfied. Thus satisfaction of Eqs. (4.3.6) also implies satisfaction of Eqs. (4.3.4).

Statically Determinate and Statically Indeterminate Reactions

Consider the example of a coplanar body supported by a smooth pin at A and by a roller at B, as shown in Fig. 4-14 and loaded by its weight and the given force of 100 lb to the right. In the free-body diagram the pin reaction is represented by two components, while the roller reaction is shown as a single vertical component. A_x represents the **magnitude** of the force, assumed acting to the left as shown. If the

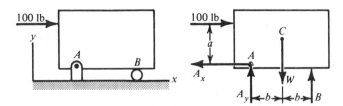

Fig. 4-14 Example of Statically Determinate Reactions

solution gives a negative value for A_x, that will indicate that we assumed the wrong direction. (Alternatively, you could let A_x and A_y represent rectangular component coefficients of the force vector $A_x \mathbf{i} + A_y \mathbf{j}$ instead of magnitudes and show each unknown component on the free-body diagram pointing in the positive coordinate direction.) Application of Eqs. (4.3.4) to the free-body diagram of Fig. 4-14 gives

$$[+\backslash\, \textstyle\sum m_A = 0]: \quad 2bB - bW - 100a = 0, \qquad B = \frac{1}{2}W + 50\frac{b}{a}$$

$$[\textstyle\sum f_y = 0]: \quad A_y + B - W = 0, \qquad A_y = \frac{1}{2}W - 50\frac{b}{a} \qquad (4.3.7)$$

$$[\textstyle\sum f_x = 0]: \quad 100 - A_x = 0, \qquad A_x = 100\,\text{lb}.$$

The sense of the positive convention for moments m_A is indicated by the curved arrow to the left of $\sum m_A$ in the equation. Notice that in this example A_x represents the magnitude of a force assumed acting to the left at A. Since it came out positive, the correct direction was shown on the diagram.

Since the support reactions can be determined by the equilibrium equations, the reactions are called *statically determinate*. If the roller at B is replaced by another pin support, there would be four unknown reactions as in Fig. 4-15 and the reactions are called *statically indeterminate* because they cannot be determined by the equilibrium equations.

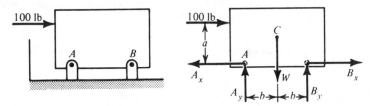

Fig. 4-15 Example of Statically Indeterminate Reactions

The only one of the three equilibrium equations of Eqs. (4.3.7) affected by the change is the last one, which now becomes

$$[\textstyle\sum f_x = 0]: \quad B_x + 100 - A_x = 0, \quad A_x - B_x = 100\,\text{lb}. \qquad (4.3.8)$$

Although we are still able to solve for A_y and B_y, we cannot determine A_x and B_x but only their difference $A_x - B_x$ by Eq. (4.3.8).

Complete and Partial Constraint. In the two cases shown in Figs. 4-14 and 4-15 the body is *completely constrained*; that is, it could not move under any system of applied loads. (The conventionally represented roller is assumed capable of exerting either upward or downward force.) An example of ***partial constraint*** is shown in Fig. 4-16 where the body is supported by two rollers. Evidently it will not be in equilibrium for arbitrary loading. In fact the 100-lb load shown will cause acceleration to the right. It becomes a problem in ***translation dynamics*** instead of equilibrium. The vertical force summation is unchanged, but both the horizontal force summation and the moment equation must be replaced by components of the translation dynamics equations; see Sec. 8.8 of Vol. II.

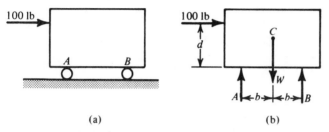

(a)   (b)

Fig. 4-16 Example of Partial Constraint; Translation Dynamics

The foregoing three examples lead us to believe that if a rigid body is to be completely constrained and have statically determinate reaction components, ***there must be as many unknown reaction components as there are equations of equilibrium***. This is in fact a correct statement, but it is only a necessary condition and ***not a sufficient condition*** for complete constraint and statical determinacy. For example, a body supported by three rollers as in Fig. 4-17 is obviously not completely constrained, although there are three reaction components. It is not statically determinate either—and in fact not even solvable by rigid-body dynamics. We would have three dynamics equations and four unknowns (three reactions and the acceleration). Such a problem can be solved only by considering the body as deformable and obtaining additional equations relating the forces to the deformations.

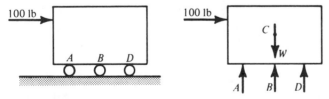

Fig. 4-17 Improper Constraint Example

Another kind of improper constraint is illustrated in Fig. 4-18. The short rigid link *BE* is a two-force member (neglecting its weight), which exerts a force in line with its ends. Hence all the support reaction components pass through *A*, and it is impossible to satisfy the moment equilibrium equation $\sum m_A = 0$ even though there are just as many unknowns as there are equations. This is statically indeterminate by rigid-body methods. What will actually happen under such a loading is that the link *BE* will stretch and the body will deform as it rotates about pin *A* until it reaches a position where the force exerted by the link is able to maintain moment equilibrium, if the link or one of the pins does not break first.

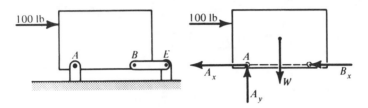

Fig. 4-18 Improper Constraint

From the examples of Figs. 4-17 and 4-18 we conclude that *a coplanar rigid body is improperly constrained whenever the supports are such that the reactions are all parallel forces or are all concurrent forces.*

Two-force body. We have already stated and several times used the result that when a body is in equilibrium under the action only of forces applied at two points then the resultant of the forces at each point must pass through the other point. The two resultants must be equal, opposite, and collinear. We shall now give a proof of this important result. Consider an arbitrary body loaded only by forces at *A* and at *B*, as illustrated in Fig. 4-19. By the parallelogram law of forces, the forces at *A* may be replaced by their vector sum without changing their effect, and the forces at *B* may likewise be replaced by their vector sum. If the body is to be in equilibrium, the first equation in Eqs. (4.3.1) requires the resultant at *B* to be equal and opposite to that at *A*, but they could still form a couple as in Fig. 4-19(b), except that the second equation in Eqs. (4.3.1) also requires the total moment to vanish. Hence the total force at each point must pass through the other point, as shown in Fig. 4-19(c). Notice that this is true even though the body is not a straight member.

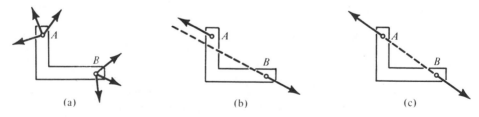

(a) (b) (c)

Fig. 4-19 Two-Force Body

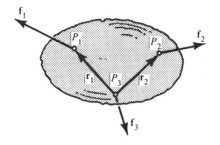

Fig. 4.20   Three-Force Body

Three-force body. When a body is in equilibrium under the action of just three forces, the forces must be coplanar and either concurrent or parallel. This useful result can be proved as follows. Suppose that forces $\mathbf{f}_1$, $\mathbf{f}_2$, $\mathbf{f}_3$ act at points P_1, P_2, P_3 as illustrated in Fig. 4-20. Choose P_3 as the origin and draw $\mathbf{r}_1$ and $\mathbf{r}_2$. For moment equilibrium about P_3,

$$\mathbf{r}_1 \times \mathbf{f}_1 + \mathbf{r}_2 \times \mathbf{f}_2 = 0.$$

Hence

$$\mathbf{r}_2 \times \mathbf{f}_2 = -\mathbf{r}_1 \times \mathbf{f}_1.$$

This shows that the plane determined by $\mathbf{r}_2$ and $\mathbf{f}_2$ has a normal parallel to the normal of the plane of $\mathbf{r}_1$ and $\mathbf{f}_1$. Since these two planes have point P_3 in common, they coincide. Hence $\mathbf{f}_1$, $\mathbf{f}_2$, and P_3 all lie in a plane. The force $\mathbf{f}_3$ must also lie in the plane in order that the vector polygon $\mathbf{f}_1 + \mathbf{f}_2 + \mathbf{f}_3$ can close as required for equilibrium. Thus we have proved that the three forces must be coplanar. If any two of them intersect at a point A, the third one must also pass through A in order that $\sum \mathbf{m}_A = 0$. This shows that they must all three be concurrent, unless they are all three parallel.

Three-dimensional problems will be treated in Sec. 4.4. Here we shall consider only coplanar problems. Of course no physical body is actually coplanar. But when the body is approximately symmetric with respect to its midplane and the forces acting on it are also symmetric with respect to the midplane then the motion or equilibrium of the body can be analyzed as though the forces all acted in the midplane. For such symmetric loading, any force components perpendicular to the plane are balanced by symmetric force components on the opposite side, and the pairs of balancing forces can be omitted.

SAMPLE PROBLEM 4.3.1

Determine the hinge reactions for each equilibrium case of Fig. 4-13(a) and (b).

SOLUTION.   The free-body diagrams shown in Fig. 4-13 are repeated in Fig. 4-21 with the weight force resolved into components parallel to the axes.

Case (a)

$$[\sum f_x = 0]: \quad A_x - W \sin \theta = 0, \qquad A_x = W \sin \theta \quad \textbf{\textit{Answer.}}$$

$$[\sum f_y = 0]: \quad A_y - W \cos \theta = 0, \qquad A_y = W \cos \theta \quad \textbf{\textit{Answer.}}$$

$$[+\searcol \sum m_A = 0]: \quad c - bW \sin \theta - aW \cos \theta = 0$$

$$c = W(b \sin \theta + a \cos \theta) \quad \textbf{\textit{Answer.}}$$

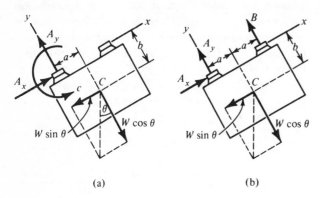

Fig. 4-21   Free-Body Diagrams for Sample Problem 4.3.1

Case (b)

$$[\textstyle\sum f_x = 0]: \quad A_x - W \sin\theta = 0, \qquad A_x = W \sin\theta \quad \textbf{\textit{Answer.}}$$

$$[+\curvearrowright \textstyle\sum m_A = 0]: \quad 2aB - bW \sin\theta - aW \cos\theta = 0$$

$$B = \frac{W}{2}\left(\frac{b}{a}\sin\theta + \cos\theta\right) \quad \textbf{\textit{Answer.}}$$

$$[\textstyle\sum f_y = 0]: \quad A_y + B - W \cos\theta = 0$$

$$A_y = \frac{W}{2}\left(\cos\theta - \frac{b}{a}\sin\theta\right) \quad \textbf{\textit{Answer.}}$$

SAMPLE PROBLEM 4.3.2

The 6-m cantilever beam AB in Fig. 4-22 is built in at A and carries a triangularly distributed downward load with intensity varying from 20 kN/m at A to zero at B. Determine the support reactions.

SOLUTION.   On the free-body diagram of Fig. 4-22(b) we have replaced the distributed load by its resultant, a concentrated downward force of 60 kN, equal to the area of the load triangle and acting through the centroid of the triangle at $x = 2$ m. (See Sample Problem 4.2.1.)

$$[\textstyle\sum f_x = 0]: \quad A_x = 0 \quad \textbf{\textit{Answer.}}$$

$$[\textstyle\sum f_y = 0]: \quad A_y - 60 = 0, \qquad A_y = 60 \text{ kN} \quad \textbf{\textit{Answer.}}$$

$$[+\curvearrowright \textstyle\sum m_A = 0]: \quad c_A - 2(60) = 0, \qquad c_A = 120 \text{ kN-m} \quad \textbf{\textit{Answer.}}$$

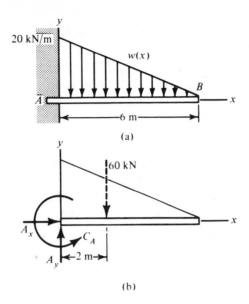

(a)

(b)

Fig. 4-22 Sample Problem 4.3.2

SAMPLE PROBLEM 4.3.3

The uniform block of weight W, height h, and base length b in Fig. 4-23 is placed on a plane inclined at angle θ to the horizontal. The coefficient of static friction is μ_s. (a) If the block is in equilibrium, determine the normal component N and frictional component F of the resultant of the distributed support reaction and the distance x_R from A to the position of the resultant. (b) If θ is slowly increased until equilibrium can no longer be maintained, will the block slide or will it tip over?

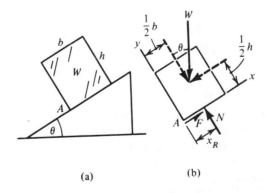

(a)   (b)

Fig. 4-23 Sample Problem 4.3.3

SOLUTION.   (a) The free-body diagram is shown in Fig. 4-23(b).

$$[\Sigma f_x = 0]: \quad F - W \sin \theta = 0, \qquad F = W \sin \theta \quad \textbf{\textit{Answer.}}$$

$$[\Sigma f_y = 0]: \quad N - W \cos \theta = 0, \qquad N = W \cos \theta \quad \textbf{\textit{Answer.}}$$

$$[+\curvearrowright \Sigma m_A = 0]: \quad Nx_R + \tfrac{1}{2}hW \sin \theta - \tfrac{1}{2}bW \cos \theta = 0$$

$$x_R = \tfrac{1}{2}(b - h \tan \theta) \quad \textbf{\textit{Answer.}}$$

(b) For impending tip, $x_R = 0$, so that $\tan \theta = b/h$. For impending slip, $F/N = \mu_s$, so that $\tan \theta = \mu_s$. Which will occur depends on the relationship of μ_s to b/h.

For $\mu_s > b/h$, tip occurs at $\tan \theta = b/h$; for $\mu_s < b/h$, slip occurs at $\tan \theta = \mu_s$ **Answer.**

EXERCISES

1. The coplanar right-angle bracket shown is supported by a smooth pin at A and leans at D against a smooth wall inclined at 60° to the horizontal as shown. It carries two applied loads, a downward 100-lb force at C and a clockwise 260-lb-ft couple. Determine the support reactions at A and D for equilibrium.

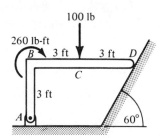

Exercise 1

2. The uniform 10-ft ladder of weight W leans against a smooth wall with the foot of the ladder 6 ft from the foot of the wall. The coefficients of friction between ladder and floor are $\mu_s = 0.5$ and $\mu_k = 0.4$. What is the friction force exerted by the floor on the ladder?

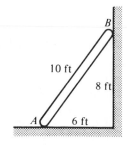

Exercise 2

3. Determine all forces acting on the coplanar bracket ABC.

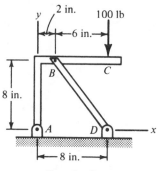

Exercise 3

4. The figure shows a uniform 100-lb boom supporting a 1000-lb load at *B*. *AB* itself is supported in a vertical plane by a smooth pin connection at *A* and a light cable *DC*. Find the cable tension and the pin reaction at *A*.

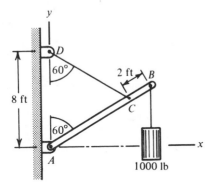

Exercise 4

5. A uniform hoop weighing 80 N is supported by a cord and a smooth incline as shown. Compute the tension in the cord.

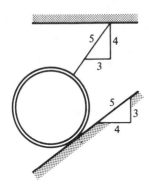

Exercise 5

6. The 20-ft ladder *AB* weighs 40 lb and leans against the **smooth** wall which is inclined at 60°. The ladder itself is at 45° and is hinged to a 100-lb block to resist slipping. If slip takes place when a 160-lb man has climbed halfway up the ladder, what is the coefficient of static friction between the block and the floor?

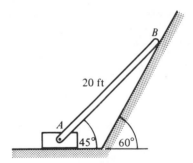

Exercise 6

7. The cantilever carries a uniform load of intensity *w* per unit length and a concentrated force *P* as shown. Determine the support reactions at *A*.

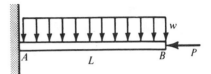

Exercise 7

8. Solve Ex. 7 if the uniform load is replaced by a triangular load distribution with intensity varying linearly from zero at *A* to w_B at *B*.

9. The light rigid right-angle L-shaped bar carries a 500-kg mass at *C* and is supported by a hinge at *B* and cable *AD*. Determine the support reactions.

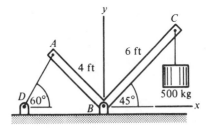

Exercise 9

10. A homogeneous hemisphere of weight *W* and radius *R* rests on a perfectly smooth horizontal table as shown. Edge *B* is tied to a point *C*

by a light vertical string such that $\theta = 30°$. Find the tension T in the string and the reaction of the table.

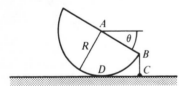

Exercise 10

11. The figure shows an arrangement for raising the uniform pole AB, which weighs 300 lb. The coefficient of friction between the pole and the ground is 0.5. If the force P is gradually increased until motion occurs, does the pole slide or does it tip about A?

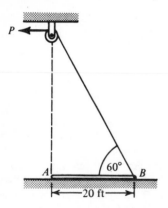

Exercise 11

12. The figure shows three different support arrangements for a uniform 10×8 ft rectangular coplanar body of weight W. Determine the support reactions for equilibrium in each

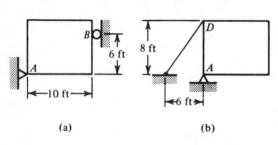

(a)   (b)   (c)

Exercise 12

case. Neglect the weights of the post in (c) and of the guy wire in (b).

13. A small truss is shown supported in four different ways by smooth pins, rollers, or light links. In each case determine (1) whether the truss is completely, partially, or improperly constrained; (2) whether the reactions are statically determinate or indeterminate; and (3) those reactions that you can compute. Neglect the weight of the truss, and answer (3) in terms of the vertical load P.

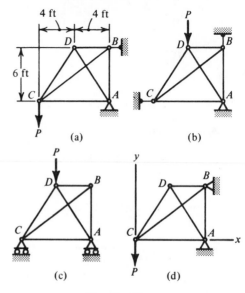

(a)   (b)

(c)   (d)

Exercise 13

14. A 16-in. diameter uniform cylinder weighing 300 lb is rolled over a 2-in.-high step as shown. Determine the force P required in each case.

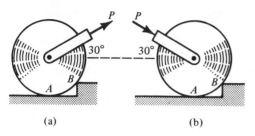

(a)   (b)

Exercise 14

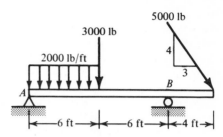

Exercises 17 and 19

15. Given $\mu_s = \frac{1}{3}$, $\mu_k = \frac{1}{4}$, (a) compute the force P which must be applied horizontally to begin to draw the homogeneous 200-kg chest up the incline. (b) What is the maximum value of h if the chest is not to tip before it slides?

20. The dolly shown has its mass center C 30 cm from the inclined plane. The uphill wheels at A are locked, while those at B are free. The coefficient of static friction is 0.4 between wheel and plane. If the mass of the dolly is 200 kg and $\theta = 8°$, determine the normal reaction forces at A and B and the friction force for equilibrium.

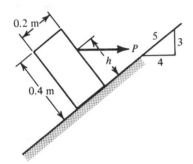

Exercise 15

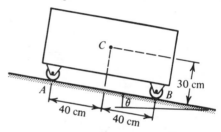

Exercise 20

16 & 17. Determine the reactions of the beam shown supported by a hinge and a roller.

18 & 19. Determine the support reactions if the roller is removed and the hinge is replaced by a built-in support at the left end.

21. For the system of Ex. 20, would it be better to have the dolly turned around with the lockable wheels downhill? What is the largest θ for which slip would be prevented with locked wheels (a) uphill? (b) downhill?

22. The center of mass of the cabinet shown is at

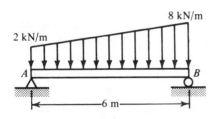

Exercises 16 and 18

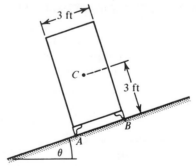

Exercise 22

distance 3 ft from the inclined plane. The coefficient of static friction is $\mu_s = 0.4$. If $\theta = 20°$, determine the normal reactions at A and B and the total frictional force at A and B for equilibrium. Answer in terms of the weight W.

23. If the angle θ in Ex. 22 is increased, for what range of values of μ_s will the cabinet (a) slip before it tips or (b) tip before it slips?

24. The uniform shaft is supported in bearings at 30° to the horizontal as shown. There is $\frac{1}{16}$ in. of motion permitted in the axial direction. If the shaft is 4 ft long, weighs 30 lb, and is loaded by a 200-lb-ft clockwise couple at its midpoint, determine the support reactions for equilibrium. (The couple acts in the vertical plane containing AB).

25. The figure shows a light airplane airfoil section at a high angle of attack $\alpha = 23°$. The aerodynamic forces on the wing at that part of the span are equivalent to a distributed lift force intensity $L = 7300$ N/m and drag $D = 2000$ N/m. (Force intensities are per unit length in the spanwise direction perpendicular to the figure.) The forces may be considered to act at the center of pressure C. For equilibrium of the airfoil element, calculate the force intensities parallel to the y-axis exerted by the airfoil on the front and rear beams and the force (per unit spanwise length) exerted parallel to the x-axis by the airfoil acting on the drag truss. Show that the force on the drag truss is actually forward for this high angle of attack.

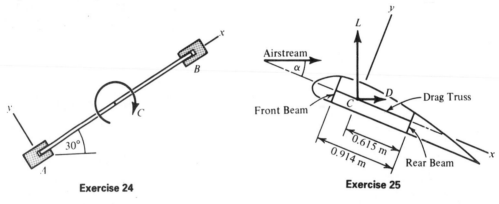

Exercise 24   Exercise 25

4.4 RIGID-BODY EQUILIBRIUM PROBLEMS IN THREE DIMENSIONS

The general three-dimensional equilibrium equations have already been given as the two vector equations of Eqs. (4.3.1) or the six component equations of Eqs. (4.3.2), but for simplicity we have up to now considered only coplanar problems where there were only three independent component equations for each free body.

Figure 4-24 summarizes *idealized support reactions for three-dimensional problems*, classified according to the number of support force and couple components that the idealized support can exert. As in the discussion of Figs. 4-12 and 4-13, we frequently omit some of the couple reactions that a hinge, bearing, or flanged wheel could exert when there are enough other supports to maintain the equilibrium without allowing enough rotation for the elastic deformations to develop significant couple reactions. When several force and couple components act at one point, it is often preferable to show the support reactions simply as a vector **f** and vector **c** instead of showing the components, in order to avoid cluttering the figure.

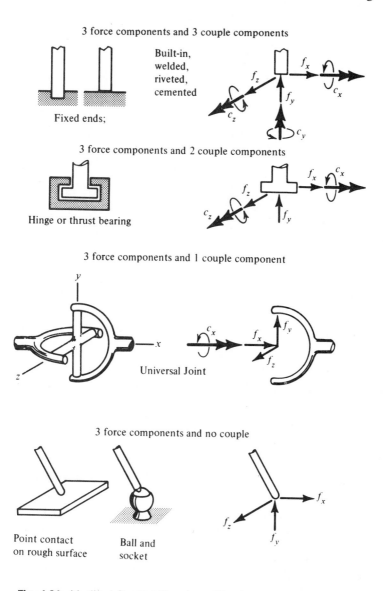

Fig. 4-24 Idealized Support Reactions; Sketch at Right Shows Reaction Components on Free Body

2 force components and 2 couple components

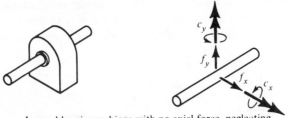

Journal bearing or hinge with no axial force, neglecting frictional couple c_z. (c_x and c_y may also be negligible.)

2 force components and 1 couple component

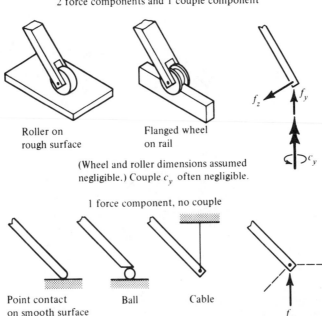

Roller on   Flanged wheel
rough surface   on rail

(Wheel and roller dimensions assumed negligible.) Couple c_y often negligible.

1 force component, no couple

Point contact   Ball   Cable
on smooth surface

Fig. 4-24   (Continued)

The discussion of free-body construction, idealization of support reactions, statically determinate and indeterminate reactions, complete, partial, and improper constraint given in Sec. 4.3 should be reviewed at this time, since with suitable modifications it applies also in three dimensions.

If the idealized supports for a three-dimensional body involve more than six unknown reactions, there are more unknowns than equations and some of the reactions are ***statically indeterminate***, although it is usually possible to solve for part of the reactions by rigid-body methods. See, for example, Sample Problem 4.4.2. If the reactions involve fewer than six unknowns, the rigid body is only **partially constrained**, so that the equations of equilibrium cannot be satisfied for arbitrary loadings. But if the loads are suitably restricted, some of the equations become trivial identities of the form 0=0, or it may happen that the six equations are not independent as in Sample Problem 4.4.1, so that the remaining equations can be solved for the unknowns. **Improper constraint** of a three-dimensional body may also occur. For example, when the support reactions must be supplied by links that all intersect one line, moment equilibrium about that line can require excessive elongation of the links as the body rotates.

Vector equilibrium equations are more often convenient to use directly (instead of component equations) in three dimensions than in coplanar problems. But even with three-dimensional problems it is often simpler to work with the component equations. Both procedures are illustrated in the sample problems of this section. Both forms of the equations are repeated below. Recall that only the **external** forces and couples shown acting on a free body enter into the equations for that body. Every force or couple shown on the free body will appear in one or more of the equations of equilibrium for the body.

Equations of Equilibrium for External Forces and Couples

$$\sum \mathbf{f} = 0 \tag{4.4.1}$$

and

$$\sum \mathbf{m}_O = 0 \tag{4.4.2}$$

or

$$\sum f_x = 0, \qquad \sum f_y = 0, \qquad \sum f_z = 0 \tag{4.4.3}$$

$$\sum m_{Ox} = 0, \qquad \sum m_{Oy} = 0, \qquad \sum m_{Oz} = 0. \tag{4.4.4}$$

SAMPLE PROBLEM 4.4.1

The uniform beam AB shown in Fig. 4-25(a) is 10 ft long and weighs 60 lb. It is supported in the horizontal plane by a ball-and-socket joint at A and by the two light cables CB and DB as shown. Determine the support reactions for equilibrium.

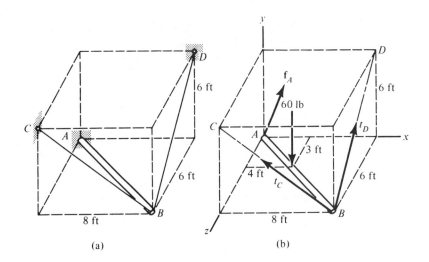

Fig. 4-25 Sample Problem 4.4.1

SOLUTION.   Two solutions will be given. One uses the component equations of equilibrium, while the other uses the vector equations. Figure 4-25(b) shows the free-body diagram with $\mathbf{f}_A$ denoting the force (of unknown direction) exerted on AB by the ball-and-socket joint at A. The cable forces $\mathbf{t}_C$ and $\mathbf{t}_D$ have known directions, as shown, and are labeled by their unknown magnitudes t_C and t_D.

We represent all vectors in standard form:

$$\mathbf{f}_A = f_{Ax}\mathbf{i} + f_{Ay}\mathbf{j} + f_{Az}\mathbf{k}$$

$$\mathbf{t}_C = t_C\hat{\mathbf{e}}_{BC} = -0.8t_C\mathbf{i} + 0.6t_C\mathbf{j}$$

$$\mathbf{t}_D = t_D\hat{\mathbf{e}}_{BD} = 0.707t_D\mathbf{j} - 0.707t_D\mathbf{k}$$

$$\text{weight:}\quad \mathbf{f}_W = -60\mathbf{j}\ \text{lb}.$$

For the component solution it is convenient to represent all forces on the free-body diagram by their rectangular components if this does not clutter the figure too much. But in doing this we must incorporate the known information about the direction of the forces. For example, we show the components of $\mathbf{t}_C$ as $0.8t_C$ acting in the negative x-direction and $0.6t_C$ acting in the positive y-direction. (Do **not** merely show unknowns t_{Cx}, t_{Cy}, t_{Cz} when you know the force direction.) The free-body diagram for the component solution is shown in Fig. 4-26. For this solution the force components at A have been denoted by A_x, A_y, A_z. We now apply the three moment equilibrium equations. Because the forces are all shown by rectangular components, the moment of each component equation about an axis can be read off as a force times a perpendicular moment arm, with algebraic sign suitably chosen for positive moments to produce counterclockwise rotation (as seen from the positive end of the axis), i.e., according to the usual right-hand rule.

$$[\Sigma \, m_x = 0]: \quad -6(0.6t_C) - 6(0.707t_D) + 3(60) = 0$$

$$[\Sigma \, m_y = 0]: \quad -6(0.8t_C) + 8(0.707t_D) = 0$$

$$[\Sigma \, m_z = 0]: \quad 8(0.6t_C) + 8(0.707t_D) - 4(60) = 0.$$

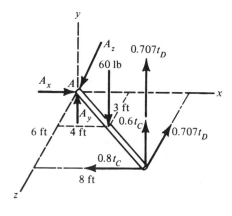

The first and last of these three are equivalent (both reduce to $1.2t_C + 1.414t_D = 60$). Simultaneous solution with the second equation then yields

$$t_C = 25 \text{ lb}, \qquad t_D = 21.21 \text{ lb} \quad \textbf{\textit{Answer.}}$$

The force equations

$$[\Sigma \, f_x = 0]: \quad A_x - 0.8t_C = 0$$

$$[\Sigma \, f_y = 0]: \quad A_y + 0.6t_C + 0.707t_D - 60 = 0$$

$$[\Sigma \, f_z = 0]: \quad A_z - 0.707t_D = 0$$

Fig. 4-26   Free-Body Diagram for Component Solution of Sample Problem 4.4.1

then furnish the results

$$A_x = 20 \text{ lb}, \qquad A_y = 30 \text{ lb}, \qquad A_z = 15 \text{ lb} \quad \textbf{\textit{Answer.}}$$

SOLUTION WITH VECTOR EQUILIBRIUM EQUATIONS.   It is usually advantageous to apply the moment equation first, choosing as moment center the point with the most unknown force components. Refer to the free-body diagram of Fig. 4-25(b), since Fig. 4-26 would not necessarily have been drawn for this second method of solution.

$$[\Sigma \mathbf{m}_A = 0]:$$

$$(4\mathbf{i} + 3\mathbf{k}) \times (-60\mathbf{j}) + (8\mathbf{i} + 6\mathbf{k}) \times [(-0.8t_C\mathbf{i} + 0.6t_C\mathbf{j}) + (0.707t_D\mathbf{j} - 0.707t_D\mathbf{k})] = 0,$$

whence

$$(-3.6t_C - 4.24t_D + 180)\mathbf{i} + (-4.8t_C + 5.656t_D)\mathbf{j} + (4.8t_C + 5.656t_D - 240)\mathbf{k} = 0.$$

This furnishes three component equations:

$$-3.6t_C - 4.24t_D + 180 = 0$$
$$-4.8t_C + 5.656t_D = 0$$
$$4.8t_C + 5.656t_C - 240 = 0.$$

These equations are of course the same as the three moment equations we obtained before by taking moments about the three coordinate axes through A. Hence, we find again $t_C = 25$ lb and $t_D = 21.21$ lb. Then the vector equation

$$[\sum \mathbf{f} = 0]: \quad \mathbf{f}_A - 0.8t_C\mathbf{i} + 0.6t_C\mathbf{j} + 0.707t_D\mathbf{j} - 0.707t_D\mathbf{k} - 60\mathbf{j} = 0$$

gives

$$\mathbf{f}_A = 20\mathbf{i} + 30\mathbf{j} + 15\mathbf{k} \text{ lb} \quad \textit{Answer.}$$

The two procedures can be mixed. We could, for example, use the vector moment equation and the three component force equations.

The reason that there were only five independent equations and five unknowns is that the supports were such that all the reactions are forces and all the forces acting on the body intersect one line, namely AB. This kind of support could not maintain equilibrium if the loads included an applied couple with a moment around the axis AB.

SAMPLE PROBLEM 4.4.2

The uniform 12×4 ft platform shown in Fig. 4-27(a) weighs 600 lb. It is supported in a horizontal position by hinges at two corners and by the light rod CD, which we model as though it were connected by ball-and-socket joints at C and D. Determine as many equilibrium support reactions as possible.

SOLUTION. We assume that the hinge couples are negligible, since rotation is prevented by other supports. Since CD is a two-force member, the force it exerts on the platform must be in line with CD. Hence

$$\mathbf{f}_C = f_C\hat{\mathbf{e}}_{CD} = -\tfrac{6}{7}f_C\mathbf{i} + \tfrac{3}{7}f_C\mathbf{j} - \tfrac{2}{7}f_C\mathbf{k}.$$

On the free-body diagram of Fig. 4-27(b) all forces have been represented by their rectangular components. The symbol f_C denotes a magnitude, and the senses of the components of $\mathbf{f}_C$ have been shown by the directions of the arrows instead of by algebraic signs. We consider component equations of equilibrium:

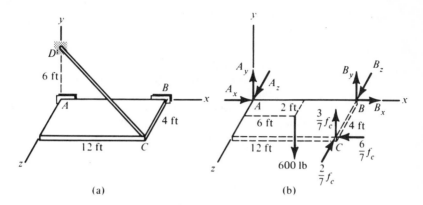

Fig. 4-27 Sample Problem 4.4.2

$$[\textstyle\sum m_x = 0]: \quad 2(600) - 4(\tfrac{3}{7}f_C) = 0, \qquad f_C = 700 \text{ lb}$$

$$[\textstyle\sum m_{BC} = 0]: \quad 6(600) - 12A_y, \qquad A_y = 300 \text{ lb}$$

$$[\textstyle\sum m_y = 0]: \quad 12(\tfrac{2}{7}f_C) - 4(\tfrac{6}{7}f_C) - 12B_z = 0$$

or (with $f_C = 700$ lb),

$$2400 - 2400 = 12B_z, \qquad B_z = 0.$$

$$[\textstyle\sum f_x = 0]: \quad A_x + B_x - \tfrac{6}{7}f_C = 0, \qquad A_x + B_x = 600 \text{ lb}$$

$$[\textstyle\sum f_y = 0]: \quad A_y + B_y + \tfrac{3}{7}f_C - 600 = 0 \quad \text{or}$$

$$300 + B_y + 300 - 600 = 0, \qquad B_y = 0$$

$$[\textstyle\sum f_z = 0]: \quad A_z + B_z - \tfrac{2}{7}f_C = 0 \quad \text{or}$$

$$A_z + 0 - 200 = 0, \qquad A_z = 200 \text{ lb}$$

Thus

$$A_x + B_x = 600 \text{ lb}, \qquad A_y = 300 \text{ lb}, \qquad A_z = 200 \text{ lb}.$$

$$B_y = 0, \qquad\qquad B_z = 0, \qquad\qquad f_C = 700 \text{ lb} \quad \textbf{\textit{Answer.}}$$

The supports are not statically determinate, since there were seven unknowns, but all of them were determinate except A_x and B_x, whose sum is known. If only one of the two hinges could support axial force, the problem would be completely determinate.

SAMPLE PROBLEM 4.4.3

Figure 4-28 shows two views of a single-wheel retractable landing gear for a small airplane (one side of the main gear of a tricycle system). The gear is shown subjected to an upward runway reaction of 70 kN acting at an angle of 12° to the oleo strut (telescoping main strut). This runway reaction is assumed transmitted to the axle at the wheel bearing E, and it is required to determine the support reaction components at A, B, and C. Couple reactions at A and B are assumed negligible, and only the bearing support at A is assumed capable of exerting a side reaction.

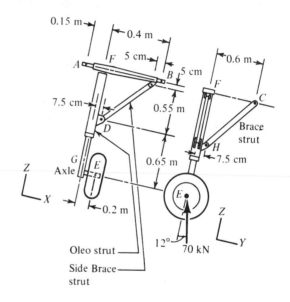

Fig. 4-28   Sample Problem 4.4.3

SOLUTION.   Figure 4-29 shows a simplified three-dimensional free-body diagram sketch for the gear with the wheel removed and each member represented by its center line. The runway reaction has been represented by its components $70 \cos 12° = 68.47$ kN and $70 \sin 12° = 14.55$ kN. HC is a two-force member of length $(0.6^2 + 0.525^2)^{1/2} = 0.797$ m. Its support force C has been resolved into components $C_z = (0.6/0.797)C = 0.753C$ and $C_y = (0.525/0.797)C = 0.658C$. We write equations of equilibrium, as follows, so that only one unknown appears in each equation and solve for the unknown components:

$$[\textstyle\sum m_{AB} = 0]: \quad 0.6(0.753C) - 1.25(14.55) = 0,$$

$$C_z = 0.753C = 30.31 \text{ kN} \quad \textit{Answer.}$$

Hence

$$C = 40.3 \text{ kN} \quad \text{and} \quad C_y = 0.658C = 26.5 \text{ kN} \quad \textbf{\textit{Answer.}}$$

$[\sum m_{By} = 0]$: $0.40(30.31) + 0.20(68.47) - 0.55A_z = 0$, $A_z = 46.9 \text{ kN}$ **Answer.**

$[\sum m_{Az} = 0]$: $0.55B_y - 0.35(14.55) + 0.15(26.5) = 0$, $B_y = 2.03 \text{ kN}$ **Answer.**

$[\sum f_x = 0]$: $A_x = 0$ **Answer.**

$[\sum f_y = 0]$: $2.03 + 26.5 - 14.55 - A_y = 0$, $A_y = 13.98 \text{ kN}$ **Answer.**

$[\sum f_z = 0]$: $68.47 - 46.9 + 30.31 - B_z = 0$, $B_z = 51.9 \text{ kN}$ **Answer.**

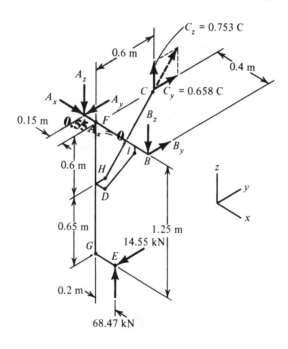

Fig. 4-29 Free-Body Diagram for Sample Problem 4.4.3

(As a check you can verify that with these results moment equilibrium is satisfied about coordinate axes through E.)

EXERCISES

1. The 6×8 ft uniform sliding door weighs 120 lb. It is supported by two flanged wheels A and B on an overhead rail, and its smooth lower edge is restrained by an unflanged roller at corner D. When the door is in such a position that D is just below B, a force P of magnitude 90 lb acts perpendicular to the plane of the door at the middle of its lower edge as shown. Determine the reactions at the wheels and the roller.

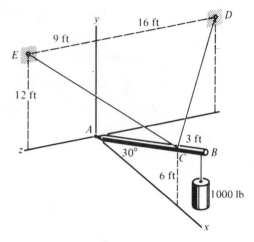

Exercise 3

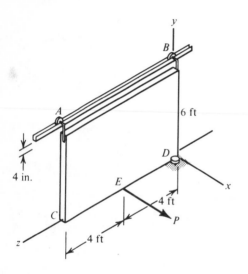

Exercise 1

2. Solve Ex. 1 if the roller is halfway between D and E.

3. The 100-lb uniform boom AB, 15 ft long, is supported in the xy-plane inclined at 30° to the x-axis by a ball-and-socket joint at A and by cables from C to points D and E in the yz-plane as shown. Determine the support reactions for equilibrium when the boom carries a 1000-lb load at B.

4. Solve Ex. 3 if the two cables are lengthened until the boom lies on the x-axis.

5. The light 15-ft bar DE is supported in the xy-plane by a ball-and-socket joint at D and by two light bars BC and AC connected by ball-and-socket joints at their ends. It carries two loads at E, as shown: 1200 lb downward and 800 lb in the negative z-direction. Determine the support reactions at A, B, and D.

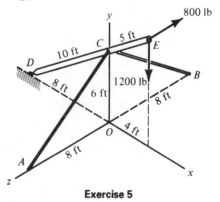

Exercise 5

6. The rigid right-angle bar is formed from a uniform bar weighing 6 lb/ft. It is built in at

A and carries two loads at *D*, a horizontal force **f** of magnitude 60 lb perpendicular to *BD* as shown and a couple *C* of 300 lb/ft acting around the axis *DB* with the sense shown. Determine the support reactions at *A* for equilibrium.

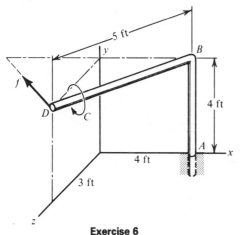

Exercise 6

7. The uniform quarter-circular curved beam *AB* of radius 6 ft is supported in the horizontal *xz*-plane by the built-in end *A*. It weighs 100 lb and carries a force **f** of magnitude 200 lb in the horizontal plane making a 40° angle with the negative *x*-direction as shown. Determine the support reactions at *A*.

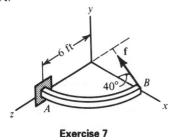

Exercise 7

8. Two semicircular plates, each of 16-in. radius and each weighing 60 lb, are welded to a vertical rod 40 in. long weighing 20 lb and supported by a thrust bearing at *A* and a journal bearing at *B*, as shown. Determine the support reactions for equilibrium.

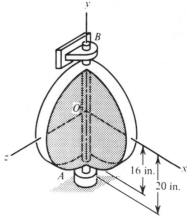

Exercise 8

9. The rigid body shown is constructed of welded tubing weighing 10 lb/ft. It consists of a vertical 8-ft post and a horizontal T-shaped portion with *BC* = 5 ft and *DE* = 4 ft. It is supported by a ball-and-socket joint at *A* and three light guy wires *BF*, *DG*, and *EH* as shown (*EH* is in line with *DE*, parallel to the *z*-axis). In addition to its own weight the structure carries a 100-lb load at *E*. Determine the support reactions for equilibrium.

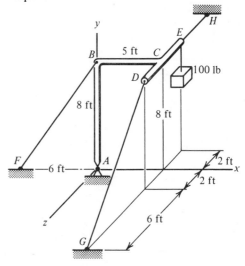

Exercise 9

10. Solve Ex. 9 if the three guy wires are removed and the end *A* is built in.

11. The light rigid Y-shaped member *AOBC* in the horizontal *xz*-plane is built in at *A* and carries three loads: a 100-lb downward force at *C*, a 600-lb-in. couple acting around the shaft *CO* as shown, and force **f** = $200\mathbf{i} - 200\mathbf{j} + 100\mathbf{k}$ lb at *B*. Determine the support reactions at *A* for equilibrium.

13. Solve Ex. 12 if the two bars *AC* and *AD* are lengthened and attached at point (12, 6, 12) ft to *OB*.

14. The 100-lb downward force on the platform *C* of the bell crank balances the tensile force *T* in the control rod attached to *D* at 30° to the horizontal in a plane parallel to the plane of the bell crank as shown. Determine the bearing reactions.

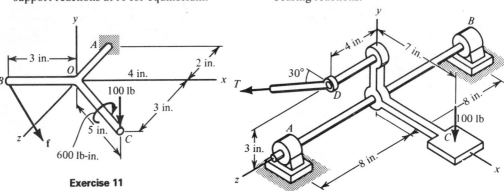

Exercise 11

Exercise 14

12. The 24-ft light boom *OAB* is supported at its midpoint *A*(8, 4, 8) ft by two light ball-and-socket-connected bars *AC* and *AD* as shown and by a ball-and-socket joint at *O*. It carries a 1000-lb load at *B*. Determine the support reactions for equilibrium.

15. Solve Ex. 14 if bearing *A* is removed and bearing *B* is the only support.

16. The uniform 1×2.4 m trapdoor has mass 200 kg. It is hinged at corners *A* and *B*, and, in the position shown, the light cable *EF* supports it at the midpoint *E* of *CD*, so that edge *CD* is 0.6 m below the horizontal *xz*-plane. Determine the support reactions insofar as possible.

17. Solve Ex. 16 if the cable is attached at *D* instead of *E* and the trapdoor is raised until it lies in the horizontal plane.

18. The T-shaped frame shown is supported by three casters that can roll freely, but the casters are locked so that they cannot swivel about their vertical pins. Caster *A* can roll in the *x*-direction, while casters *B* and *C* can roll in the *z*-direction. Determine the support reactions for equilibrium if the frame carries a 100-lb weight *W* as shown and a 20-lb force *P* acts in the *z*-direction at the center of the T. If the coefficient of static friction is 0.5 between rollers and floor, will equilibrium be maintained?

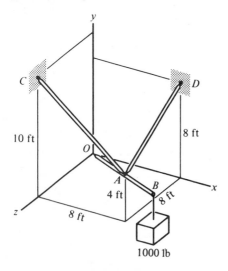

Exercise 12

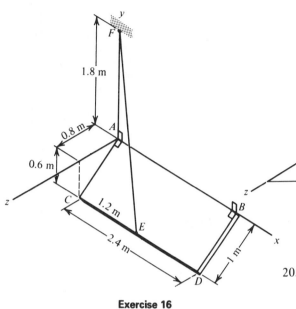

Exercise 16

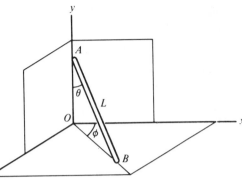

Exercise 19

20. If in Ex. 19, $\theta = 26.6°$, $\varphi = 45°$, and a *horizontal force* of magnitude $W/4$ acts perpendicular to the beam at its midpoint in a sense tending to increase φ, (a) determine the support reactions for equilibrium and, (b) determine the minimum coefficient of static friction at B for equilibrium.

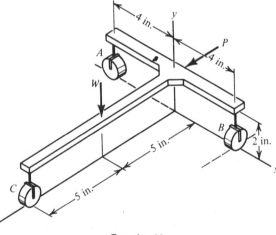

Exercise 18

19. The uniform pole AB of length L and weight W leans at angle θ to the vertical against the corner of two smooth walls as shown and rests on a rough horizontal floor at B. (OB is at angle φ to the x-axis.) (a) Determine the support reactions for equilibrium. (b) What is the largest possible angle θ for equilibrium if the coefficient of static friction at B is μ?

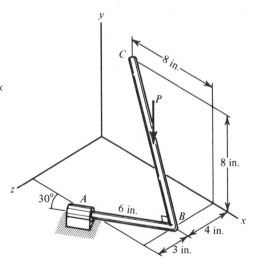

Exercise 20

21. The right-angle L-shaped light arm *ABC* is formed with *AB* = 6 in., *BC* = 12 in. It is supported by a thrust bearing at *A* and leans against a smooth wall at *C* as shown. Determine the support reactions if the arm carries a downward load *P* = 100 lb at the midpoint of *BC* as shown. Express your answers as vectors in terms of **i**, **j**, and **k**.

22. Determine the force acting at *D* and the couple component c_{Fz} and force components exerted at *F* on the oleo strut by the member *AB* in Sample Problem 4.4.3. The fitting that connects *AB* to the oleo strut can transmit only the one couple component (torque preventing rotation of the oleo strut about its axis). (*Hint*: Draw a free-body diagram of the oleo unit *FGE* with member *AB* removed.)

4.5 SUMMARY

The most general system of forces **f** and couples **c** acting on a body is equipollent (i.e., for rigid-body mechanics, equivalent) to a ***resultant force*** $\mathbf{f}^R$ acting at an assigned point *A* plus a ***resultant couple*** $\mathbf{c}_A^R$,

$$\mathbf{f}^R = \sum \mathbf{f} \qquad \mathbf{c}_A^R = \sum \mathbf{c} + \sum (\mathbf{r}' \times \mathbf{f}) \tag{4.5.1}$$

where **r**′ is the vector from *A* to any point on the line of action of **f**. These can be combined into a ***wrench resultant*** (see Fig. 4-2) at another point.

Special cases considered in Sec. 4.1 included

1. The resultant is a couple $\mathbf{c}^R$ when $\sum \mathbf{f} = 0$.
2. ***Concurrent forces*** at *A*. The resultant is $\mathbf{f}^R$, since $\mathbf{c}_A^R = 0$.
3. ***Coplanar system.*** $\mathbf{f}^R$ in the plane and $\mathbf{c}_A^R$ perpendicular to it (represented by two force components and one couple component) can be easily combined into a single force at another point if $\mathbf{f}^R \neq 0$.
4. ***Parallel force system.*** Resultant $\mathbf{f}^R$ and $\mathbf{c}_A^R$ can be easily combined into a single force at another point if $\mathbf{f}^R \neq 0$.

In Sec. 4.2 we considered distributed ***parallel force systems*** and defined ***center of mass***, Eqs. (4.2.2), and (4.2.9); ***center of gravity***, Fig. 4-9; and ***centroid***, Eqs. (4.2.10) with ρ = const. In a uniform parallel gravitational force field the distributed weight forces acting on a body are equipollent to a single force, the total weight, acting at the center of mass. When the density is uniform (ρ = const.), the center of mass is the centroid of the body. Methods of locating centroids are given in Secs. 7.3–7.5. Tables A3 and A4 of Appendix A give the locations for a number of common geometric shapes of arcs, areas, and volumes.

Rigid-body equilibrium methods were presented in Sec. 4.3 for coplanar problems and in Sec. 4.4 for three dimensions. The most important technique in the solutions is the careful construction of the ***free-body diagram***, isolated from its supports and showing all the known and unknown ***external forces that act on the body***, including support reactions.

The vector *equations of equilibrium*

$$\sum \mathbf{f} = 0, \qquad \sum \mathbf{m}_o = 0 \tag{4.5.2}$$

give at most six independent component equations per free body (three independent equations for coplanar problems). In certain cases the number of useful equations is further reduced (e.g., all forces concurrent in a point, or all forces intersecting a line in a three-dimensional problem). If the free-body diagram is correctly drawn, the equations of equilibrium should contain no forces or couples not shown on the diagrams, and every force or couple shown should appear in at least one equilibrium equation.

If there are more unknowns than equations, the equilibrium problem is statically indeterminate (see, e.g., Fig. 4-15).

For a rigid body to be *completely constrained* and have *statically determinate* reaction components there must be as many unknown reaction components as there are equations of equilibrium. In certain cases of improper constraint (e.g., Figs. 4-17 and 4-18) the reactions may still be *indeterminate*, or the body may not be completely constrained, even when there are as many unknowns as equations.

Equilibrium solutions are frequently simplified by recognizing each *two-force body* (see Fig. 4-19). Also for equilibrium of a *three-force body* (see Fig. 4-20), the forces must be coplanar and either concurrent or parallel. Other simplifications are made possible by the *support reaction idealizations*, summarized in Figs. 4-12 and 4-24. Before proceeding to the following chapters, the student is advised to study those two figures again and reread the discussion accompanying them on techniques of constructing and using a free-body diagram.

In Chapter 5 the power of the free-body diagram technique will become fully apparent. There we shall consider connected systems of rigid bodies. When we take them apart and draw in a correct systematic fashion the interaction forces which now become external forces on the separate free bodies we are able to formulate in a straightforward manner problems which would otherwise seem hopelessly involved. We shall also consider in Chapter 5 the separation of one rigid body into two parts in the discussion of the stress resultants, including bending moments and shear forces transmitted from one part of a beam to another one across an internal cross section. A section on flexible cables whose weight is not negligible is also included in Chapter 5.

CHAPTER FIVE

Structures and Machines

5.1 CONNECTED BODIES; STATICALLY DETERMINATE FRAMES AND MACHINES

The procedure for simplifying and analyzing a structure or machine by separating it into component free bodies will be presented by examples. We shall illustrate it first by two simple coplanar examples.

EXAMPLE 5.1.1

Consider the coplanar frame formed by two uniform members, each weighing 50 lb, connected to each other and to the supports by smooth pins and loaded as shown in Fig. 5-1(a). The free-body diagram of the whole structure is shown in Fig. 5-1(b). Since there are four unknown reactions and only three equations of equilibrium available we cannot determine all the unknown reactions on this free body, although we could determine the vertical reactions.

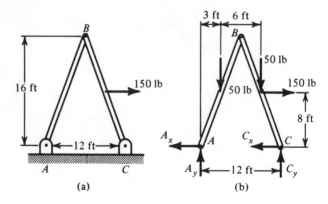

Fig. 5-1 Coplanar Frame of Example 5.1.1

We now separate the structure into two separate free bodies as shown in Fig. 5-2, supposing that the pin at *B* is part of one of the bodies, say of *AB*. The force that this pin exerts on *BC* now becomes an ***external force*** in the free-body diagram of Fig. 5-2(b). Note that it does not appear in the free-body diagram of the whole structure in Fig. 5-1 because there it is an internal force. We have represented this force by two component magnitudes B_x and B_y assumed acting to the left and up at point *B* on *BC*.

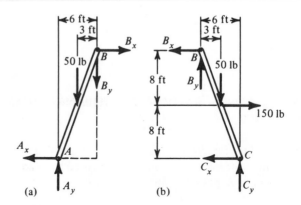

Fig. 5-2 Free-Body Diagrams of the Separate Members for Example 5.1.1

If we have guessed a direction wrong, the solution will give a negative value for the magnitude of the component guessed wrong. By Newton's third law an equal and opposite force acts on pin *B* of *AB*. This fact has been expressed by drawing the components acting at *B* on *AB* **in the opposite directions but labeled with the same symbols** B_x **and** B_y. Be very attentive to the convention just stated. A consistent procedure here is the key to successful analysis of connected bodies. When only a single body is involved you may either try to guess the correct directions for support reactions B_x and B_y or you may always show them pointing in the positive coordinate directions. But the second alternative is not available for the interaction forces on both bodies when you draw separate diagrams of the parts. If you show B_x to the right on AB, you *must show it to the left on BC.*

Now although each of the two free bodies in Fig. 5-2 has four unknown components acting on it, there are only six unknowns altogether because two of the unknowns appear on both free bodies. Since we have three equations for each body, the problem is now statically determinate.

When you write equilibrium equations for more than one free body, be sure to label the equations in some way to indicate what body they apply to. For example, we may write the following equations:

For AB:

$$[+\backslash) \sum m_A = 0]: \quad -16B_x - 6B_y - 3(50) = 0 \tag{5.1.1a}$$

$$[\sum f_x = 0]: \quad -A_x + B_x = 0 \tag{5.1.1b}$$

$$[\sum f_y = 0]: \quad A_y - B_y - 50 = 0 \tag{5.1.1c}$$

For BC:

$$[+\backslash) \sum m_C = 0]: \quad 16B_x - 6B_y + 3(50) - 8(150) = 0 \tag{5.1.1d}$$

$$[\sum f_x = 0]: \quad -B_x - C_x = 0 \tag{5.1.1e}$$

$$[\sum f_y = 0]: \quad B_y + C_y - 50 = 0. \tag{5.1.1f}$$

Simultaneous solution of these six equations determines the six unknowns. We can begin by adding Eqs. (1.5.1a) and (1.5.1d) to eliminate B_x and find $B_y = -100$ lb. Hence B_y is actually 100 lb *down on BC* and 100 lb *up on AB*. The equations then yield

$$A_x = B_x = 27.5 \text{ lb}, \qquad A_y = -50 \text{ lb}$$
$$C_x = -25 \text{ lb}, \qquad C_y = 150 \text{ lb}.$$

As a check we can verify that the values obtained for the reactions at A and C satisfy the following equilibrium equations for the free-body diagram of the whole structure, Fig. 5-1(b). Although the body ABC does not appear to be a rigid body when it has been removed from its supports, we can still use rigid-body equilibrium equations for the reactions of the supports, since when these reactions are exerted by the supports, no motion is occurring.

For *ABC*:

$$[+\curvearrowright \sum m_A = 0]:\quad 12C_y - 3(50) - 9(50) - 8(150) = 0,\qquad C_y = 150\,\text{lb}$$

$$[\sum f_y = 0]:\quad A_y + C_y - 50 - 50 = 0,\qquad A_y = -50\,\text{lb}\quad (5.1.2)$$

$$[\sum f_x = 0]:\quad -A_x - C_x = 0,\qquad A_x = -C_x.$$

Actually it would have been a little quicker to solve the problem by using as the two free bodies the whole body *ABC* and body *AB* and then check with body *BC*, but the main purpose of the example was to illustrate the sign conventions on the interaction forces. It is usually best to consider the whole body first.

Another possible procedure is to put unit vectors on the components and label them, for example, as $B_x\mathbf{i}$ on *AB* and $-B_x\mathbf{i}$ on *BC*, in Fig. 5-2, since one vector is the negative of the other. Or, if components are not used, you can show a vector $\mathbf{f}_B$ on one at an arbitrary direction and label $-\mathbf{f}_B$ on the other, drawn in the opposite direction. The important thing is to use a consistent procedure. The component method illustrated in Example 5.1.1 is usually the most efficient, unless it clutters up the figures too much.

For the forces at *B* it was suggested that the pin at *B* could be considered a part of one of the members. When only two members are connected to a pin, it does not matter which one is assumed to have the pin. The free-body diagrams look the same for either choice. But, as the second example will illustrate, it can make a difference in the following three cases:

Pin Location in Free Body

1. When more than two members are connected by one pin,
2. when two members are pinned to one support, or
3. when an applied load acts on a pin connecting two members, it may make a difference where the pin is assumed to be.

It is usually more efficient to assume that the pin is part of a multiforce member instead of a two-force member.

EXAMPLE 5.1.2

The coplanar structure in Fig. 5-3(a) is formed of light pin-connected bars and carries a downward load *P* as shown. We wish to determine the forces acting on member

AB. Figure 5-3(b) shows a free-body diagram of the whole structure except for the two-force member *CD*, which exerts the horizontal force C_x. Moment equilibrium about *A* shows that $C_x = P$. Force summations then show that $A_x = P$ and $A_y = P$ also.

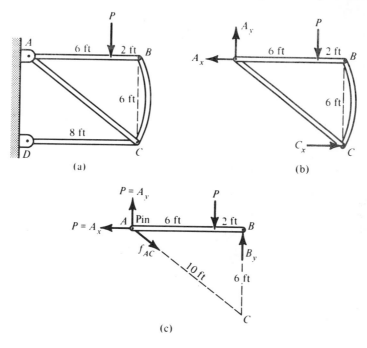

Fig. 5-3   Example 5.1.2

Two different possibilities will be illustrated. We shall first consider the pin at *A* to be part of the multiforce member *AB* as shown in Fig. 5-3(c). The free-body diagrams of *AC* and *BC* are not shown, since they are recognized as two-force members whose actions at *A* and *B* are in line with *AC* and *BC*, respectively. (f_{BC} has been labeled B_y because it is a vertical force on pin *B*.) We have shown $A_y = P$ and $A_x = P$, as previously determined from the analysis of the whole structure *ABC*.

For *AB*:

$$[+\circlearrowleft \sum m_A = 0]: \quad 8B_y - 6P = 0, \qquad\qquad B_y = \tfrac{3}{4}P$$

$$[\sum f_x = 0]: \quad -P + \tfrac{4}{5}f_{AC} = 0, \qquad\qquad f_{AC} = \tfrac{5}{4}P$$

$$[\sum f_y = 0]: \quad P - P + B_y - \tfrac{3}{5}f_{AC} = 0, \qquad B_y = \tfrac{3}{4}P.$$

The second procedure is illustrated in Fig. 5-4, where pin *A* is assumed part of member *AC*. The interaction forces between the pin and member *AB* have been labeled A'_x and A'_y, since they are not the same as the forces A_x and A_y exerted on the pin by the support.

For AB in Fig. 5-4:

$$[+\text{\Large)}\, \textstyle\sum m_A = 0]: \quad 8B_y - 6P = 0, \qquad B_y = \tfrac{3}{4}P$$

$$[\textstyle\sum f_y = 0]: \quad A'_y + B_y - P = 0, \qquad A'_y = \tfrac{1}{4}P$$

$$[\textstyle\sum f_x = 0]: \qquad\qquad\qquad\qquad A'_x = 0.$$

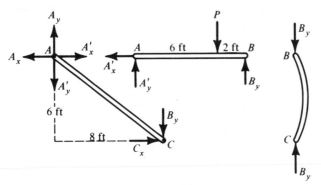

Fig. 5-4   Free Bodies with Pin A in AC

We can check that these results satisfy the equations of equilibrium for AC. Note that we have labeled the vertical force at C by B_y because of the two-force member BC.

For AC:

$$[+\text{\Large)}\, \textstyle\sum m_A = 0]: \quad 6C_x - 8B_y = 0, \qquad 6P - 8(\tfrac{3}{4}P) = 0$$

$$[\textstyle\sum f_x = 0]: \quad A'_x - A_x + C_x = 0, \qquad 0 - P + P = 0$$

$$[\textstyle\sum f_y = 0]: \quad A_y - A'_y - B_y = 0, \qquad P - \tfrac{1}{4}P - \tfrac{3}{4}P = 0.$$

The second procedure gave the same force B_y as the first one, but the forces acting on AB at A are now different, since they are forces exerted by the pin on AB instead of forces exerted by the foundation and member AC acting on the pin. The two sets of forces have the same resultant, because the pin itself is in equilibrium. The first procedure is usually preferable, but the important thing is to adopt a procedure, make clear what the procedure is by appropriate labeling of the free-body diagrams, and then follow it consistently.

Degrees of Freedom of a System

In complicated machines the parts may move relative to each other. To characterize the **configuration** of such a system of bodies we specify a certain number of **generalized coordinates** defining the positions of its parts. A "generalized coordinate" may be an actual x, y, or z coordinate of a point on the body, or it may be the angle that a bar makes with another bar, or it may be the perpendicular distance from a point on one bar to another bar. We shall not attempt to list all the possibilities but shall discuss a few simple examples. If there are a total of N generalized coordinates for a system and they can change independently without violating any constraints placed on the system, we say that the system has N **degrees of freedom**.

For example, a free particle in space has three degrees of freedom, a particle constrained to move on a given surface has only two degrees of freedom, and a particle constrained to move on a given curve has only one degree of freedom. (The single independent generalized coordinate for the last case might be the path-length variable s measured from some origin on the curve.)

A rigid body constrained to move in a plane has three degrees of freedom. The generalized coordinates might be chosen as the x- and y-coordinates of one point A on the body and the angle θ_x for the line AB to any other point on the body.

A rigid body free to move in space has six degrees of freedom. The generalized coordinates might be the x, y, z-coordinates of one point A of the body; two direction cosines, say θ_x and θ_y, of a line AB on the body (note that θ_z is not independent of θ_x and θ_y, since the sum of the squares of the three cosines equals unity); and an angle φ through which the body rotates about AB. The general displacement of a rigid body will be considered further in Chapter 10 of Vol. II.

For connected systems of rigid bodies the possibilities can become quite complicated. In this chapter, however, we shall be concerned mainly with systems that are completely constrained (zero degrees of freedom) or in which the complete configuration of a given system is specified by only one variable parameter (one degree of freedom).

Structures; Statically Determinate Rigid Frames

By the word "structure" we usually mean a system of connected bodies supported in such a way that it is completely constrained against motion. For rigid-body analysis this is a system with no degrees of freedom. We sometimes, however, speak of aircraft or aerospace structures, for example, which are not completely constrained. Structures constructed of slender members are classified into two categories: (1) **frames** and (2) **trusses**. A truss is a structure than can be analyzed as though all its members were two-force members, with the loads applied to the structure only at the joints. Truss analysis will be treated in Sec. 5.2.

A frame is a structure constructed of slender members, such that some of the members are multiforce members (not two-force members). We have discussed two simple examples of structural frames at the beginning of this section. Example 5.1.1 had two multiforce members, while Example 5.1.2 had only one, the bar *AB*. The loads on a multiforce member tend to bend it and to shear it as well as to stretch or shorten it. Bending moment and shear force distributions in a beam will be discussed in Sec. 5.4. Here we are concerned only with the interaction forces between the members where they are joined together and with the support reactions.

In most of our examples the frames will be coplanar, and the connections will be assumed to be by smooth pins. Extension to three dimensions is not difficult in principle, but in practice the problems quickly become very complicated. Sample Problem 5.1.2 is a three-dimensional example that can be treated fairly easily by making judicious choices of moment equations. More complicated problems can be analyzed by using advanced techniques of systems analysis (beyond the scope of this book) to formulate the problems and digital computers to solve the problems.

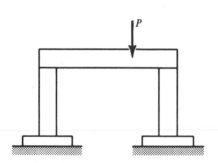

In practical frame design most connections are not made by single pins, but instead the members are welded together or fastened together by several rivets. This makes statically indeterminate even structures that are very simple in appearance, for example, the welded portal frame of Fig. 5-5. The analysis of such frames is beyond the scope of this book.

Fig. 5-5   Welded Portal Frame

Statically Determinate Frames

Even when the frame is pin-connected it may still be statically indeterminate. To study this question for any given frame we first recognize the multiforce members and draw a separate free-body diagram of each multiforce member. We then compare the number of unknowns appearing on these diagrams with the number of independent equations available for these free bodies. When the number of unknowns and equations turns out to be equal (as in Example 5.1.1) the system may be statically determinate and completely constrained. (It might not be if some of the members are improperly constrained. See the discussions of improper constraint in Secs. 4.3 and 4.4.) If there are more unknowns than equations, the system is statically indeterminate. If there are fewer unknowns than equations, the system is not completely constrained; such a structure is called *nonrigid*, but it can still sometimes be analyzed by rigid-body statics for suitably restricted loadings.

Machines or Mechanisms

A partly constrained frame may become a *mechanism*, a structure ordinarily intended to move and to transmit input force or input work from one point of the structure to another point where it becomes output force or work. The most important problems in machines are dynamics problems rather than equilibrium problems. Since machine dynamics usually involves rotational dynamics of rigid bodies, we defer such problems until Chapter 12 of Vol. II.

The analysis of the equilibrium of a mechanism with one degree of freedom is quite similar to that of a frame; problems of the two kinds are mixed together at the end of this section. One feature appearing in some of the mechanism problems that does not enter the frame problems is that for given loads it may be possible for equilibrium to occur only for certain configurations of the body (specified by certain values of the single parameter defining the configuration of the one-degree-of-freedom system). Part of the problem is to determine the possible values of the parameter for equilibrium. An example of this is given in Sample Problem 5.1.1. Another example is in the discussion of stability in connection with Fig. 5-6 below, where the determination of the equilibrium positions is trivial.

Stability of Equilibrium

When a body is not completely constrained **unstable equilibrium** may occur. Unstable equilibrium means that although under the assigned external forces the body is in equilibrium in a certain position, any small accidental displacement from the equilibrium position gives rise to an unbalanced force system tending to increase the displacement. Three trivial examples are shown in Fig. 5-6, where a uniform bar of weight W is supported by a single pin, either at one end or at its center. The upper figures show the initial equilibrium positions, while the free-body diagram below each is after a small rotation about A. The actual analysis of the rotational dynamics following the small perturbation is beyond the scope of this chapter, but it is intuitively clear that case (a) is **stable**, since after a small displacement the moment of W tends to return the bar to its initial position, while case (b) is **unstable**, because the moment of W now tends to increase any small displacement. Case (c) is *neutral*, since there is neither a moment tending to return the bar to the original equilibrium configuration nor one tending to increase the displacement. For this case any position is a possible equilibrium position. Methods of finding the equilibrium configuration of a mechanism and of deciding the stability characteristics of an equilibrium configuration by potential energy methods will be discussed in Sec. 6.2.

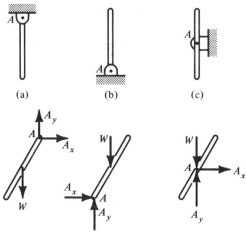

Fig. 5-6 Examples of Stability and Instability of Equilibrium

In Sec. 5.2 we shall consider truss analysis. We shall close Sec. 5.1 with several sample problems on analysis of pin-connected frames and machines.

SAMPLE PROBLEM 5.1.1

The coplanar "four-bar linkage" mechanism schematically represented in Fig. 5-7 consists of three identical light rigid bars each of length $2L$, connected to each other and to the foundation (the "fourth bar") by smooth pins. For what angles θ can the mechanism be in equilibrium, and what are the forces acting on bars AB and BC for equilibrium when given equal forces P act vertically at the midpoint of BC and horizontally at the midpoint of CD as shown?

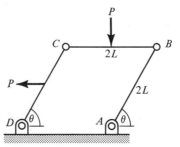

Fig. 5-7   Sample Problem 5.1.1

SOLUTION.   Free-body diagrams for the whole mechanism ABC (three bars) and for bars BC and CD are shown in Fig. 5-8. Since AB is a two-force member, we have represented the force it exerts on BC by the same letter A as for the force the support exerts on AB.

We count six unknowns: A, C_x, C_y, D_x, D_y, and θ on the two free bodies BC and CD. Hence we can hope for statical determinacy. We begin with the whole structure and use two moment equations for each body in order to keep the simultaneous solution simple. We obtain the following six equations.

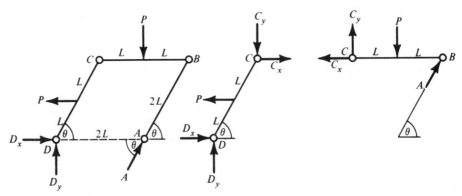

Fig. 5-8   Free-Body Diagrams for Sample Problem 5.1.1

For the whole mechanism $ABCD$:

$$[+\rangle \sum m_D = 0]: \quad (2L \sin \theta)A + (L \sin \theta)P - (L + 2L \cos \theta)P = 0 \qquad \text{(a)}$$

$$[+\rangle \sum m_A = 0]: \quad -2LD_y + (L \sin \theta)P - (2L \cos \theta - L)P = 0 \qquad \text{(b)}$$

$$[\sum f_x = 0]: \quad A \cos \theta + D_x - P = 0 \qquad \text{(c)}$$

For bar BC:

$$[+\rangle \sum m_B = 0]: \quad LP - 2LC_y = 0 \qquad \text{(d)}$$

$$[+\rangle \sum m_C = 0]: \quad 2L(A \sin \theta) - LP = 0 \qquad \text{(e)}$$

$$[\sum f_x = 0]: \quad -C_x + A \cos \theta = 0. \qquad \text{(f)}$$

We solve Eq. (e) for $A = P/(2 \sin \theta)$ and substitute the result into Eq. (a) to obtain

$$LP + LP \sin \theta - LP - 2LP \cos \theta = 0,$$

whence $\tan \theta = 2$. There are thus two possible equilibrium angles:

$$\theta_1 = 63°26' \quad \text{and} \quad \theta_2 = \theta_1 + 180° = 243°26'$$

(assuming that motion is possible down to the second configuration). The equations give the following forces for each configuration:

$$\text{for } \theta = 63°26': \quad \sin \theta = 0.8944, \quad \cos \theta = 0.4472$$
$$A = 0.559P, \quad C_x = \tfrac{1}{4}P, \quad C_y = \tfrac{1}{2}P$$
$$D_x = \tfrac{3}{4}P, \quad D_y = \tfrac{1}{2}P.$$

$$\text{for } \theta = 243°26': \quad \sin \theta = -0.8944, \quad \cos \theta = -0.4472$$
$$A = -0.559P, \quad C_x = \tfrac{1}{4}P, \quad C_y = \tfrac{1}{2}P$$
$$D_x = \tfrac{3}{4}P, \quad D_y = \tfrac{1}{2}P.$$

The negative value for A in the second case means that bar AB is in tension instead of in compression as was assumed in drawing the free-body diagrams. The results can be checked by showing that the results satisfy Eq. (a) and also the equilibrium equations for bar CD.

For *CD*:

$$[\textstyle\sum f_x = 0]: \quad D_x + C_x - P = 0$$

$$[\textstyle\sum f_y = 0]: \quad D_y - C_y = 0$$

$$[+\rangle \textstyle\sum m_D = 0]: \quad -(2L\cos\theta)C_y - (2L\sin\theta)C_x + (L\sin\theta)P = 0.$$

SAMPLE PROBLEM 5.1.2

The horizontal symmetrical A-frame shown in Fig. 5-9(a) is formed of light members and carries a vertical downward load *P* attached to bar *AB* at its midpoint *D*. The connections at *B*, *D*, *E*, *F*, and *G* may be treated as ball-and-socket joints. The frame is supported at *A* and *C* by hinges with axis along *AC*. Only the hinge at *A* can support axial load. Determine the support reactions and the interaction forces between members.

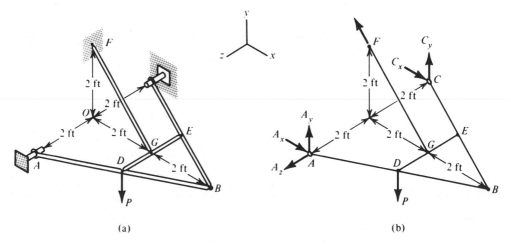

(a)   (b)

Fig. 5-9   Sample Problem 5.1.2

SOLUTION.   Figure 5-9(b) shows the free-body diagram of the whole structure. Note that *FG* is a two-force member and that it has been assumed that the bearing at *C* supports no axial force and that neither bearing exerts any couple.

For the whole structure:

$$[\textstyle\sum m_{Az} = 0]: \quad 2\left(\frac{1}{\sqrt{2}}F\right) - 2P = 0, \qquad\qquad F = P\sqrt{2}$$

$$[\textstyle\sum m_{Ay} = 0]: \quad -4C_x + 2\left(\frac{1}{\sqrt{2}}F\right) = 0, \qquad\qquad C_x = \frac{1}{2}P$$

$$[\Sigma f_x = 0]: \quad A_x + C_x - \frac{1}{\sqrt{2}}F = 0, \qquad A_x = \frac{1}{2}P$$

$$[\Sigma m_{Ax} = 0]: \quad 4C_y - (1)P + 2\left(\frac{1}{\sqrt{2}}F\right) = 0, \qquad C_y = -\frac{1}{4}P$$

$$[\Sigma f_y = 0]: \quad A_y + C_y + \frac{1}{\sqrt{2}}F - P = 0, \qquad A_y = +\frac{1}{4}P$$

$$[\Sigma f_z = 0]: \qquad\qquad\qquad\qquad\qquad\qquad A_z = 0.$$

The separate free-body diagrams for the three multiforce members are shown in Fig. 5-10 with the results of the above analysis for the whole body incorporated. The force $F = P\sqrt{2}$ has been represented by its two components at G, each equal to P.

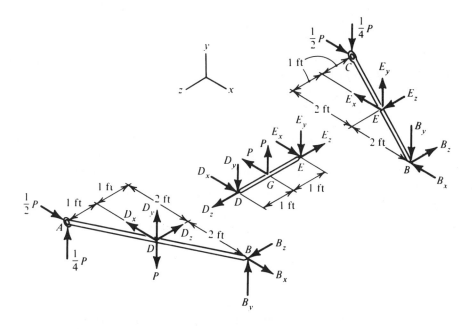

Fig. 5-10 Free-Body Diagrams of Separate Members for Sample Problem 5.1.2

For *DE*:

$$[\Sigma m_{Gx} = 0]: \quad D_y - E_y = 0$$
$$[\Sigma m_{Gy} = 0]: \quad D_x - E_x = 0.$$

Force summations then give

$$D_x = E_x = \tfrac{1}{2}P \quad \text{and} \quad D_y = E_y = \tfrac{1}{2}P.$$

For AB:

$$[\textstyle\sum f_x = 0]:\ \tfrac{1}{2}P + B_x - D_x = 0, \qquad\qquad B_x = 0$$

$$[\textstyle\sum f_y = 0]:\ \tfrac{1}{4}P + D_y - P + B_y = 0, \qquad\quad B_y = \tfrac{1}{4}P$$

$$[\textstyle\sum m_{D_y} = 0]:\ (1)(\tfrac{1}{2}P) - (1)B_x - (2)B_z = 0, \qquad B_z = \tfrac{1}{4}P$$

$$[\textstyle\sum f_z = 0]:\ D_z - B_z = 0, \qquad\qquad\qquad D_z = \tfrac{1}{4}P.$$

For DE:

$$[\textstyle\sum f_z = 0]:\ E_z = D_z, \qquad\qquad\qquad E_z = \tfrac{1}{4}P.$$

You can check by writing the equilibrium equations for bar BC and verifying that they are identically satisfied by these results.

We have used component summations throughout instead of vector moments in order to be able to use different moment axes and thereby to eliminate as many unknowns as possible from each moment equation. And we have applied the equilibrium equations in a sequence that enabled us to avoid solving a complicated system of simultaneous equations. This is a desirable procedure when it is possible. For very large complicated structures, it may not be so easy to choose a suitable sequence.

EXERCISES

1. Determine the support reactions for the co-planar frame, connected by smooth pins at A, B, and C and loaded by the 150-lb downward load shown. Bar weights are negligible.
2. Draw free-body diagrams for the coplanar frame shown, formed from light bars pin-connected except at B and E where smooth contact is assumed. Draw one diagram for the whole structure and one for each member.
3. Determine the components of all forces acting on member ACE of Ex. 2.
4. The plane frame is formed by four light bars AB, BC, DH, HE pin-connected to each

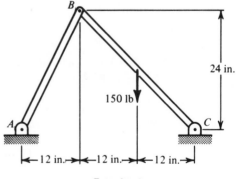

150 lb

24 in.

|←12 in.→|←12 in.→|←12 in.→|

Exercise 1

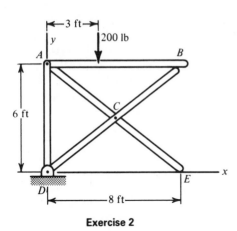

Exercise 2

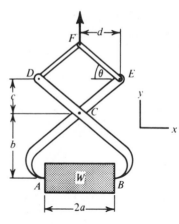

Exercise 6

other and to the supports at *A* and *C* as shown. Draw free-body diagrams of the pin at *H*, member *AB*, member *BC*, and the whole structure.

8. The pin-connected plane frame carries two loads at *G*, as shown. Determine the horizontal and vertical components of all forces acting on member *ACE*.

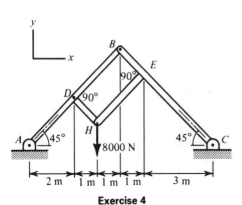

Exercise 4

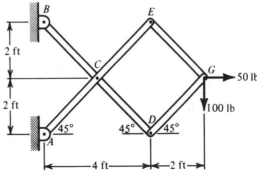

Exercise 8

5. Determine the support reactions for the frame of Ex. 4.

6. The tongs shown are formed by two continuous members *ACE* and *BCD* pin-connected at *C* and to the members *DF* and *EF*. Draw free-body diagrams of the pin *F*, member *ACE*, and member *BCD*. Member weights are negligible.

7. Determine the force components acting on *BCD* at *B* and *C* in Ex. 6 in terms of *a, b, c, d, θ,* and *W*.

9. The 100-lb weight is suspended by a flexible cable that goes over a small frictionless pulley at *C* and is attached to member *DB* at *B*. Determine the minimum coefficient of friction at *D* to prevent collapse of the coplanar frame. Figure is on following page.

10. The symmetrical coplanar A-frame shown rests on a smooth horizontal surface. Determine the force exerted on member *AB* at *A* when a 5-kN downward force acts at *F*. Figure is on following page.

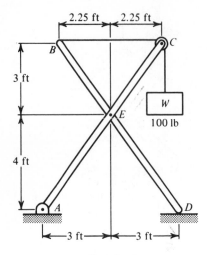

Exercise 9

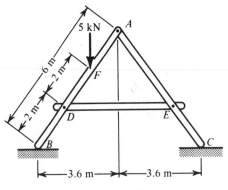

Exercise 10

11. Solve Ex. 10 if the structure is pin-supported at B and a horizontal force of 5 kN acts to the right at F in addition to the vertical force.

12. Determine the support reactions at A and C. Assume frictionless pins, and neglect weights of the pulley and of the two angle members AB and BC.

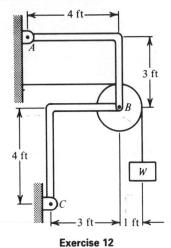

Exercise 12

13. A schematic diagram of a movable overhead hoist is shown. Assume smooth pins, negligible member weights, and flexible cable attached at E. Determine the counterweight C required to prevent tipping and the components of the pin force acting on AD at B when $W = 2000$ lb and the system is in equilibrium.

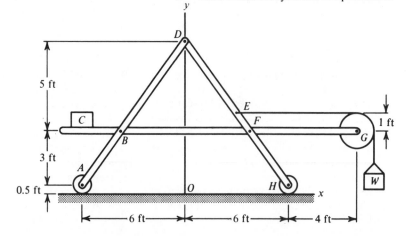

Exercise 13

14. The coplanar frame is built in at *C* and other-
    wise pin-connected. Determine all the
    actions on member *AC*.

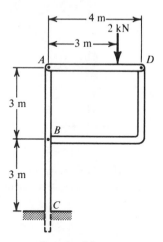

Exercise 14

15. For the pin-connected plane frame shown,
    determine all actions on member *BD*, ne-
    glecting member weights.

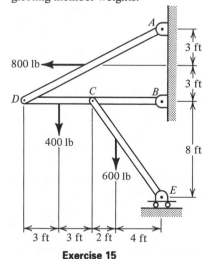

Exercise 15

16. Determine the couple *M* which must be
    transmitted to the crank *OA* to equilibrate a
    piston force *P*, as a function of crank angle
    θ. The crank arm and connecting rod have
    lengths *a* and *b*, respectively.

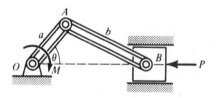

Exercise 16

17. Solve Sample Problem 5.1.1 if the downward
    force on the horizontal member is increased
    to $2P$, while the other force is unchanged.

18. For the toggle vise shown, for given vertical
    force *P* and lengths *L*, determine the horizon-
    tal gripping force *H* at *A* for equilibrium, as a
    function of θ, neglecting friction and treating
    the problem as coplanar. Sketch a plot of *H*
    versus θ, and discuss the result as θ
    approaches zero from a practical point of
    view.

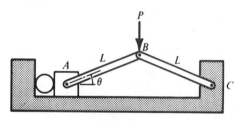

Exercise 18

19. For the quick-return mechanism shown, and
    given vertical load *P* at *D*, express the couple
    M that must be transmitted to the arm *AB* by
    the shaft at *A* for equilibrium, as a function of
    the angle θ. Neglect friction and member
    weights.

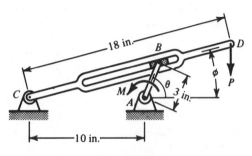

Exercise 19

20. The space frame shown consists of three 16-ft bars *AD*, *BD*, *CD* joined at their midpoints by three 8-ft bars *EF*, *FG*, *GE* and resting on a smooth horizontal floor at the vertices of equilateral triangle *ABC* of side length 16 ft. All bars are uniform and weigh 5 lb/ft, and the structure is loaded only by its own weight. The joints may be treated as ball-and-socket joints. Draw free-body diagrams of (a) the whole frame, (b) member *EF*, and (c) the A-frame *ABD* with *EF* attached but the other members removed. Determine all forces exerted on the A-frame. (*Hints*: Use the symmetry of the problem. It may help to draw a top view of the structure in order to clarify the dimensions. The origin *O* shown is directly below the midpoint of *EF*.)

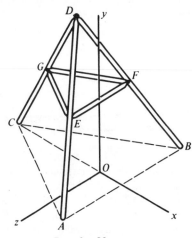

Exercise 20

21. Determine as a function of *θ* the required vertical force *P* and all reactions on member *AB* to support the smooth slider at *D* of weight *W*. The light members *AB* and *BC* are of length 2*a* and *a*, respectively. Assume smooth pins at *B*, *C*, and *D* and treat the problem as coplanar.

22. Neglecting all weights and treating the problem as coplanar, determine as a function of *θ* the frictional force at *A* required for the symmetrical gripping tool to hold in equilibrium the cylinder of radius *a*. (The applied

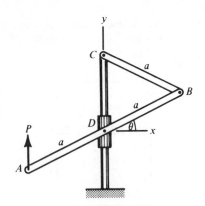

Exercise 21

forces *P* and the dimensions *b* and *L* are given.) If the coefficient of static friction is *μ*, what is the largest angle *θ* for which equilibrium can be maintained?

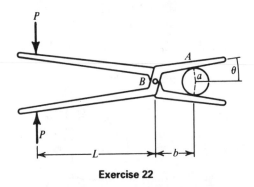

Exercise 22

23. The two heavy horizontal uniform members *AB* and *BC* weigh 30 lb/ft. They are supported by the light bar *BD* as shown. All joints may be treated as ball-and-socket joints. Determine the support reactions at *A* and *C* for equilibrium.

24. Solve Ex. 23 if the light bar *BD* is replaced by another heavy uniform bar weighing 30 lb/ft.

25. The shaft *BC* carries a 2-ft-diameter pulley and is supported by journal bearings at *B* and *C*. Shaft and pulley together weigh 100 lb, and the three 10-ft members of the symmetrical frame each weigh 100 lb. Joints at *A*, *D*, *E*, and *F* may be treated as ball-and-socket joints. Determine the support reaction components at *D*, *E*, and *F*.

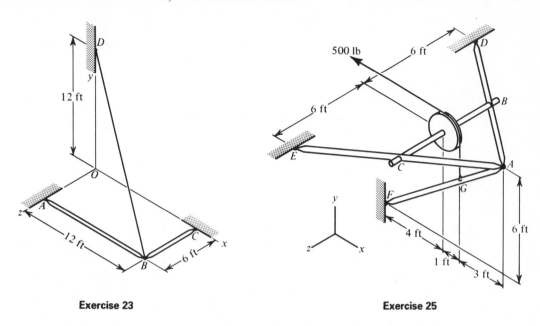

Exercise 23   Exercise 25

5.2 STATICALLY DETERMINATE PLANE TRUSSES; METHOD OF JOINTS AND METHOD OF SECTIONS

A *truss* is a structure composed of slender straight members connected only at their ends by joints and loaded only at the joints. It is one of the most important types of engineering structure, used in buildings, bridges, and aerospace vehicles. Practical structures are formed from several trusses joined together to form a space framework. Since the members are slender, they can support little lateral load. When a concentrated load must be applied between two joints or when a distributed load is to be supported by the truss (for example, a bridge floor), an auxiliary structure must be provided, which uses stringers and floor beams to transmit the load to the joints. In this section we shall consider only the analysis of a plane truss and not of the auxiliary structure.

Figure 5-11 shows a small plane truss. In most practical trusses the joints are formed with gusset plates riveted or welded to the members, as illustrated schematically in Fig. 5-11(a). These joints must be carefully designed so that the center lines of the members meeting at a joint are concurrent at a point in order that large couples are not developed and applied to the ends of the members. The primary stress analysis of the truss is then carried out by modeling the actual joints as frictionless pin connections, as illustrated in Fig. 5-11(b). The conventional representation of Fig. 5-11(c) is used for the pin-connected model of Fig. 5-11(b). Because some small

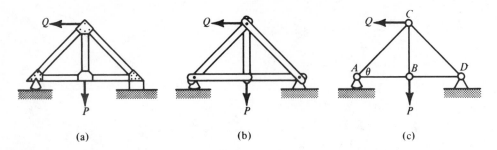

(a)   (b)   (c)

Fig. 5-11   Plane Truss

amount of bending is inevitably induced by the actual joints and by the member weights, a secondary stress analysis is often necessary. Analysis of the stress and deformation in the gusset plates of the joints may also be needed, especially of the stress concentrations around rivet holes, for example. Here we shall consider only the primary stress analysis. The term "stress analysis" is usually used for the determination of the internal forces in the bars of the truss (equal in magnitude to the forces exerted by the bars on the pins). The word "stress," however, means force per unit area.

For the primary stress analysis, a plane truss is considered to be composed of straight two-force members pin-connected only at their ends and loaded only at the pins. No member is continuous through a joint. If the member weights are included at all, half the member weight is assumed to act on the pin at each end.

A *simple plane truss* is defined as one that can be formed by beginning with a pin-connected triangle (e.g., *ABC* in Fig. 5-12) and then adding successively two members at a time, in such a way that the truss is "rigid"; that is, it cannot deform as a mechanism in its own plane if the members are rigid bars. Most simple plane trusses appear to be formed of triangles, but, as the last example in Fig. 5-12 shows, this need not be the case, since *CDEF* is not a triangle. A simple truss is rigid even when not supported. This is not the only way to construct a rigid truss, but it is the most common way. In the original triangle *ABC* there are three members and three joints. For every additional joint in a simple truss there are two additional members. Hence, if *m* is the total number of members and *j* is the total number of joints, we have $m - 3 = 2(j - 3)$. Thus **in a simple plane truss**

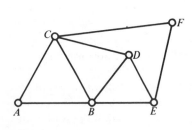

Fig. 5-12   Simple Plane Truss

$$m = 2j - 3. \tag{5.2.1}$$

We shall see that $m = 2j - 3$ is a necessary condition for a plane truss to be statically determinate when it is supported in a suitable manner.

Analysis of the forces in a statically determinate truss is commonly performed by (1) the ***method of joints*** or (2) the ***method of sections***.* In both methods suitable free bodies are chosen by isolating various portions of the truss. In the method of joints, each free body consists of only one joint, while in the method of sections a larger part of the truss is isolated by passing imaginary sections through several members in order to cut out the free body. Figure 5-13 shows a free-body diagram of the whole truss of Fig. 5-11 and also shows the truss completely dismembered with a free-body diagram of each member and one of each pin. Since all the members are two-force members, the analysis of a member is trivial, and it is customary to omit the members from the drawing as in Sample Problem 5.2.1. But notice in Fig. 5-13 how the awareness of the two-force nature of each member enables us to show consistent senses for the force *AC* acting on pin *A* and the force *AC* acting on pin *C* (***both labeled with the same symbols identifying the member that exerts the forces*** but drawn in opposite directions). Notice also that ***when a member force is shown pushing on a pin the member exerting that force is in compression, while when the member force pulls on the pin the member doing the pulling is in tension***. We can begin by guessing which members are in tension and which are in compression and drawing the joint diagrams accordingly. If the analysis gives a negative value for the magnitude, that tells us we have guessed wrong. [Alternatively, you could guess them all in tension (positive); then your wrong guesses would come out negative (meaning compression).]

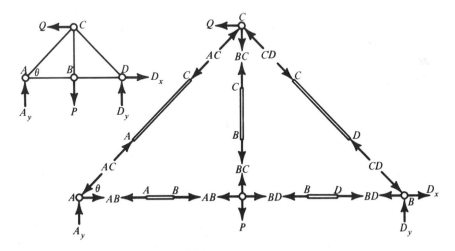

Fig. 5-13   Method of Joints (Member Free-Bodies May Be Omitted)

---

* The method of analysis by joints was introduced by an American engineer, S. Whipple, in 1847. The method of sections was used by J. W. Schwedler in 1851 and A. Ritter in 1862.

Joint Analysis

Since all forces acting on a pin are concurrent, the moment equilibrium equation for the joint free body is trivially satisfied and useless. We are left with two force equilibrium equations for each joint, $2j$ equations for j joints with $m + 3$ unknowns (the m member forces and the three support reactions for statically determinate support of the whole truss as a rigid body). Since $m = 2j - 3$ in a simple truss, we see that there are just enough joint equilibrium equations to solve for all the unknowns in a simple plane truss with three support reactions. The three additional equations for equilibrium of the whole truss as a rigid body are not independent but furnish a check.

In practice it is often more convenient to find the support reactions first from the free-body diagram of the whole truss. Then three of the joint equations are available as checks. In a simple truss, when the support reactions are known it is always possible to find a joint where there are at most two unknowns. Then you can proceed from joint to joint solving at most two simultaneous equations at each joint instead of setting up the whole system of $2j$ equilibrium equations to solve simultaneously, as may be necessary for a statically indeterminate truss where deformation equations must also be added to the system.

The equation $m = 2j - 3$ is a necessary condition for statical determinacy in any rigid truss, but it may fail to be sufficient in cases of improper support or improper arrangement of the members. Also some trusses are not rigid unless they are supported. See, for example, Ex. 6 at the end of this section, where two pin supports are necessary to hold the truss in position instead of the pin-and-roller simple support. Thus there are four support conditions, and the number of unknowns is $m + 4$. **When there are more than $2j$ unknowns the plane truss is statically indeterminate**, and deformation analysis must be included along with the force analysis.

Method of Sections

Figure 5-14 shows the truss of Fig. 5-11 again and also shows two free-body diagrams formed by passing a section through members AC, BC and BD. The section is shown as the dashed curve SS in the first figure. Whenever possible choose a section cutting at most three members, since three equations of equilibrium are usually

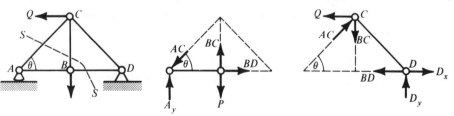

Fig. 5-14 Method of Sections

available (including the moment equation since the three unknown bar forces are not usually concurrent in one point). The members that are cut by the section are omitted from the free bodies and replaced by the forces that they exert on the free bodies. (You may either try to guess the correct sense or alternatively show them all as tension.)

If the support reactions have already been determined from a free-body diagram of the whole truss, then either of the free bodies of Fig. 5-14 may be used to solve for *AC, BC,* and *BD.* The remaining bar forces *AB* and *CD* can be determined by choosing two additional free bodies obtained by cutting *AB* and *CD,* respectively. In this case *AB* can be determined by analysis of joint *B* and *CD* by joint *D.*

Complete analysis of a large truss by the method of sections will require many different sections, and the total computational effort is about the same as that for complete analysis by the method of joints. The method of sections possesses two advantages, however. It can often be used to determine one critical unknown in the interior of a large truss without the step-by-step analysis of working in from one end by the method of joints. See Sample Problem 5.2.2. The method of sections also minimizes the accumulation of roundoff errors in large trusses and can provide a check on the analysis by joints before the analysis by joints is complete.

Kips. In many of the following examples and exercises the loads are given in kips (thousands of pounds). For example, a load $P = 3000$ lb may be given as $P = 3$ kips or $P = 3$ k, sometimes written $P = 3^k$.

Zero-Force Members; Special Loading Conditions

For certain loading conditions it is possible to determine some of the member forces by inspection. Consider, for example, the joints *K, B,* and *H* of Fig. 5-15, whose free bodies are shown at the right. By summing horizontal and vertical forces on each of the free bodies we see immediately that $KJ = 0$, $KA = 0$, $AB = BC$, $BJ = 10$ k, $GH = HJ$, and $CH = 0$. The ***zero-force members*** are especially easy to spot and

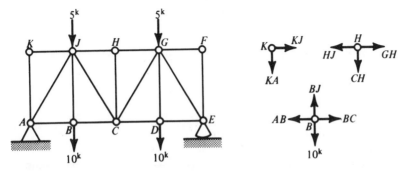

Fig. 5-15 Joints with Special Loading Conditions

greatly simplify the analysis of the rest of the truss. These results could have been written down immediately without actually drawing the three free bodies, since we can imagine these very simple free bodies just by looking at the truss. An unloaded two-member joint can be in equilibrium only if the bar forces are equal and opposite or are both zero as at K where they cannot be equal and opposite. When only four members or three members and a load meet at a joint in such a way that they are oppositely directed in pairs they must be equal in pairs as at B and H. (H is an especially useful and obvious case where the load opposing CH is zero.) Beware, however, of trying to extend this to joints such as J and G where other members are available to carry part of the vertical load.

Zero-force members are not useless in a truss, since for a different loading of the same truss they may be needed.

Since there are no horizontal loads on the truss of Fig. 5-15, the support reaction $A_x = 0$. Hence *the symmetric truss is loaded symmetrically, and symmetrically placed members will have equal forces* ($AJ = EG$, $CJ = CG$, etc.). *Beware of nonsymmetrical loading, however*, as in Fig. 5-11, where force AC is *not* equal to force CD.

It is not necessary for the special loading joints such as K, B, and H to involve perpendicular members. For example, if free bodies like those of Fig. 5-16 can be visualized in a truss, then $f_1 = f_3$, $f_2 = f_4$, $f_5 = f_6$, $f_7 = 0$, $f_8 = 0$, and $f_9 = 0$, as you can verify by summing first vertical components and then horizontal components.

Section 5.3 deals with space trusses.

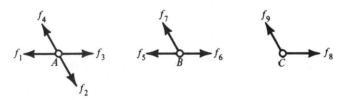

Fig. 5-16   Joints that Can Be Analyzed by Inspection

SAMPLE PROBLEM 5.2.1

Analyze the truss of Fig. 5-17 by the method of joints. Label answers T for tension or C for compression.

SOLUTION.   For the whole truss $\sum m_E = 0$ gives $C_x = 4$ k. Force summations then give $E_x = 4$ k and $E_y = 3$ k, which we show on the free body of joint E. Hence $EF = 4$ k T and $ED = 3$ k T *Answer.*

Next we go to joint A where there are only two unknowns. The force triangle for pin A is shown to the right of the free body. Since $\tan \theta = \frac{3}{4}$, we obtain $AF = 4$ k T and $AB = 5$ k C *Answer.*

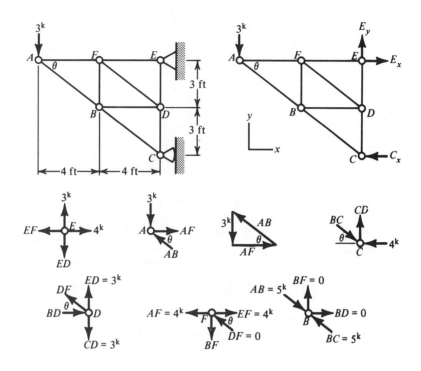

Fig. 5-17 Sample Problem 5.2.1

Joint C could be treated similarly. Alternatively, force component summations give $BC \cos \theta - 4 = 0$ and $CD - BC \sin \theta = 0$, whence $BC = 5 \text{ k}$ C and $CD = 3 \text{ k}$ T **Answer.**

Joint D now has only two unknowns. Vertical force summation followed by horizontal force summation shows that $DF = 0$ and $BD = 0$ **Answer.**

Joint F now has only one unknown BF. Vertical force summation shows that $BF = 0$ **Answer.**

One free body remains with no unknowns, joint B, whose equilibrium is a check on the previous calculations. Since all answers came out positive, correct senses had been guessed on the free bodies. Hence those shown pulling are tensile, and those shown pushing are compressive.

SAMPLE PROBLEM 5.2.2

Determine the forces in members CG, CD, and DG of the truss whose free-body diagram is shown in Fig. 5-18. (A hinge support at A and a roller at E have been removed.)

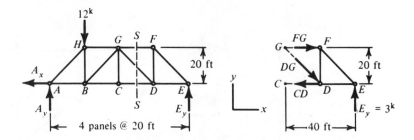

Fig. 5-18 Sample Problem 5.2.2

SOLUTION. CG is obviously a zero-force member. For the free body of the whole truss, moment equilibrium about A gives $E_y = 3$ k. Section SS is then used to isolate the free body on the right. For the isolated body, we have

$$[+\curvearrowright \Sigma m_G = 0]:\quad 40(3) - 20(CD) = 0,\qquad CD = 6\ k\ T\quad \textbf{\textit{Answer.}}$$

$$[\Sigma f_y = 0]:\quad 3 - DG \sin 45° = 0,\qquad DG = 4.24\ k\ C\quad \textbf{\textit{Answer.}}$$

EXERCISES

1. Determine the bar forces of the truss of Fig. 5-11 in terms of P, Q, and θ by the method of joints.
2. Analyze the truss of Sample Problem 5.2.2 by the method of joints.
3. Determine bar forces CD and CG in Fig. 5-15 if the truss is 12 ft high and has four 8-ft panels horizontally.
4. What are the forces in members BC and BG of the truss of Sample Problem 5.2.2.

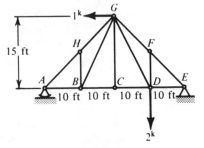

Exercise 6

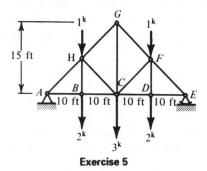

Exercise 5

Exercise 7

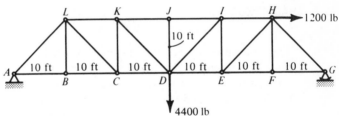

Exercise 11

5–10. Use the method of joints to analyze each of the trusses in the figures shown. Indicate which members are in tension and which are in compression.

11. Determine forces in *KD, CD, DJ,* and *DI.*
12. Determine forces in *CG* and *CF.*
13. Determine forces in *AH* and *JH* for truss of Ex. 12.
14. Determine bar forces *KJ* and *KN.*
15. Determine bar forces *JI* and *IO* in the truss of Ex. 14.

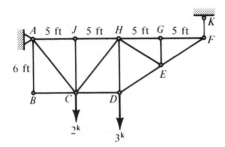

Exercise 8

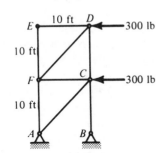

Exercise 9

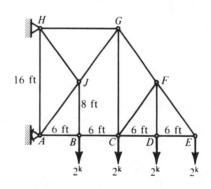

Exercise 12

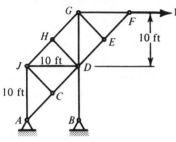

Exercise 10

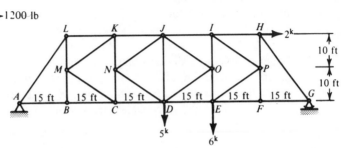

Exercise 14

*5.3 SPACE TRUSSES

The analysis of a three-dimensional truss is in principle no different from that of a plane truss, but it very quickly becomes more complicated. The primary stress analysis of a space truss models it as formed of light straight members connected by frictionless ball-and-socket joints at their ends.

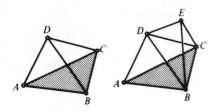

A **simple space truss** is defined as one that begins with a just-rigid four-joint tetrahedron and proceeds by adding successively three bars and one joint at a time, as illustrated in Fig. 5-19 where the starting tetrahedron is *ABCD*.

The starting tetrahedron has six members and four joints. We then add three members for each additional joint. Hence, if m is the total number of members and j the total number of joints, we have

Fig. 5-19   Simple Space Truss

$m - 6 = 3(j - 4)$, or $m = 3j - 6$, in a simple space truss.

For analysis by the method of joints we have three equilibrium equations for each joint, and for statically determinate support of the whole truss as a rigid body there are six support reactions and m unknown bar forces. Hence there are $3j$ equations for $m + 6$ unknowns, just enough equations to analyze completely a simple space truss with six support reactions.

The six equilibrium equations for the truss as a whole are not independent of the $3j$ joint equations but may be used in place of six of them. The six unused equations can then be used as a check. If there are more than $3j$ unknowns, the space truss is statically indeterminate. In a simple space truss, if the support reactions are known, it will always be possible at each stage of the analysis to find a joint where there are only three unknowns for the next step of the analysis. Hence it will never be necessary to solve more than three simultaneous equations at a time.

The method of sections may be used to cut out a free body of a larger portion of the truss. There will be six equilibrium equations for each part cut out.

The first step in analysis of a joint is to express each unknown bar force in terms of its unknown magnitude and its known direction cosines. The direction is known (except possibly for sense) because each member is a two-force member. We consider here only a few small space trusses in order to avoid long calculations. In larger problems a digital computer analysis is advisable. Many of the following exercises can be simplified by recognizing the symmetry that exists. But be careful to check that both the loads and the truss are symmetric.

In Sec. 5.4 we shall consider stress resultants in beams.

---

* This section may be omitted.

EXERCISES

1. Determine the forces in all members connected to joints A and B of the space truss
shown. The support at D is a ball-and-socket
joint, and C is supported by two short links and
E by one link. The truss carries downward
loads of 3 kips at A and 2 kips at B.

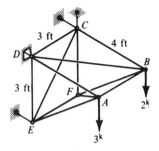

Exercise 1

2. The truss shown is a regular tetrahedron
formed with six equal bars. Determine the
force in each bar when joint D carries a downward load P, while joints ABC rest on the
smooth horizontal floor in the xz-plane.

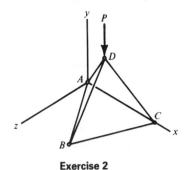

Exercise 2

3. Determine the force in each member of the
simple truss shown formed with nine bars of
equal length and loaded by two oppositely
directed loads as shown. (The upper "load"
could be a cable support reaction.)

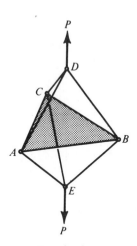

Exercise 3

4. The truss shown is symmetric with respect to
the xy-plane. It is formed by four equal-length
bars joining E to supports A, B, C, and D at
the vertices of a 4×8 ft rectangle in the xz-
plane and then three added members joining
B, C, and E to joint F. If the vertical load P is
1000 lb, determine the force in each bar.

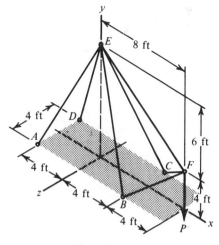

Exercise 4

*5.4   STRESS RESULTANTS IN BEAMS; BENDING MOMENT AND SHEAR FORCE DIAGRAMS

Beams are slender structural elements that offer resistance to bending by applied loads. Most applications involve straight prismatical bars long in comparison to their cross-section dimensions, loaded by bending couples or forces perpendicular to the bars. But curved beams also exist, and straight or curved beams sometimes carry loads that tend to elongate or twist the member as well as to bend it.

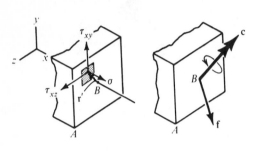

Internal forces in members must be analyzed in order to determine the load-carrying capacity of the members and to predict their deformations. The internal forces acting across a cross section are distributed forces whose intensity (force per unit area) is called ***stress***. Figure 5-20 shows schematically in the left figure the stress on an element of area dA, resolved into three rectangular components: the ***normal stress*** σ (perpendicular to the cross section) and two components τ_{xy} and τ_{xz} of tangential or ***shear stress*** acting in the y- and z-directions, respectively.

Fig. 5-20   Vector Stress Resultants on Beam Cross Section

Stress Resultants in Beams

We now define stress resultants for the most general possible traction distribution (internal force per unit area) over the cross section of a beam. We know from the results of Sec. 4.1 that in the most general case the distributed forces on the cross section are equipollent to a couple and a single concentrated force at an arbitrarily chosen point, as illustrated by the couple **c** and the force **f** at B in Fig. 5-20.

If we knew the traction vector **t** (internal force per unit area),

$$\mathbf{t} = \sigma\mathbf{i} + \tau_{xy}\mathbf{j} + \tau_{xz}\mathbf{k}, \tag{5.4.1}$$

as a function of position on the cross section A of the beam shown in Fig. 5-20 then we could calculate the resultant at B of the distributed tractions on the area A by evaluating the integrals

$$\mathbf{f} = \int_A \mathbf{t}\, dA = \int_A (\sigma\mathbf{i} + \tau_{xy}\mathbf{j} + \tau_{xz}\mathbf{k})\, dA$$

$$\mathbf{c} = \int_A \mathbf{r}' \times \mathbf{t}\, dA. \tag{5.4.2}$$

* This section may be omitted or postponed.

Since we do not usually know the traction distribution in advance, the **stress resultants f** and **c** must be determined by some other method besides evaluating the defining integrals of Eq. (5.4.2). The free-body equilibrium methods of Chapter 4 will suffice to determine the stress resultants at an arbitrary section of a beam with statically determinate support reactions. When the supports are statically indeterminate, deformation analysis must be used simultaneously with the equilibrium analysis. We shall illustrate the procedure for the statically determinate case by an example.

EXAMPLE 5.4.1

The uniform cantilever of Fig. 5-21(a) is built in at the end $x = L$. In addition to its own weight of w lb per unit length, it carries an axial tensile load P and a horizontal transverse load Q at the free end $x = 0$. We imagine the beam cut at section A and analyze the free-body diagram of the portion of the beam of length x to the left of A, Fig. 5-21(b). The weight wx of this portion of the beam has been shown as a concentrated force at its midpoint (at distance $\frac{1}{2}x$ from 0), and the six stress resultant components have been shown acting on the right end. A trivial application of the six equations of equilibrium to this free body gives

$$f_x = P, \qquad f_y = wx, \qquad f_z = -Q$$
$$c_x = 0, \qquad c_y = -Qx, \qquad c_z = \tfrac{1}{2}wx^2. \tag{5.4.3}$$

The couple components are best obtained by moment equations about axes through the moment center in the cross section, so that only one unknown appears in each equation. For example,

$$[\textstyle\sum m_{Az} = 0]: \quad (\tfrac{1}{2}x)(wx) + c_z = 0. \tag{5.4.4}$$

The determination of the stress resultant components for the cantilever thus uses exactly the same equilibrium analysis as we would use to determine the support reactions at the end $x = L$. Indeed, substituting $x = L$ into Eq. (5.4.3) gives the support reactions at $x = L$.

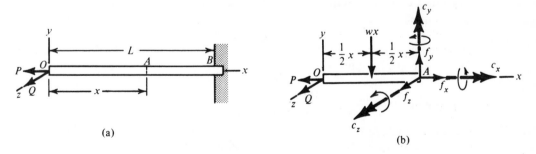

(a)

(b)

Fig. 5-21 Stress Resultants on Cross Section Area A of Cantilever

For the statically determinate uniform beam with no concentrated loads except at the ends, a single free body of "typical length" x suffices to determine the stress resultants at all sections. When there are concentrated loads in the interior, additional free bodies must be used, as the coplanar examples treated later in this section will illustrate.

Of course the stress resultants at a section are only ***equipollent*** to the actual traction distribution on the section, that is, equivalent for rigid-body analysis of the free-body. They are not strictly equivalent for deformation analysis near the cross section. But for long slender members, experience indicates that elastic beam deflections can be analyzed quite successfully by means of simplified theories using the stress resultants as the variables in the governing equations instead of the actual stresses. Since the stress resultants are variable functions of only one position coordinate x, while the stresses are in general variable functions of x, y, and z, a tremendous simplification is possible when the problem can be solved in terms of stress resultants. Actual deformable body analysis of beams will not be treated here,* but some of the terminology and the methods of determining the stress resultants by equilibrium methods are given here.

Because the stress resultants can be used as ***generalized stresses*** for the design and analysis of structural elements, the determination of the stress resultants as functions of position along the beam is usually called ***stress analysis***, even when the actual stresses are not determined.

From Fig. 5-21(b) we can intuitively attribute different kinds of deformational effects to the different components of the stress resultants. Thus c_x would be a twisting couple, while c_y would bend the beam in the horizontal plane and c_z would bend it in the vertical plane. Force f_x would stretch the beam, while f_y and f_z would shear it in the y- and z-directions, respectively. All this presumes some resisting loads or support reactions on the free body so that the resultants on A will not merely translate and rotate the body. The very same stress resultants at one section can produce quite different effects on the free body: (1) external effects (motion and/or support reactions) and (2) internal effects (deformation). It is customary to use ***different sign conventions*** for the stress resultants when they are considered as ***generalized stress*** variables in the deformation analysis from the conventions used when they are considered as force or couple components in the study of motion. This fact frequently causes confusion for the beginner, who, for example, may fail to distinguish between the ***bending moment*** and the ***turning moment*** at a section because they are the same stress resultant. Be very attentive to the distinction in the following coplanar illustrations.

Bending Moment and Shear Force

For the remainder of this section we shall consider only beams for which the xy-plane is a plane of symmetry for both the geometry of the beams and the applied

---

* This topic is treated in books on mechanics of materials. See, for example, E. P. Popov, *Introduction to Mechanics of Solids* (Englewood Cliffs, N.J.: Prentice-Hall, Inc., 1968).

loads. Bending and stretching take place in the xy-plane, and the whole problem can be treated as coplanar. The only couple component is the z-component, and this will usually be represented by a curved arrow in the xy-plane instead of as a vector. (The couple vector load representation would not appear to be symmetric.)

Consider first the cantilever of negligible weight loaded only by a bending couple of couple-moment magnitude M at one end as shown in Fig. 5-22(a). Free-body diagrams are shown in Fig. 5-22(b) and (c) for the portions of length x and $L - x$, respectively, to the left and to the right of A. The other stress resultants at A have been omitted since a trivial equilibrium analysis shows that all are zero except c_z. The vector stress resultant is $\mathbf{c} = M\mathbf{k}$ at A on OA, while by Newton's third law it is $\mathbf{c} = -M\mathbf{k}$ at A on AB. Thus, as a turning moment, $c_z = +M$ at A on OB, while $c_z = -M$ at A on AB. But the bending moment is considered to be $+M$ on both. In fact, for this example the bending moment is $+M$ everywhere in the beam.

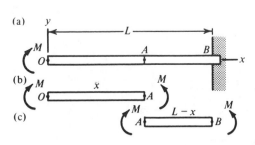

Fig. 5-22 Bending Moments

For a symmetrically loaded symmetric beam there are only three stress resultants: the axial force, denoted by P; one transverse shear force in the plane of the loads, denoted by V; and the bending moment M. *We adopt the sign conventions of Fig. 5-23* for positive M, V, and P. The sign conventions for M and P are well standardized. Some authors use the opposite sense for positive V, but the one given here is the one most widely used in American engineering practice.

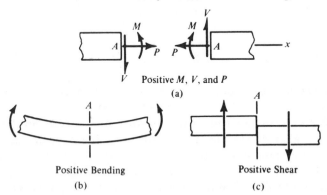

Fig. 5-23 Positive Bending Moment, Shear, and Axial Force

As a memory device it is sometimes said that *positive bending bends the beam into a shape that would hold water* [concave upward, as in Fig. 5-23(b)], while *positive shear is caused by a loading that causes the right-hand end of the beam to shear off downward*, as in Fig. 5-23(c). (To apply this memory device to the shear you must view the beam from the side that has the x-axis pointing to the right.)

Shear and Bending Moment Diagrams

An important tool in the analysis and design of structures formed of beams is the bending moment diagram, a graph of bending moment versus position coordinate along the beam. A graph of the shear force is also useful, called a shear diagram. Construction of such graphs will be illustrated by examples of beams with statically determinate supports. Statically indeterminate beams require deformation analysis along with the force analysis.

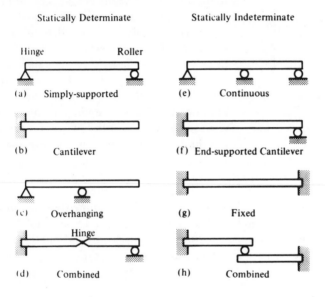

Fig. 5-24   Some Types of Beams

Figure 5-24 shows **_conventional symbols_** and names for several types of beams supported in various ways. The **_simply supported_** beam, for example, has a hinge at one end and a roller at the other. The conventionally represented roller is assumed capable of exerting either upward or downward force. Many beams are in fact mounted in such a way that they can expand axially when the temperature rises. This means that at least one end must be free to move axially; in practice it may not actually be mounted on a roller but is restrained axially only by friction negligible in comparison to the loads. For simplicity the hinge support is represented conventionally by a small triangle and the roller by a circle. The combined beams joined by a hinge in Fig. 5-24(d) form a statically determinate system, as you can verify by drawing separate free-body diagrams of the two beams and counting unknowns. Other types of continuous beams may have more than the one intermediate support shown in Fig. 5-24(e).

Differential Relationships

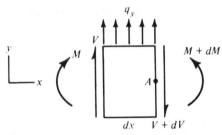

Fig. 5-25 Beam Element for Derivation of Differential Relationships

Figure 5-25 shows a free-body diagram of a beam element of infinitesimal length dx loaded by a continuous distribution of upward load $q_y(x)$ per unit length that may be considered constant q_y for the infinitesimal length dx. We assume that there is no axial force, and write two equilibrium equations for the free-body of Fig. 5-25:

$$V - (V + dV) + q_y\, dx = 0$$

(5.4.5)

$$[+\curvearrowleft) \sum m_A = 0]:\ (M + dM) - M - V\, dx - q_y\, dx(\tfrac{1}{2}dx) = 0.$$

(5.4.6)

The last term vanishes in the limit (after division by dx). We obtain the two *differential equilibrium relationships*

$$\frac{dV}{dx} = q_y(x)$$

(5.4.7)

$$\frac{dM}{dx} = V(x),$$

(5.4.8)

which together imply

$$\frac{d^2M}{dx^2} = q_y(x)$$

(5.4.9)

The algebraic signs occurring in these equations result from the sign conventions adopted for positive M, V, and q_y. Many authors use a different sign convention for V and/or q_y. You can always check the algebraic signs by repeating this simple derivation with all quantities shown in the direction they would have if they were positive according to the sign convention used.

The differential relationships of Eqs. (5.4.7)–(5.4.9) are helpful for sketching the shear and bending moment diagrams. *In particular, when there is no distributed load, they imply that V is constant and that M varies linearly with x between any two concentrated-load or support points.*

We shall close this section with several examples of the construction of shear and bending moment diagrams. Notice that in each example the slope of the moment diagram is equal to the ordinate to the shear diagram at the same position x, except at a concentrated force load, where the slope of the moment diagram and the magnitude of the shear force are both undefined.

"Summation methods" of determining shear and moment diagrams, based on integrating Eqs. (5.4.7) and (5.4.8) for given $q_y(x)$, are given in books on mechanics of materials and on structural analysis. The student should, however, master the free-body "method of sections" presented here, which is less subject to gross errors and which is a valuable check on the summation methods.

Method of Sections

The basic free-body method used in the following examples is to isolate a portion of the beam by passing an imaginary cutting section perpendicular to the beam at a *typical position* x between two concentrated-load or support points. *Never choose the section at a support or at a concentrated load.* The free body isolated (of length x or $L - x$) is bounded between one end of the beam and the section plane at position x. On the free body we show all external forces and couples. These include applied forces, support reactions, and the stress resultants M, V, and P at the cut section. (The axial force P is often omitted when there are no axial loads.) Equilibrium equations for the free body then give $M(x)$, $V(x)$, and $P(x)$ at the position x. It is necessary to repeat the process for each span between two points of support or concentrated load.

Discontinuities. The shear diagram has a jump discontinuity at each concentrated force (load, or support reaction) of magnitude equal to the concentrated force. This produces a slope discontinuity in the moment diagram. The moment diagram has a jump discontinuity at a concentrated couple of magnitude equal to the magnitude of the couple. The concentrated couple does not affect the shear diagram. All these discontinuities are automatically taken care of by the free-body procedure stated above, if it is correctly applied.

The diagrams should close. For example, if $M \neq 0$ at an end, draw a vertical segment on the moment diagram at the end, so that it starts and ends at $M = 0$.

Maximum bending moment. The most critical value for beam design is usually the maximum bending moment magnitude. If the bending moment diagram has a relative maximum or minimum in the interior of a span, the shear force will be zero there. This helps to locate the critical maximum moment point, but *do not forget to check the end points of each span for possible moment magnitudes greater than at the interior relative maximum and minimum*.

Units. In the SI system of units the shear force should be given in newtons (N), or more conveniently in kilonewtons or meganewtons (kN or MN), and length in meters (m). Some European engineering writers have used the kilogram force unit, but this is not proper in the SI system. In the British system force may be given in pounds, tons, or kips and length usually in feet or inches. Moment units are frequently written as lb-ft to distinguish them from work (ft-lb), but ft-lb is also used for moments.

Historical note. The use of bending moment and shear force as well as Eq. (5.4.8) and its use to locate the maximum bending moment are attributed to J. W. Schwedler in 1851 by S. Timoshenko, *History of Strength of Materials* (New York: McGraw-Hill Book Company, 1953).

EXAMPLE 5.4.2

The simply supported beam *AB* of length *L* carries a concentrated downward force *Q* at $x = a$. The beam weight is neglected. Figure 5-26 shows four free-body diagrams: the whole beam and the portions *AC*, *AD*, and *DB*. (The last two are not both needed but illustrate two alternative procedures.) The unknown *M* and *V* are drawn at each cut section as though they were positive. *M* is determined in each case by taking moments about a center in the cut section. The equations for free bodies *AC* and *DB* are the most convenient to use; these equations are shown on the figure. Alternatively to the equations for *DB*, we may use for *AD*

$$[+\backslash \sum m_D = 0]: \quad M + Q(x - a) - \frac{b}{L} Qx = 0$$

$$M = \frac{b}{L} Qx - Q(x - a).$$

Since $b = L - a$, this is equal to the moment

$$M = aQ\left(1 - \frac{x}{L}\right)$$

determined for the same section by equilibrium analysis of free body *DB*. The shear and moment diagrams are shown plotted at the bottom of the figure. The most significant item on the diagrams is usually the position and magnitude of the maximum bending moment. For a single concentrated load on a simply supported beam, we see that this occurs under the load at $x = a$ and has magnitude

$$M_{\max} = Qa\left(1 - \frac{a}{L}\right).$$

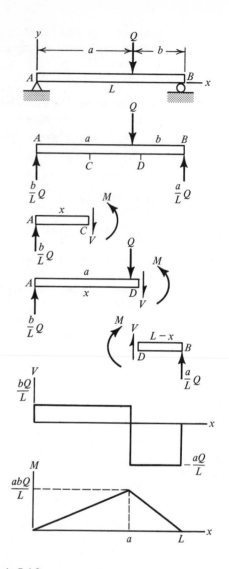

For *AC*:

$$[+\curvearrowright \sum m_C = 0]: \quad M = \frac{b}{L}Qx$$

$$[\sum f_y = 0]: \quad V = \frac{b}{L}Q.$$

For *DB*:

$$[+\curvearrowright \sum m_D = 0]: \quad M = aQ\left(1 - \frac{x}{L}\right)$$

$$[\sum f_y = 0]: \quad V = -\frac{a}{L}Q.$$

Shear Diagram

Moment Diagram

Fig. 5-26   Example 5.4.2

EXAMPLE 5.4.3

The 16-ft cantilever of Fig. 5-27 carries an upward load of 100 lb at the left end and a horizontal 500-lb load applied through a 2-ft rigid arm at the midpoint as shown. Since we are interested in the beam *AB* instead of the arm, we replace the horizontal load by an equipollent force-couple system at the midpoint of the free body *AB*. The analysis is shown beside the figure. Note the jump of 1000 lb-ft on the moment diagram at the point where the concentrated couple acts on *AB*. It is left as an exercise to show that the same results can be obtained for $x > 8$ by a free body of *AE*.

For *AB*:

$$[+\curvearrowleft \textstyle\sum m_B = 0]: \quad C + 1000 - 16(100) = 0$$

$$C = 600 \text{ lb-ft}$$

$$B_x = 500 \text{ lb}, \qquad B_y = 100 \text{ lb}\downarrow.$$

For *AD*:

$$M = 100x, \qquad V = 100 \text{ lb}, \qquad P = 0.$$

For *EB*:

$$[+\curvearrowright \textstyle\sum m_E = 0]: \quad M = 600 - 100(16 - x)$$

$$V = 100 \text{ lb}, \qquad P = 500 \text{ lb}.$$

Shear Diagram

Moment Diagram

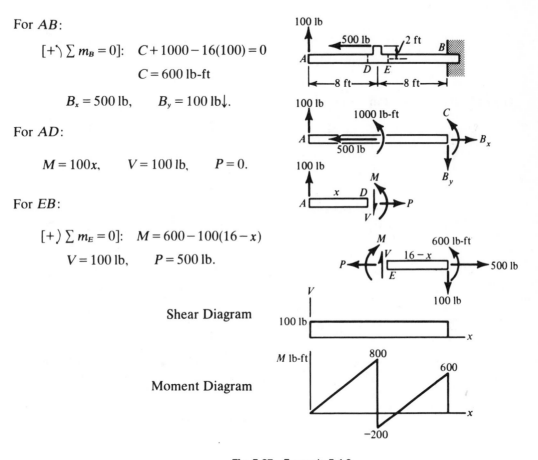

Fig. 5-27   Example 5.4.3

EXAMPLE 5.4.4

An overhung beam with a uniform load is shown in Fig. 5-28.

For *AC*: Replace uniform load by resultant 720 lb:

$$[+\rangle\,\textstyle\sum m_A = 0]:\quad B = 450\,\text{lb}$$

$$[\textstyle\sum f_y = 0]:\quad A = 270\,\text{lb}.$$

For *AD*: Replace uniform load by resultant $48x$:

$$[+\rangle\,\textstyle\sum m_D = 0]:\quad M = 270x - 24x^2$$

$$V = 270 - 48x.$$

For *EC*: Replace uniform load by $48(15 - x)$:

$$[+\rangle\,\textstyle\sum m_E = 0]:\quad M = -24(15 - x)^2$$

$$V = 48(15 - x).$$

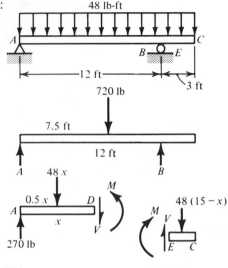

Shear Diagram

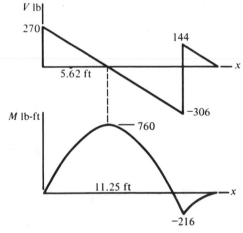

Moment Diagram

Fig. 5-28   Example 5.4.4

EXAMPLE 5.4.5

We shall now consider an example of a nonuniform distributed load in order to illustrate the general integration procedure that can be used for any distributed load. If the distribution is such that the integrals cannot be evaluated explicitly, they can always be evaluated numerically with a variable upper limit, a routine available in any digital computer center.

The cantilever of Fig. 5-29 of length $2a$ carries a parabolically distributed load $w(x)$ lb/ft on the portion $0 < x < a$ with maximum value w_m at $x = a/2$. The portion $a < x < 2a$ is not loaded, so that $w(x) = (4w_m/a^2)x(a - x)$ on the portion $0 < x < \frac{1}{2}a$ only. Free-body diagrams are drawn for two typical portions of length x, with sections cut at C and D, respectively.

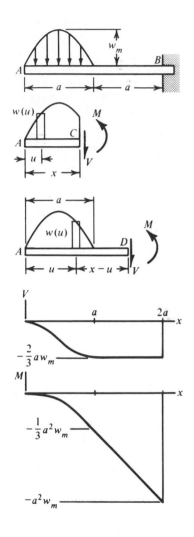

Fig. 5-29 Example 5.4.5

For AC:

$$[\textstyle\sum f_y = 0]: \quad V = -\int_0^x w(u)\, du$$

$$[+\textstyle\rangle \sum m_C = 0]: \quad M = -\int_0^x (x - u)w(u)\, du,$$

where u is a dummy integration variable representing the abscissa of the element of load $w(u)\, du$ and $x - u$ is the perpendicular moment arm from the moment center in the cut section at the variable upper limit x to the element of load.

Thus, for AC

$$V = -\frac{4w_m}{a^2}\int_0^x (au - u^2)\, du$$

$$= -\frac{4w_m}{a^2}\left[\frac{1}{2}ax^2 - \frac{1}{3}x^3\right]$$

$$= -\frac{2}{3}aw_m\left[3\left(\frac{x}{a}\right)^2 - 2\left(\frac{x}{a}\right)^3\right]$$

$$M = -\frac{4w_m}{a^2}\int_0^x (x - u)(au - u^2)\, du$$

$$= -\frac{4w_m}{a^2}\int_0^x [axu - (a + x)u^2 + u^3]\, du$$

$$= -\frac{4w_m}{a^2}\left[\frac{1}{2}ax^3 - \frac{1}{3}(a + x)x^3 + \frac{1}{4}x^4\right]$$

$$= -\frac{1}{3}a^2 w_m\left[2\left(\frac{x}{a}\right)^3 - \left(\frac{x}{a}\right)^4\right].$$

For $x > a$, we use free body AD:

$$[\Sigma f_y = 0]: \quad V = -\int_0^a w(u)\, du$$

$$[+\circlearrowleft \Sigma\, m_D = 0]: \quad M = -\int_0^a (x-u)w(u)\, du.$$

The only difference is that the integrals are now evaluated with upper limit a instead of x, since there is no load for $x > a$. Thus for $a < x < 2a$,

$$V = -\frac{4w_m}{a^2}\left[\frac{1}{2}a^3 - \frac{1}{3}a^3\right] = -\frac{2}{3}aw_m$$

$$M = -\frac{4w_m}{a^2}\left[\frac{1}{2}a^3x - \frac{1}{3}(a+x)a^3 + \frac{1}{4}a^4\right]$$

$$= -\frac{1}{3}a^2 w_m\left[2\left(\frac{x}{a}\right)-1\right].$$

Hence, for $a < x < 2a$, V is constant, and M is linear with x.

EXAMPLE 5.4.6

The integration procedure illustrated in Example 5.4.5 can usually be avoided by replacing a distributed load on each free body by one or more concentrated loads equipollent to the given loads as in Solution 2 of Sample Problem 4.2.1.

Consider the 12-m simply supported beam of Fig. 5-30, loaded by a distributed load varying linearly from 1 kN/m at the left end to 5 kN/m at the right end.

For the whole beam the triangular part of the load is equipollent to 24 kN at $x = 8$ m, while the rectangular part is equipollent to 12 kN at $x = 6$ m, as shown in the second figure, where the replacement loads have been shown by dashed arrows.

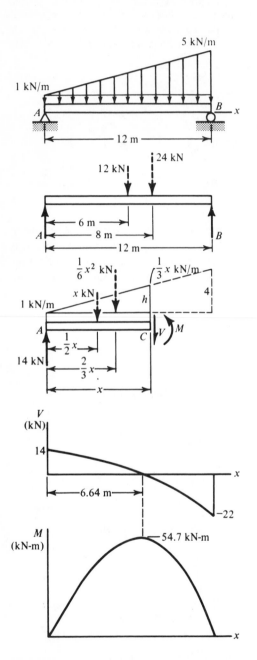

Fig. 5-30   Example 5.4.6

Reactions:

$$[+\!\curvearrowleft m_B = 0]:\quad 12A - 6(12) - 4(24) = 0$$

$$A = 14 \text{ kN}, \qquad B = 22 \text{ kN}.$$

The 12- and 24-kN concentrated loads cannot be used for calculating the internal force resultants. The third figure shows the free body AC of typical length x. By similar triangles the load intensity h at C is given by

$$\frac{h}{x} = \frac{4}{12},$$

whence

$$h = \tfrac{1}{3}x \text{ kN/m}.$$

Now we replace the triangular part of the load on AC by the "area" $\frac{1}{2}(x)(\frac{1}{3}x) = \frac{1}{6}x^2$ kN at distance $\frac{2}{3}x$ from A and the rectangular part $(1)(x) = x$ kN at distance $\frac{1}{2}x$ as indicated by the dashed arrows. For AC

$$[+\!\uparrow \Sigma f_y = 0]:\quad 14 - x - \tfrac{1}{6}x^2 - V = 0$$

$$V = 14 - x - \tfrac{1}{6}x^2 \text{ kN}$$

$$[+\!\curvearrowleft \Sigma m_C = 0]:\quad M + (\tfrac{1}{6}x^2)(\tfrac{1}{3}x) + (x)(\tfrac{1}{2}x) - (14)(x) = 0$$

$$M = 14x - \tfrac{1}{2}x^2 - \tfrac{1}{18}x^3.$$

The maximum moment occurs where $V = 0$:

$$x^2 + 6x - 84 = 0$$

$$x = 6.64 \text{ m}, \qquad M_{max} = 54.7 \text{ kN-m}.$$

Closure

This completes our introduction to stress resultants in beams and to shear and moment diagrams, topics that will be developed further in courses in mechanics of materials and in engineering structural analysis. In Sec. 5.5 we shall consider the shapes of flexible cables carrying distributed loads.

EXERCISES

Unless otherwise instructed, use free-body methods to determine expressions for the bending moment and shear at each typical section x, and sketch the loaded beam and the bending moment and shear diagrams with points of interest labeled, including maximum moment and shear points.

1. Simply supported beam with uniform downward load w per unit length.
2. Cantilever with transverse concentrated load Q at left end (free end).
3. Simply supported beam with two equal transverse concentrated loads Q at $x = L/3$ and $x = 2L/3$.
4. Cantilever with triangular distributed downward load $w(x)$ varying from zero at $x = 0$ to w_m at the support $x = L$.
5. Simply supported beam with the load of Ex. 4.
6. Cantilever with the triangular load of Ex. 4 reversed so that maximum occurs at free end.
7. Simply supported beam with half-sine distributed upward load $q_y(x) = q_m \sin(\pi x/L)$.
8. Overhanging beam of Fig. 5-24(c) with roller at $x = a$ and concentrated downward load Q at right end $x = L$.
9. Same as Ex. 8 except with uniform distributed weight w per unit length included and with $a = \frac{1}{2}L$.
10. Combined beam of Fig. 5-24(d) if each span has length 3 m and each carries a 4-kN downward concentrated load at midspan.
11. The combined beam of Fig. 5-24(d) if each span is 8 ft long and carries a uniform downward load of 300 lb/ft.
12. Obtain the results of Example 5.4.3, Fig. 5-27, for $x > 8$ ft by using a free-body diagram AE.
13–22. Beam AB shown: The symbol k denotes kip (1000 lb).

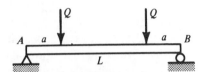

Exercise 13

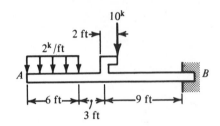

Exercise 14

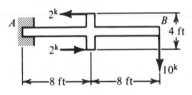

Exercise 15

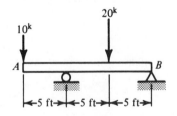

Exercise 16

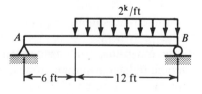

Exercise 17

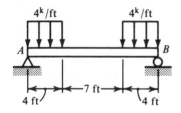

Exercise 18

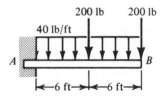

Exercise 21

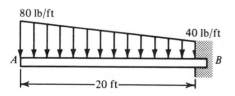

Exercise 19

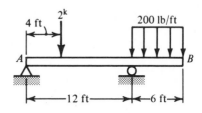

Exercise 22

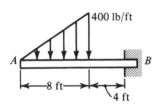

Exercise 20

23. For a uniform quarter-circular cantilever of radius R, lying in a horizontal plane and loaded only by its own weight w per unit length; define stress resultants as in Fig. 5-21 for a local coordinate system with the x-axis tangent to the beam center line. Express these resultants as functions of a central-angle coordinate θ measured from the free end.

---

*5.5 FLEXIBLE CABLES

A flexible cable has negligible resistance to bending, so that the resultant of the internal forces on a cross section is a tensile force tangent to the center line of the cable. In this section we shall consider cables whose weight is not negligible or which may be subjected to other loads distributed along their lengths. An important example is one of the main cables of a suspension bridge, where the significant loads are the bridge deck and pavement, which may sometimes be modeled as a loading of the cable that is uniformly distributed along the horizontal. Another example is a cable freely hanging under its own weight, which is uniformly distributed along the arc length instead of along the horizontal. We shall consider first cables that are slack enough to have a horizontal tangent at the lowest point.

* This section may be omitted.

The engineer needs to know the relations among the span, sag, and length of the cable and the maximum tension for given loads. He may also wish to know the exact shape of the arc. Figure 5-31(a) shows schematically a cable whose lowest point is taken as the origin O. Figure 5-31(b) is a free-body diagram of an arc OP of the cable from O to a typical point whose horizontal distance from O is x and whose distance along the arc is s.

W represents the resultant of the distributed weight and any other (vertical) distributed load on the arc OP. H is the horizontal tension at O, and T is the tension at P (at angle θ to the horizontal). For equilibrium we clearly have

$$T \sin \theta = W \quad \text{and} \quad T \cos \theta = H, \tag{5.5.1}$$

whence

$$T^2 = H^2 + W^2 \quad \text{and} \quad \tan \theta = \frac{W}{H}. \tag{5.5.2}$$

These results are independent of the way in which the load is distributed. We shall consider two special cases: the **parabolic cable**, with load distributed uniformly along the horizontal ($W = wx$, for constant w), and the **catenary**, with loads distributed uniformly along the arc ($W = ws$ for constant w).

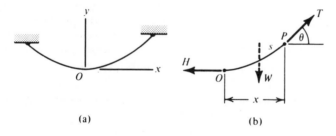

(a)   (b)

Fig. 5-31   (a) Hanging Cable; (b) Free-Body Diagram

Parabolic Cable

If the load is uniformly distributed along the horizontal, then Eq. (5.5.2) gives

$$W = wx, \quad T = [H^2 + w^2 x^2]^{1/2}, \quad \text{and} \quad \frac{dy}{dx} = \tan \theta = \frac{wx}{H}. \tag{5.5.3}$$

The last equation can be integrated immediately to give

$$y = \frac{w}{2H} x^2 \tag{5.5.4}$$

for the equation of the parabola. (The integration constant is zero, since $y = 0$ when $x = 0$.)

If x_m is chosen as the abscissa of the higher support and y_m is the sag below the support, then

$$\left. \begin{array}{c} H = \dfrac{wx_m^2}{2y_m} \quad \text{and} \quad T_m = wx_m\left[1 + \dfrac{1}{4}\left(\dfrac{x_m}{y_m}\right)^2\right]^{1/2} \\[4mm] \text{or} \quad T_m = H\left(1 + \dfrac{4y_m^2}{x_m^2}\right)^{1/2} \end{array} \right\} \tag{5.5.5}$$

give the horizontal force H at O and the maximum tension T_m (at the support) in terms of prescribed horizontal distance x_m and sag y_m between O and the support. The equation of the curve may be written as

$$y = \frac{y_m x^2}{x_m^2}. \tag{5.5.6}$$

The length s of the arc from O to the support is

$$s = \int_0^{x_m} \left[1 + \left(\frac{dy}{dx}\right)^2\right]^{1/2} dx = \int_0^{x_m} \left(1 + \frac{4y_m^2 x^2}{x_m^4}\right)^{1/2} dx. \tag{5.5.7}$$

Changing the integration variable to

$$u = \frac{2y_m x}{x_m^2} \qquad \text{with} \qquad dx = \frac{x_m^2}{2y_m} du$$

transforms Eq. (5.5.7) to

$$s = \frac{x_m^2}{2y_m} \int_0^{2y_m/x_m} (1 + u^2)^{1/2} \, du.$$

From integral tables

$$\int (1 + u^2)^{1/2} \, du = \frac{u}{2}(1 + u^2)^{1/2} + \frac{1}{2} \ln[u + (1 + u^2)^{1/2}].$$

For the definite integral, dropping the subscripts m to obtain the arc length OP to a typical point (which could be a support), we find

$$s_{OP} = \frac{x}{2}\left(1 + 4\frac{y^2}{x^2}\right)^{1/2} + \frac{x^2}{4y} \ln\left[\frac{2y}{x} + \left(1 + 4\frac{y^2}{x^2}\right)^{1/2}\right]. \tag{5.5.8}$$

Catenary

If the vertical load is the cable's own uniformly distributed weight w per unit length, then for the length s of OP in Fig. 5-31, we have $W = ws$, and, by Eq. (5.5.2),

$$W = ws, \qquad T = [H^2 + w^2 s^2]^{1/2}, \qquad \frac{dy}{dx} = \tan\theta = \frac{ws}{H}. \qquad (5.5.9)$$

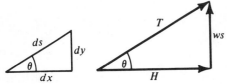

The last equation cannot be integrated immediately for y as a function of x because it contains the variable arc length s. But from the similar triangles of Fig. 5-32, we see that

Fig. 5-32 Similar Triangles for Eq. (5.5.10)

$$\frac{dx}{ds} = \cos\theta = \frac{H}{(H^2 + w^2 s^2)^{1/2}}. \qquad (5.5.10)$$

We introduce the parameter $c = H/w$ and obtain

$$H = cw, \qquad dx = \frac{c\,ds}{\sqrt{c^2 + s^2}}. \qquad (5.5.11)$$

The last form may be integrated with respect to s to obtain (after setting the integration constant equal to zero because $s = 0$ at $x = 0$)

$$x = c \sinh^{-1}\left(\frac{s}{c}\right), \qquad (5.5.12)$$

where $\sinh^{-1}$ denotes the inverse hyperbolic sine function (see any calculus textbook). Equation (5.5.12) is the inverse form of

$$s = c \sinh\left(\frac{x}{c}\right) \equiv \frac{c}{2}[e^{x/c} - e^{-x/c}]. \qquad (5.5.13)$$

The Cartesian equation of the curve may now be obtained by substituting Eq. (5.5.13) into the last equation in Eqs. (5.5.9), with $H = cw$, and integrating with respect to x:

$$\frac{dy}{dx} = \sinh\left(\frac{x}{c}\right)$$

$$y = c \cosh\left(\frac{x}{c}\right) + \text{const.}$$

The integration constant equals $-c$ if $y = 0$ at $x = 0$, since $\cosh(0) = 1$. Thus

$$y = c \, \cosh\left(\frac{x}{c}\right) - c \tag{5.5.14}$$

is the **equation of a catenary** with vertex at the origin.

If x and y are the coordinates of the upper support, Eq. (5.5.14) is a transcendental equation which may be solved numerically for the parameter c in terms of prescribed horizontal distance x and sag y between O and the support. A more convenient form for numerical solution is obtained by dividing through by x, namely,

$$\frac{y}{x} = \frac{\cosh z - 1}{z}, \qquad \text{where } z = \frac{x}{c}. \tag{5.5.15}$$

For prescribed x and y the first of these equations determines z, and hence $c = x/z$. Then, with the additional prescription of w, the first equation of Eqs. (5.5.11) gives H, and Eqs. (5.5.9) and (5.5.13) give

$$T = wc\left[1 + \sinh^2\left(\frac{x}{c}\right)\right]^{1/2}$$

or after use of the identity

$$1 + \sinh^2 u = \cosh^2 u \tag{5.5.16}$$

$$\frac{T}{wx} = \frac{c}{x} \cosh\left(\frac{x}{c}\right). \tag{5.5.17}$$

Equations (5.5.13) and (5.5.14) can be combined by use of the identity of Eq. (5.5.16) to give the following convenient formula for the arc length from the origin at the vertex of the catenary to the point (x, y) on the curve:

$$s^2 = (y + c)^2 - c^2. \tag{5.5.18}$$

Equations (5.5.9), (5.5.11), (5.5.13), and (5.5.14) furnish the following simple results for use after c has been determined:

$$H = wc, \qquad W = ws, \qquad T = w(y + c). \tag{5.5.19}$$

For design purposes a graph or table can be constructed from Eq. (5.5.15) to determine z in terms of prescribed y/x. A few values are given in Table 5.5.1.

Table 5.5.1 z in Terms of y/x

y/x	0	0.1	0.2	0.3	0.4	0.5	0.6
z	0	0.1993	0.395	0.583	0.762	0.931	1.088

Note: $c = x/z$.

Additional values and other tabulations for the catenary determining z for given T/wx or given s/x may be found in handbooks.*

For small sag ratios y/x, the catenary is quite closely approximated by a parabola (the power series expansion of the hyperbolic cosine terminated with the quadratic term). For $y_m/x_m < 0.2$ the formulas given earlier for the parabolic curve are quite satisfactory even when the weight is really distributed uniformly along the arc instead of along the horizontal. The maximum tension formula of Eq. (5.5.5) underestimates the catenary tension less than 1.5% with $y_m/x_m = 0.2$, but the error increases to more than 14% at $y_m/x_m = 0.6$ and increases rapidly for still higher ratios.

SAMPLE PROBLEM 5.5.1

A uniform cable weighing 5 lb/ft (Fig. 5-33) is suspended between two points 400 ft apart at the same height. If the sag is 90 ft, determine (a) the total length of the cable and (b) the maximum and minimum tension.

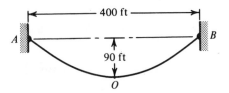

Fig. 5-33   Catenary of Sample Problem 5.5.1

SOLUTION.   With the origin at the vertex, we have

$$\frac{y}{x} = \frac{90}{200} = 0.45 \qquad \text{for the support } B.$$

Equation (5.5.15) then becomes

$$1 + 0.45z = \cosh z \qquad\qquad (5.5.20)$$

---

* See, e.g., T. Baumeister, ed., *Marks' Standard Handbook for Mechanical Engineers*, 7th ed. (New York: McGraw-Hill Book Company, 1967), Part 2, p. 57.

to solve numerically (or graphically) for z. A trial-and-error solution is indicated below, beginning with $z = 0.84$ (about halfway between the values corresponding to y/x values of 0.4 and 0.5 in Table 5.5.1):

Trial z	$1 + 0.45z$	$\cosh z$
0.84	1.3780	1.3740
0.85	1.3825	1.3835
0.846	1.3807	1.3797
0.848	1.3816	1.3816

We accept $x/c = z = 0.848$. Then

$$c = \frac{x}{z} = \frac{200}{0.848} = 236.$$

(a) By Eq. (5.5.18), for arc OB:

$$s^2 = (y + c)^2 - c^2 = 326^2 - 236^2$$

$$\text{half-length } s = 225 \text{ ft}$$

$$\text{total length} = 450 \text{ ft} \quad \textbf{Answer.}$$

(b) By Eqs. (5.5.19),

$$T = w(y + c) = 5(y + 236)$$

$$T_{max} = 1630 \text{ lb}, \qquad T_{min} = H = 1180 \text{ lb} \quad \textbf{Answer.}$$

SAMPLE PROBLEM 5.5.2

The uniform cable of weight w per unit length in Fig. 5-34(a) is stretched so taut that no interior part is horizontal. If the slope at O is m, derive equations that can be used to determine the tension.

SOLUTION.   From the free-body diagram, Fig. 5-34(b), of arc OP to a typical point P, we see that

$$T \sin \theta = ws + V, \qquad T \cos \theta = H, \tag{5.5.21}$$

where H and V are the horizontal and vertical support reactions at O. Thus

$$\frac{dy}{dx} = \tan \theta = \frac{ws + V}{H} \quad \text{and} \quad m = \frac{V}{H}. \tag{5.5.22}$$

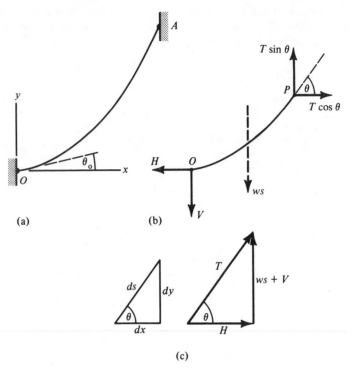

Fig. 5-34 (a) Catenary of Sample Problem 5.5.2; (b) Free-Body Diagram of *OP*; (c) Similar Triangles at *P*

From the triangles of Fig. 5-34(c), we obtain

$$\frac{dx}{ds} = \cos \theta = \frac{H}{[H^2 + (ws + V)^2]^{1/2}}. \tag{5.5.23}$$

We let $c = H/w$ and proceed in a manner similar to that following Eq. (5.5.10). Thus

$$H = cw, \qquad dx = \left[1 + \left(\frac{s}{c} + m\right)^2\right]^{-1/2} ds \tag{5.5.24}$$

$$x = c\left[\sinh^{-1}\left(\frac{s}{c} + m\right) - k\right], \tag{5.5.25}$$

where

$$k = \sinh^{-1} m \tag{5.5.26}$$

is known. Thus

$$s = c\left[\sinh\left(\frac{x}{c}+k\right)-m\right].$$ (5.5.27)

From Eqs. (5.5.22) and (5.5.27), with $H/w = c$, we obtain

$$\frac{dy}{dx} = \frac{s}{c}+m = \sinh\left(\frac{x}{c}+k\right),$$ (5.5.28)

whence

$$y = c\left[\cosh\left(\frac{x}{c}+k\right)-\cosh k\right].$$ (5.5.29)

Evaluating for $y = y_m$ and $x = x_m$, with $x_m/c = z$, we obtain a transcendental equation

$$\frac{y_m}{x_m} = \frac{\cosh(z+k)-\cosh k}{z}$$ (5.5.30)

to solve numerically for z when x_m, y_m, k are known. With z determined and with $c = x_m/z$ we determine s by Eq. (5.5.27). Then with $H = cw$, the last equation of Eqs. (5.5.22) determines V, and, by Eqs. (5.5.21) and (5.5.28),

$$T^2 = (ws + V)^2 + H^2 = H^2\left[\left(\frac{s}{c}+m\right)^2+1\right] \quad \textbf{\textit{Answer.}}$$

By Eqs. (5.5.27) and (5.5.16), this may be written as

$$T = cw\,\cosh\left(\frac{x}{c}+k\right) \quad \textbf{\textit{Answer.}}$$

EXERCISES

1. A uniform cable of total mass 74.5 kg is stretched between two supports 100 m apart at the same level. If the cable sag is 2 m at the center, determine the maximum tension.

2. The span of a cable is 80 ft between two supports at the same level. What is its weight if the maximum tension is 500 lb and the sag at the middle is 1 ft?

3. What percentage error in the maximum tension would result if Sample Problem 5.5.1 were solved by assuming that the total weight of 2250 lb is uniformly distributed along the horizontal?

4. A suspension bridge cable spans 3000 ft with a sag of 250 ft. If the horizontally distributed load carried by the cable is 2000 lb/ft, determine the maximum and minimum tensions.

5. Determine the length of the cable of Ex. 4.
6. Unstayed telephone poles are not intended to resist any net horizontal force at the top. If the cable across the 150-ft span BC sags 7.5 ft, what must be the sag in the 225-ft span AB for no net horizontal force on the top of pole B? (Assume parabolic cables with the same weight per horizontal foot in each span.)

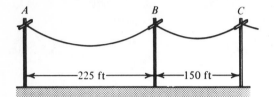

Exercise 6

7. A uniform cable of mass 3 kg/m, suspended between two supports at the same level 150 m apart, sags 30 m at the middle. Determine (a) the cable length and (b) the maximum tension.
8. Determine the length and the maximum tension in a uniform cable of weight w per unit

length, hanging between two supports 200 ft apart, if the sag is (a) 28 ft, (b) 34 ft, (c) 42 ft, and (d) 55 ft.
9. Show for the catenary of Fig. 5-31 that if s and x from O to the support are given, the parameter $z = x/c$ is determined by the transcendental equation $s/x = (\sinh z)/z$.
10. (a) Use the result of Ex. 9 to determine the sag of a catenary cable of total length 30 m and span 20 m between supports at the same level. (b) What is the maximum tension if the total mass of the cable is 60 kg?
11. For a parabolic cable with distributed load w per unit horizontal length, stretched as in Sample Problem 5.5.2, determine the maximum tension and the support reactions at O if $\tan \theta_0 = \frac{1}{2} \tan \theta_A$ and the coordinates of A are (x_m, y_m).
12. A cable weighing 60 lb is stretched between two supports 100 ft apart at the same level. It carries a concentrated load of 200 lb at the middle with a sag of 3 ft. (a) Determine the maximum cable tension. (b) Compare this with the tension estimated by treating the cable as weightless but adding a fictitious additional load of 30 lb (half the cable weight) at the middle (still located 3 ft below the supports).

5.6 SUMMARY

The methods of rigid-body equilibrium analysis developed in Chapter 4 have been applied in this chapter to the determination of interaction forces in structures and machines formed of pin-connected rigid members and to the determination of internal moments and forces in beams and cables.

The most significant application of the free-body method is to frames and machines containing multiforce members. If you thoroughly understand the technique of taking the body apart and the sign conventions for labeling the interaction forces on the free-body diagrams of the multiforce members as illustrated in the first two simple examples of Sec. 5.1, you should have no difficulty in principle with extending the method to more complicated examples. A review of these two examples at this time should be profitable.

Trusses are modeled as structures formed of two-force members joined by smooth pins (or by ball-and-socket joints for space trusses) and loaded only at the pins. Member weights are either neglected, or else half the member weight is assumed to act as a load on each end pin. The basic ***method of joints*** then analyzes a free-body diagram of each pin. The ***method of sections*** provides a check and also a quicker method for finding the force in one member without a complete analysis.

Stress resultants in beams, introduced in Sec. 5.4, are basic to the analysis of internal forces in structural and machine elements and will be extensively used in courses in mechanics of deformable solids. In Sec. 5.5 we considered flexible cables.

The free-body techniques will also be basic in dynamics, where equations of motion are written for each isolated body instead of equations of equilibrium. Alternative methods of work and energy will also be useful in dynamics.

In Chapter 6 we shall consider the concepts of work and potential energy and their use to determine the stability or instability of an equilibrium configuration.

CHAPTER SIX

Work, Energy, and the Stability of Equilibrium

6.1 WORK OF A FORCE; CONSERVATIVE FORCE FIELDS AND POTENTIAL ENERGY

The work done by a force is a cumulative measure of the effect of the force as the particle it acts on moves through a certain distance. In dynamics we shall see that the total work done on a rigid body is equal to the change of its kinetic energy. This fact furnishes one method of finding the body's velocity change as it moves through a certain distance. The definition of the work done by a force is another important application of the scalar product of Sec. 3.2.

Work by a Force: Definition

The increment dU of work done by a force $\mathbf{f}$ acting on a particle that moves through an infinitesimal vector displacement $d\mathbf{r}$ is defined to be

$$dU = \mathbf{f} \cdot d\mathbf{r} = f \cos \theta \, ds, \tag{6.1.1}$$

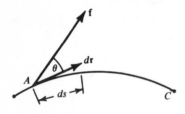

Fig. 6-1   Work $dU = \mathbf{f} \cdot d\mathbf{r}$

where θ is the angle between $\mathbf{f}$ and $d\mathbf{r}$ and where f and ds are the magnitudes of $\mathbf{f}$ and $d\mathbf{r}$ as in Fig. 6.1 *In general both the magnitude of $\mathbf{f}$ and the angle θ will vary with position* of the moving material point A of application of the force, so that to calculate the work U done by the force in moving a finite distance along a path C it is necessary to evaluate the line integral

$$U = \int_C \mathbf{f} \cdot d\mathbf{r} = \int_C (f \cos \theta) \, ds \qquad (6.1.2)$$

along the curve C. This can be a difficult problem, but it is often possible to express all varying quantities in terms of a single parameter varying along the path and evaluate the integral. A simple example will be treated in Example 6.1.1. We usually use the letter U for work instead of the more natural W in order to avoid confusion when we calculate the work done by the weight W. When confusion will not arise, the work may be denoted by W.

The dimensions of work are

$$[U] = [FL], \qquad (6.1.3)$$

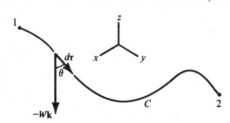

Fig. 6-2   Constant Force $\mathbf{f} = -W\mathbf{k}$

since $\cos \theta$ is dimensionless, but only when the force is constant and acts parallel to a straight-line motion is it possible to evaluate the work done by the force simply as the arithmetic product of force times distance traveled. When the force is constant, however, it is easy to evaluate the work even when the motion is not straight-line motion. For example, Fig. 6-2 shows the constant weight force $\mathbf{f} = -W\mathbf{k}$ acting on a particle moving along curve C from 1 to 2 in a uniform gravitational field. By Eq. (6.1.1),

$$dU = -W\mathbf{k} \cdot d\mathbf{r}$$
$$= -W\mathbf{k} \cdot (dx\mathbf{i} + dy\mathbf{j} + dz\mathbf{k}) \qquad (6.1.4)$$
$$= -W \, dz.$$

Hence

$$U = \int_{z_1}^{z_2} (-W) \, dz = W(z_1 - z_2). \qquad (6.1.5)$$

Thus ***the work done by the weight in a uniform gravitational field equals the weight times the loss of elevation,*** no matter what path is followed. This is an example of a ***conservative force field***; see Eq. (6.1.13).

The tangential component f_t of the force

$$f_t = f \cos \theta = \mathbf{f} \cdot \hat{\mathbf{e}}_t, \tag{6.1.6}$$

where $\hat{\mathbf{e}}_t$ is a tangential unit vector in the direction of the motion, is called the ***working component*** of the force, since the work increment is $dU = f_t \, ds$, while the normal component of the force (perpendicular to the path) does no work during the motion. ***Notice that the work done by the force is negative when*** $90° < \theta \leq 180°$, ***where the force opposes the motion.***

EXAMPLE 6.1.1

A particle moves along one quadrant of the ellipse

$$\frac{x^2}{a^2} + \frac{y^2}{b^2} = 1 \tag{6.1.7}$$

shown in Fig. 6-3. As it moves it is acted upon by a force field

$$\mathbf{f} = -kxy\mathbf{i} - kxy\mathbf{j} \text{ lb}, \tag{6.1.8}$$

where k is a given constant having units of lb/in². Determine the work done by $\mathbf{f}$ as the particle moves from $(a, 0)$ to $(0, b)$ in.

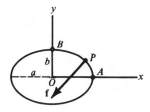

Fig. 6-3 Example 6.1.1

SOLUTION.   The parametric equations of the ellipse are

$$x = a \cos u, \qquad y = b \sin u, \tag{6.1.9}$$

and point P traverses the quadrant AB as u increases from 0 to $\pi/2$. Hence, by Eq. (6.1.2),

$$U = \int_{AB} \mathbf{f} \cdot d\mathbf{r}$$

$$= \int_{AB} (-kxy\mathbf{i} - kxy\mathbf{j}) \cdot (dx\mathbf{i} + dy\mathbf{j})$$

$$= \int_{AB} (-kxy\,dx - kxy\,dy) \qquad\qquad (6.1.10)$$

$$= k \int_0^{\pi/2} (a^2 b \sin^2 u \cos u - ab^2 \cos^2 u \sin u)\,du$$

$$= k[\tfrac{1}{3}a^2 b \sin^3 u + \tfrac{1}{3}ab^2 \cos^3 u]_0^{\pi/2}$$

$$U = \tfrac{1}{3}kab(a-b) \text{ in.-lb} \quad \textbf{\textit{Answer.}}$$

Remark. In this case the work done is evidently not independent of the path, since if the particle had moved along the axes from A to O and then from O to B, we would have had $\mathbf{f} = 0$, whence $U = 0$.

Conservative Force Fields and Potential Energy

The work U done by a force $\mathbf{f}$ acting on a particle that moves along a path C from point P_1 to point P_2 can be calculated, according to Eq. (6.1.2), by the *work integral*

$$U = \int_C \mathbf{f} \cdot d\mathbf{r} = \int_C f_t\,ds \qquad\qquad (6.1.11)$$

if the *working component*

$$f_t = \mathbf{f} \cdot \hat{\mathbf{e}}_t \qquad\qquad (6.1.12)$$

can be expressed as a function of the path coordinate s.

Even when the active force $\mathbf{f}$ is a given function of position in space (independent of velocity and time) the work integral between two positions in space will in general depend on the path that is followed from the first point to the second point. In certain important exceptional cases the work done by a force field is independent of the path, making the work integral especially easy to evaluate.

A *conservative force* field

1. Is one that depends only on position, and
2. is such that the work done by it on a particle that moves between   (6.1.13) any two positions is independent of the path followed.

We have already seen in Eq. (6.1.5) one example of a conservative force field, namely, a uniform gravitational field in which a weight W does work equal to W times the loss in altitude, no matter what path is followed by the descending weight. Any other constant force field $\mathbf{f}_0$ does work equal to f_0 times the net displacement component in the direction of $\mathbf{f}_0$, no matter what path is followed. *Friction is not a conservative force* since its direction depends on the direction of motion, and viscous friction also depends on the velocity.

When the work of a force field $\mathbf{f}$ is independent of the path between any two points in a region of space, we can choose any one fixed point in the region as the *datum point* P_0 from which the work done in moving to any other point P is then a function only of the position of P. The negative of this function is defined as the *potential energy* of the conservative force field $\mathbf{f}$, denoted $V(P)$, or $V(x, y, z)$.

Potential Energy

$$V(P) = -\int_{P_0}^{P} \mathbf{f} \cdot d\mathbf{r}, \tag{6.1.14}$$

---

where the integral is independent of the path (provided it does not go outside the region where $\mathbf{f}$ is conservative). *The potential energy is defined only up to an arbitrary additive constant, since the initial datum position P_0 is arbitrary.*

Then the work done by the force field $\mathbf{f}$ as its point of application moves from P_1 to P_2 is given by

---

Work Done by Conservative Force Field

$$U_{1\to2} = -\Delta V = -(V_2 - V_1), \tag{6.1.15}$$

---

where V_2 means $V(P_2)$ and V_1 means $V(P_1)$. This important result can be established by choosing a particular path from P_1 to P_2 that passes through P_0. Then

$$U_{1\to2} = \int_{P_1}^{P_0} \mathbf{f}\cdot d\mathbf{r} + \int_{P_0}^{P_2} \mathbf{f}\cdot d\mathbf{r} = V(P_1) - V(P_2), \qquad (6.1.16)$$

by the definition of Eq. (6.1.14). This establishes Eq. (6.1.15).

When several conservative forces act on a moving particle, their potential energies can be added algebraically, since the work is additive, to give a total potential energy V for the conservative forces acting. To evaluate the work integral of Eq. (6.1.11) by means of potential energy change, as in Eq. (6.1.15), we need formulas for the potential energy functions of various kinds of conservative forces. Three simple examples are given here.

Uniform gravitational field. We already know that the work done by a weight W in a uniform gravitational field is W times the loss in elevation; see Eq. (6.1.5). Hence

---

Uniform Gravitational Field

$$V_g = Wh, \qquad (6.1.17)$$

---

where V_g denotes the potential energy and where h is the elevation above an arbitrary datum level. Then $U_{1\to2} = -\Delta V = -W\Delta h$. Notice that as the weight descends Δh is negative; hence $U_{1\to2} = W|\Delta h|$ in descending. The work done in descending from elevation h to elevation 0 is Wh, equal to the potential energy before descent. This, in fact, is the origin of the name "potential energy," the energy potentially convertible into work.

Linear Spring. Figure 6-4 illustrates a linear spring, assumed always to remain straight and to exert a restoring force $\mathbf{f}$ of magnitude $k|\delta|$, where δ is the elongation of the spring from the unstretched length L_0.

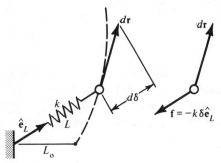

Fig. 6-4 Linear Spring

$$\mathbf{f} = -k\delta\hat{\mathbf{e}}_L, \qquad (6.1.18)$$

where

$$\delta = L - L_o, \qquad (6.1.19)$$

and where $\hat{\mathbf{e}}_L$ is a unit vector directed from the fixed end toward the movable end. It is assumed that a compressive force can be exerted by the spring when it is shortened.

For an arbitrary infinitesimal displacement $d\mathbf{r}$ of the movable end, we have an incremental elongation

$$d\delta = d\mathbf{r}\cdot\hat{\mathbf{e}}_L.$$

Hence the work done *by the spring* is

$$dU = \mathbf{f}\cdot d\mathbf{r} = -k\delta\hat{\mathbf{e}}_L\cdot d\mathbf{r}$$
$$= -k\delta\, d\delta.$$
$$U_{1\to 2} = -k\int_{\delta_1}^{\delta_2} \delta\, d\delta = -[\tfrac{1}{2}k\delta_2^2 - \tfrac{1}{2}k\delta_1^2].$$

Thus, since $U_{1\to 2} = -\Delta V$ by Eq. (6.1.15), we obtain

$$V_2 - V_1 = \tfrac{1}{2}k\delta_2^2 - \tfrac{1}{2}k\delta_1^2. \tag{6.1.20}$$

If we choose the arbitrary datum position to be where $\delta = 0$, then the elastic spring potential energy V_e in any other position is

---

Linear Elastic Spring with Datum at $\delta = 0$

$$V_e = \tfrac{1}{2}k\delta^2. \tag{6.1.21}$$

Equation (6.1.21) is often written as $V_e = \tfrac{1}{2}kx^2$, but this form should be used only for rectilinear motion along the x-axis for a spring unstretched at $x = 0$.

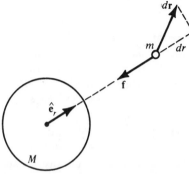

Inverse square gravitational force field. Figure 6-5 shows a particle of mass m in the gravitational field of the earth. The earth is idealized as a spherically symmetric mass M. Then, by Newton's law of gravitation, Eq. (1.3.2), the gravitational force $\mathbf{f}$ on the particle is

$$\mathbf{f} = -\frac{GMm}{r^2}\,\hat{\mathbf{e}}_r. \tag{6.1.22}$$

For a vector displacement $d\mathbf{r}$, we have

Fig. 6-5 Motion in Inverse Square Gravitational Field

$$d\mathbf{r}\cdot\hat{\mathbf{e}}_r = dr$$

Hence

$$dU = \mathbf{f} \cdot d\mathbf{r} = -\frac{GMm}{r^2} \hat{\mathbf{e}}_r \cdot d\mathbf{r} = -\frac{GMm}{r^2}\, dr, \qquad (6.1.23)$$

so that for any motion from P_1 to P_2

$$U_{1\to 2} = -GMm \int_{r_1}^{r_2} \frac{dr}{r^2} = +\left[\frac{GMm}{r_2} - \frac{GMm}{r_1}\right] \qquad (6.1.24)$$

independent of the path actually followed. The potential energy change $V_2 - V_1 = -U_{1\to 2}$ is therefore

$$V_2 - V_1 = -\left[\frac{GMm}{r_2} - \frac{GMm}{r_1}\right]. \qquad (6.1.25)$$

Hence the gravitational potential energy for the force on a particle of mass m is

---

Inverse Square Gravitational Force

$$V(r) = -\frac{GMm}{r} + \text{const.} \qquad (6.1.26)$$

The arbitrary constant will be zero if we choose the datum at infinity. Thus, $+GMm/r$ is the work done by the earth's gravitational field on a particle that moves in from infinite distance to distance r from the earth's center. (The gravitational *potential* is frequently defined as the *potential energy per unit mass of the particle* $-GM/r$.) This calculation treated the earth as an inertial reference. Where the motion of the earth is significant some modification may be needed.

Potential energy methods will greatly simplify many problems in dynamics, as we shall see in Chapter 9. We shall also use the potential energy concept to determine the equilibrium configurations of a system and to examine the stability of an equilibrium configuration in Sec. 6.2. Applications of the potential energy calculation formulas will be found in the examples and exercises of Sec. 6.2 and of Chapters 9, 12, 13 and 14 in Vol. II.

EXERCISES

1. What is the work done by the force $\mathbf{f} = k(x^2\mathbf{i} + xy\mathbf{j} + y^2\mathbf{k})$ lb on a particle moving as follows (coordinates are in feet)?
   (a) Along the x-axis from the origin to $A(2, 0, 0)$.
   (b) Along the vertical straight line from $A(2, 0, 0)$ to $B(2, 2, 0)$.
   (c) Along the straight line from the origin to B. Is this the same as the work of part (a) plus the work of part (b)?
   (d) Along the line $x = 2$, $y = 2$ parallel to the z-axis from $B(2, 2, 0)$ to $C(2, 2, 2)$.
   (e) Along the straight line from the origin to $C(2, 2, 2)$.

2. Calculate the work done by the constant $\mathbf{f} = -mg\mathbf{j}$ acting on a particle which moves in the xy-plane along the semicircular arc defined by $x = R \cos \theta$, $y = R \sin \theta$ as the parameter θ goes from $-\pi/2$ to $\pi/2$. Compare your answer with the work done by the same force on the particle if the particle follows the straight-line path between the same two points.

3. Find the work done by the force $\mathbf{f} = k(-y\mathbf{i} + x\mathbf{j})$ on the particle moving along the semicircular path of Ex. 2.

4. Find the work done by the following force field of constant magnitude f N.

$$\mathbf{f} = f\left[\frac{-y}{x^2+y^2}\mathbf{i} + \frac{x}{x^2+y^2}\mathbf{j}\right]$$

acting on a particle as it moves in the xy-plane along a square of side length 2 m.
(a) From $A(1, -1, 0)$ to $B(1, 1, 0)$. (b) From $B(1, 1, 0)$ to $C(-1, 1, 0)$. (c) All the way around the square $ABCD$ from A back to the starting point A.

6.2 POTENTIAL ENERGY CONDITIONS FOR EQUILIBRIUM AND FOR STABILITY

The equilibrium configurations of a mechanism or the equilibrium positions for an incompletely constrained body can be determined by a consideration of the potential energy of the system if friction and other dissipative forces are sufficiently negligible that all forces may be considered conservative. For such **conservative systems** the potential energy consideration also enables us to determine whether an equilibrium position is stable, unstable, or neutral.

Part 1.   One Degree of Freedom

We shall treat first a simple example that illustrates the principles involved.

Figure 6-6 illustrates four different possible equilibrium configurations for a small block of weight W constrained to move on a curved smooth guide in a vertical plane. The gravitational potential energy of the weight W is $V = Wh$, where h is the height above an arbitrary datum; hence the graph of V versus x for the weight W is

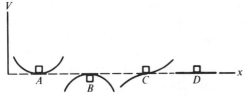

Fig. 6-6 Stability of Equilibrium (*A* is stable; *B* and *C* are unstable; and *D* is neutral)

(to a suitable scale) coincident with the shape of the guide. All four of the equilibrium positions illustrated are points where the curve has a horizontal tangent, illustrating the following important result.

An equilibrium position of a system subject to conservative forces occurs at a point where the potential energy of the system has a stationary value. Thus, if V is a differentiable function of one variable, x,

$$\frac{dV}{dx} = 0 \qquad\qquad (6.2.1)$$

at equilibrium.

In any **one-degree-of-freedom** system of connected bodies, the configuration is completely specified by a single parameter, say by θ. If the system is conservative, then the potential energy is also a function $V = V(\theta)$, and it can be shown that at an equilibrium position

$$\frac{dV}{d\theta} = 0. \qquad\qquad (6.2.2)$$

When the system is composed of several connected bodies it is usually much easier to determine the equilibrium configurations by solving Eq. (6.2.2) for the values of θ at equilibrium than it would be to draw free-body diagrams of all the separate members and write equilibrium equations for each of them as in Sec. 5.1. The simplification arises because in the conservative system the workless constraints and workless interconnection forces do not appear in Eq. (6.2.2), while spring supports and spring interconnections are easily included in the potential energy. The only difficulty might be in using trigonometry or kinematics to express all varying positions and lengths in terms of the one generalized coordinate θ.

Stability of equilibrium. Figure 6-6 also illustrates how the potential energy function can be used to determine the stability characteristics of an equilibrium position. An equilibrium position is called **stable** if every possible small displacement from the equilibrium position brings into action forces tending to return the system to the equilibrium position. Point A in Fig. 6-6 is a stable equilibrium position, since displacement either to the right or to the left moves the block to a point where the weight force has a tangential downhill component tending to return the block to A. Point B is clearly unstable. Point C is also **unstable**, since displacement to the left gives rise to a force tending to cause further displacement. For that matter, although displacement to the right from C brings about a force that returns it to C, it will return with some velocity that will cause it to overshoot and enter the region to the left of C, where the force tends to keep it going. Point D is **neutral**, since if the block is displaced slightly and then released *from rest*, it will tend to stay where it was released, neither returning as required for stability nor being displaced further by the forces brought into play.

Examination of Fig. 6-6 shows that the stable position A occurs where the potential energy V has a relative minimum and that the unstable positions B and C occur where V has a relative maximum or at a *minimax* horizontal inflection point on the curve. The neutral condition occurs where V is locally constant, neither increasing nor decreasing for a small displacement.

Potential Energy and Stability in a Conservative System	
Equilibrium Position	Potential Energy
Stable	Relative minimum
Unstable	Relative maximum or a minimax
Neutral	Locally constant

(6.2.3)

For a **one-degree-of-freedom** system, with $V = V(\theta)$, the question of whether a stationary point is a maximum or minimum may be decided by examining the second derivative if the second derivative does not vanish at the stationary point. If

$$V'(\theta_0) = 0 \text{ and } V''(\theta_0) > 0, \quad \theta_0 \text{ is a minimum point} \quad \text{(stable)}$$

$$V'(\theta_0) = 0 \text{ and } V''(\theta_0) < 0, \quad \theta_0 \text{ is a maximum point} \quad \text{(unstable)}.$$

(6.2.4)

If both the first and second derivatives vanish at a point, it is necessary to examine the higher derivatives. At a minimax V'' vanishes, but V'' may also vanish at a minimum or a maximum. (Consider, for example, $V = 10x^4$, where the origin $x = 0$ is a minimum.)

If the potential energy can be expanded in a Taylor's series in the neighborhood of θ_0, then

$$V(\theta) - V(\theta_0) = V'(\theta_0)(\theta - \theta_0) + \frac{V''(\theta_0)}{2!}(\theta - \theta_0)^2 + \cdots$$

$$+ \frac{V_{(\theta_0)}^{(n)}}{n!}(\theta - \theta)^n + \cdots$$

(6.2.5)

The shape of the curve $V = V(\theta)$ in the neighborhood of θ_0 is controlled by the first nonvanishing derivative at θ_0. If θ_0 is a stationary point, the first derivative necessarily vanishes. If the first nonvanishing derivative at a stationary point is of odd order, the point is a minimax inflection point and the equilibrium is unstable (e.g., $V = 10\theta^3$ at $\theta_0 = 0$). If the first nonvanishing derivative is of even order, say $V^{(2n)}$, then the stationary point θ_0 is either a minimum or a maximum, according to whether $V^{(2n)}(\theta_0)$ is positive or negative.

In an ideal system without any dissipation at all, a small displacement from a stable equilibrium configuration would initiate a free vibration, an oscillation back and forth about the equilibrium position. Since every real system has some dissipation, the oscillation will in fact be damped out unless it is forced to continue by continued disturbing forces. *The equilibrium positions of a real system with a relatively small amount of friction can be located and their stability analyzed by neglecting the friction.* Of course the block of Fig. 6-6, case *B*, could stay in position after a small displacement if there is enough friction, even though *B* would be called an unstable position in the analysis neglecting friction.

Notice that in all this discussion we have been considering only *small displacement from the equilibrium positions.* It is quite possible that a system could be stable for small disturbances but that a large displacement from a potential energy minimum point might take it to or even beyond a maximum point. This is apparent from the introductory example of the block on a smooth guide. In applying these results to real systems some engineering judgment should be used to decide whether there is any likelihood that large displacements would lead to instability.

EXAMPLE 6.2.1

We shall consider by potential energy methods the equilibrium and stability of the system shown in Fig. 6-7. A uniform bar *AB* of weight *W* and length *b* is supported by a smooth hinge at *A*. The end *B* is connected through a linear spring of spring constant *k* and unstretched length L_0 to point *D* above *A* at distance *d*.

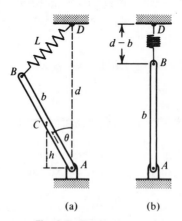

(a)   (b)

Fig. 6-7   Example 6.2.1

The total potential energy $V = V_g + V_e$ is

$$V = Wh + \frac{1}{2}k\delta^2 + \text{const.}$$

$$= W\frac{b}{2}\cos\theta + \frac{1}{2}k(L - L_0)^2 + \text{const.} \tag{a}$$

By the law of cosines

$$L^2 = b^2 + d^2 - 2bd\cos\theta, \tag{b}$$

whence

$$2L\frac{dL}{d\theta} = 2bd\sin\theta, \qquad \frac{dL}{d\theta} = \frac{bd}{L}\sin\theta. \tag{c}$$

Thus

$$\frac{dV}{d\theta} = -W\frac{b}{2}\sin\theta + k(L - L_0)\frac{dL}{d\theta}$$

$$= \sin\theta\left[kbd\left(1 - \frac{L_0}{L}\right) - W\frac{b}{2}\right] \tag{d}$$

$$\frac{d^2V}{d\theta^2} = \cos\theta\left[kbd\left(1 - \frac{L_0}{L}\right) - W\frac{b}{2}\right] + \sin\theta\left[kbd\left(\frac{L_0}{L^2}\right)\frac{dL}{d\theta}\right]$$

$$= \cos\theta\left[kbd\left(1 - \frac{L_0}{L}\right) - W\frac{b}{2}\right] + kb^2d^2\frac{L_0}{L^3}\sin^2\theta. \tag{e}$$

The equilibrium configurations are found by setting $dV/d\theta = 0$. Two configurations are found by setting the factor $\sin\theta = 0$, namely,

$$\theta_1 = 0° \quad \text{and} \quad \theta_2 = 180°.$$

If the other factor has a root θ_3 between 0° and 180°, where $\sin\theta_3 \neq 0$, then θ_3 will be a stable configuration, since $d^2V/d\theta^2$ will be positive there. [The first term on the right-hand side of Eq. (e) vanishes there, and the last term is positive for $\sin\theta \neq 0$.] To locate θ_3, we substitute for L in Eq. (d) and solve for $\cos\theta$:

$$\cos\theta = \frac{1}{2bd}\left\{b^2 + d^2 - \left[\frac{L_0}{1 - (W/2kd)}\right]^2\right\}.$$

If this gives a number between -1 and $+1$, then there is one root θ_3 between $0°$ and $180°$ and another root $\theta_4 = -\theta_3$, both stable equilibrium configurations (assuming that the construction of the system permits rotation in either direction).

We shall now examine the stability of the equilibrium configurations defined by θ_1 and θ_2. The configuration $\theta_1 = 0$ is illustrated in Fig. 6-7(b).

$$\theta = 0°, \quad L = d - b, \quad \text{and} \quad \cos\theta = 1$$

$$\frac{d^2V}{d\theta^2} = kbd\left(1 - \frac{L_0}{d-b}\right) - \frac{Wb}{2}.$$

$$\text{stable} \left(\frac{d^2V}{d\theta^2} > 0\right) \quad \text{if } k > \frac{W/2d}{1 - [L_0/(d-b)]} \equiv k_1 > 0$$

$$\text{unstable} \quad \text{if } k < k_1.$$

If $d - b > L_0$ and the spring is sufficiently strong, $\theta_1 = 0$ is a stable configuration. For $k = k_1$, we would have to examine higher derivatives to decide the stability character.

For

$$\theta = 180° \quad L = d + b, \quad \text{and} \quad \cos\theta = -1$$

$$\frac{d^2V}{d\theta^2} = -kbd\left(1 + \frac{L_0}{d+b}\right) + \frac{Wb}{2}.$$

$$\text{stable} \quad \text{for } k < \frac{W/2d}{1 + [L_0/(d+b)]} \equiv k_2, \quad \text{since} \frac{d^2V}{d\theta^2} > 0.$$

$$\text{unstable} \quad \text{for } k > k_2.$$

The downward position will be unstable if the spring is too strong ($k > k_2$).

Part 2. Systems with More than One Degree of Freedom*

We shall consider briefly only the case of two degrees of freedom. If the configuration of the system is completely determined by the specification of two *generalized coordinates*, say θ_1 and θ_2, the system is said to have two degrees of freedom (see Sec. 5.1). The generalized coordinates may be lengths or angles or any parameters fixing the configurations (θ_1 and θ_2 need not have the same dimensions).

* This part may be omitted.

The potential energy V is then a function of θ_1 and θ_2, which we assume can be expanded in a Taylor's series in the neighborhood of an equilibrium configuration (θ_1^0, θ_2^0).

$$
V(\theta_1, \theta_2) = V(\theta_1^0, \theta_2^0) + \left[\Delta\theta_1 \frac{\partial V}{\partial\theta_1} + \Delta\theta_2 \frac{\partial V}{\partial\theta_2} \right]_0
$$

$$
+ \frac{1}{2!} \left[(\Delta\theta_1)^2 \frac{\partial^2 V}{\partial\theta_1^2} + 2\Delta\theta_1\Delta\theta_2 \frac{\partial^2 V}{\partial\theta_1\partial\theta_2} \right.
$$

$$
+ (\Delta\theta_2)^2 \frac{\partial^2 V}{\partial\theta_2^2} \Big]_0 + \cdots \tag{6.2.6}
$$

$$
+ \frac{1}{n!} \left[\left(\Delta\theta_1 \frac{\partial}{\partial\theta_1} + \Delta\theta_2 \frac{\partial}{\partial\theta_2} \right)^n V \right]_0 + \cdots,
$$

where we have used the **subscript zero on the square brackets to indicate that the partial derivatives are to be evaluated at** (θ_1^0, θ_2^0) and where $\Delta\theta_1 = \theta_1 - \theta_1^0$ and $\Delta\theta_2 = \theta_2 - \theta_2^0$.

Since the equilibrium configuration is a point where the potential energy has a stationary value, the sum of the linear terms in the first bracket must be zero for arbitrary changes $\Delta\theta_1$ and $\Delta\theta_2$. Hence,

at equilibrium

$$
\frac{\partial V}{\partial\theta_1} = 0, \qquad \frac{\partial V}{\partial\theta_2} = 0. \tag{6.2.7}
$$

When V is a known function of θ_1 and θ_2, Eqs. (6.2.7) are two simultaneous equations for the determination of the equilibrium positions (θ_1, θ_2). Unfortunately the two equations are in general nonlinear. **For systems limited to very small motions about** $\theta_1 = 0$, $\theta_2 = 0$, **the potential energy may sometimes be approximated as a quadratic in** θ_1 **and** θ_2; **then Eqs. (6.2.7) would be linear.**

The stability of an equilibrium configuration is governed by the second-order terms in Eq. (6.2.6) unless the second derivatives all vanish at (θ_1^0, θ_2^0). If the second derivatives all vanish, it is necessary to consider higher derivatives.

If the second derivatives are not all zero at (θ_1^0, θ_2^0), the shape of the surface $V = V(\theta_1, \theta_2)$ in the immediate neighborhood of the equilibrium point is controlled by the second-order terms. For stability, the equilibrium point must be a strict relative minimum, so that $\Delta V = V - V_0$ will be positive for any small change $(\Delta\theta_1, \Delta\theta_2)$. A homogeneous quadratic function $f(x, y) = Ax^2 + 2Bxy + Cy^2$ is *positive-definite* (or *negative-definite*), that is, strictly positive (or strictly negative) for all real values of x and y (except $x = y = 0$) if and only if the *discriminant* $B^2 - AC$ is negative. [This means that the equation $f(x, y) = 0$ has only complex roots when solved as a quadratic for x/y or y/x.] It will be *positive-definite* if the discriminant is negative while A and C are both positive.

Hence a configuration where all the derivatives up to the second order are continuous and satisfy simultaneously

$$\frac{\partial V}{\partial \theta_1} = 0, \qquad \frac{\partial V}{\partial \theta_2} = 0$$

$$\left(\frac{\partial^2 V}{\partial \theta_1 \, \partial \theta_2}\right)^2 - \frac{\partial^2 V}{\partial \theta_1^2}\frac{\partial^2 V}{\partial \theta_2^2} < 0, \qquad \frac{\partial^2 V}{\partial \theta_1^2} > 0, \qquad \frac{\partial^2 V}{\partial \theta_2^2} > 0$$

(6.2.8)

will be a *stable equilibrium configuration*. If the discriminant is negative, while $\partial^2 V/\partial \theta_1^2 < 0$ and $\partial^2 V/\partial \theta_2^2 < 0$. the equilibrium point is unstable, since the second-order terms will be negative-definite, representing a maximum point of the potential energy $V(\theta_1, \theta_2)$.

If the discriminant is positive or zero, the stationary point is a saddle-point minimax representing unstable equilibrium, unless all three second derivatives are zero. If all three are zero, it is necessary to examine the higher derivatives. If all derivatives of all orders are zero, the function is locally constant, and the equilibrium is neutral.

In Sec. 6.3 the principle of virtual displacements will be presented. This principle is an alternative formulation of the equilibrium conditions; it is, like the potential energy method, an easier method for finding the equilibrium configuration for a system of connected bodies than free-body analysis of all the separate members of the system when the constraints of the system reduce the number of degrees of freedom to one or two and when the constraints do no work. Unlike the potential energy method, however, the method of virtual displacements gives no information about stability.

SAMPLE PROBLEM 6.2.1

Figure 6-8 shows a column formed from two identical light rigid pin-connected links that can rotate independently in the xy-plane but are restrained by four identical springs of spring constant k. Determine an upper bound on the downward load P for stability of the vertical equilibrium position using small-displacement approximations.

SOLUTION.   We choose the two angles θ_1 and θ_2 made by the centerlines of bars AB and BC, respectively, with the vertical as generalized coordinates to define the configuration of the system. For small displacements from the vertical position the spring elongations may be approximated by the horizontal displacements of B and

C. The constant downward load P has potential energy Py_C (for datum at A). Hence the total potential energy is

$$V = 2(\tfrac{1}{2}kx_B^2) + 2(\tfrac{1}{2}kx_C^2) + Py_C$$

$$= kL^2 \sin^2 \theta_1 + kL^2 (\sin \theta_1 + \sin \theta_2)^2$$

$$+ PL(\cos \theta_1 + \cos \theta_2).$$

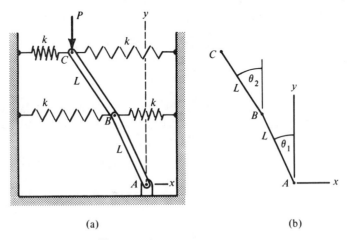

(a)   (b)

Fig. 6-8   Sample Problem 6.2.1

We linearize the problem by using small-angle approximations to give a quadratic potential energy: $\sin \theta \approx \theta$ and $\cos \theta \approx 1 - \tfrac{1}{2}\theta^2$. Then

$$V = kL^2 \theta_1^2 + kL^2(\theta_1 + \theta_2)^2 + PL(2 - \tfrac{1}{2}\theta_1^2 - \tfrac{1}{2}\theta_2^2)$$

$$\frac{\partial V}{\partial \theta_1} = 2kL^2 \theta_1 + 2kL^2(\theta_1 + \theta_2) - PL\theta_1$$

$$\frac{\partial V}{\partial \theta_2} = 2kL^2(\theta_1 + \theta_2) - PL\theta_2$$

$$\frac{\partial^2 V}{\partial \theta_1^2} = 4kL^2 - PL > 0 \qquad \text{for } P < 4kL$$

$$\frac{\partial^2 V}{\partial \theta_2^2} = 2kL^2 - PL > 0 \qquad \text{for } P < 2kL$$

$$\frac{\partial^2 V}{\partial \theta_1\, \partial \theta_2} = 2kL^2.$$

The discriminant D is [see Eq. (6.2.8)]

$$D = \left(\frac{\partial^2 V}{\partial\theta_1\,\partial\theta_2}\right)^2 - \left(\frac{\partial^2 V}{\partial\theta_1^2}\right)\left(\frac{\partial^2 V}{\partial\theta_2^2}\right) = 4k^2L^4 - (4kL^2 - PL)(2kL^2 - PL)$$

$$D = -[P^2L^2 - 6PkL^3 + 4k^2L^4].$$

Hence $D < 0$ when $(P/kL)^2 - 6(P/kL) + 4 > 0$, that is, when P/kL is outside the interval separating the two roots of the quadratic equation $(P/kL)^2 - 6(P/kL) + 4 = 0$. Thus $D < 0$ when

$$\frac{P}{kL} < 3 - \sqrt{5} \quad\text{or}\quad \frac{P}{kL} > 3 + \sqrt{5}.$$

For stability we must have the discriminant negative and the two second derivatives $\partial^2 V/\partial\theta_1^2$ and $\partial^2 V/\partial\theta_2^2$ positive. All three conditions are satisfied for

$$P < P_1 = (3 - \sqrt{5})kL = 0.764kL.$$

Hence an upper bound on the load P for stability of the column against lateral buckling is

$$P_1 = 0.764kL \quad \textbf{\textit{Answer.}}$$

SAMPLE PROBLEM 6.2.2

A uniform block of depth $2d$ and length $2L$ is balanced on a half-cylinder of radius R as shown in Fig. 6-9. Discuss the stability of the equilibrium configuration, assuming that in any disturbance it tips without slipping.

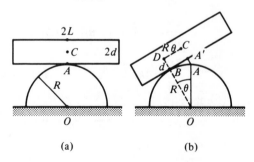

(a)   (b)

Fig. 6-9 Sample Problem 6.2.2

SOLUTION.   For tip without slip the new point of contact B will be at distance BA' along the block equal to the arc $AB = R\theta$. Hence the elevation of the mass center above the datum at O will be

$$h = (R + d) \cos \theta + R\theta \sin \theta.$$

Since $V = Wh$, we obtain

$$V(\theta) = W[(R + d) \cos \theta + R\theta \sin \theta]$$
$$V'(\theta) = W[-(R + d) \sin \theta + R \sin \theta + R\theta \cos \theta]$$
$$= W[-d \sin \theta + R\theta \cos \theta],$$

vanishing at the equilibrium position $\theta = 0$, and

$$V''(\theta) = W[(R - d) \cos \theta - R\theta \sin \theta].$$
$$V''(0) = W(R - d).$$

Hence the equilibrium is stable when $R > d$ and unstable for $R < d$. For $R = d$, $V''(0) = 0$, and we must examine the higher derivatives. For $R = d$,

$$V'''(\theta) = -RW(\sin \theta + \theta \cos \theta)$$

and

$$V^{(iv)}(\theta) = -RW(2 \cos \theta - \theta \sin \theta).$$

Hence, for $R = d$, $V'''(0) = 0$, and $V^{(iv)}(0) < 0$, so that the equilibrium is also unstable for $R = d$. Thus we have stability only for $R > d$.   **Answer.**

EXERCISES

1. The weight W is attached to the unstretched linear spring (of spring constant k) and released from rest. Use the potential energy method to determine the distance d_e down to the equilibrium position and to show that the equilibrium is stable.

2. Given that $W = \frac{1}{2}kL$, determine the values of θ for equilibrium of the system shown, and

Exercise 1

examine the stability of the equilibrium positions. Bar AB has negligible weight and is connected by smooth pins to the block A of weight W, which moves in a smooth vertical slot, and to block B, which moves in a smooth horizontal slot. The spring, of spring constant k, is unstretched when B is at D.

The weight W is 100 N. The four light 0.5-m bars are pin-connected, and W moves in a smooth vertical guide.

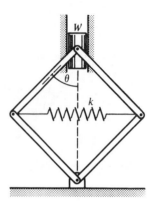

Exercise 4

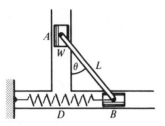

Exercise 2

3. The two drums of radii 2 ft and 3 ft and negligible mass are rigidly attached to each other and rotate together about the smooth pin support. The linear spring is unstretched in the position shown and has a spring constant of 30 lb/ft. The spring and the 80-lb weight are attached to the drums by light flexible ropes wrapped around each drum. Use the potential energy method to determine through what angle θ the drums rotate before an equilibrium position is reached. Is the equilibrium stable, unstable, or neutral?

5. The ball A, of given weight W lb, is attached to the light rod OA, of length b ft, which rotates about the smooth pin support O. The ball is attached to B(b ft above O) by a linear spring with spring constant k lb/ft. The spring's unstretched length L_0 equals b (i.e., $\theta_0 = 60°$). (a) What must be the value of k (in terms of W and b) in order that the horizontal position OA' will be an equilibrium position (where $\theta = 90°$)? (b) If k has this value, is the equilibrium stable?

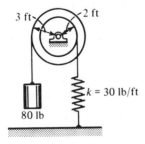

Exercise 3

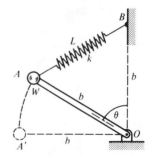

Exercise 5

4. Determine all values of θ for which the co-planar system shown is in equilibrium positions. The spring is unstretched when θ is zero and has a spring constant $k = 200$ N/m.

6. For the initial conditions of Ex. 5, (a) show that $\theta = 180°$ is an equilibrium position, and discuss the relationship between W and kb

for stability. (b) For given k and b, determine an upper bound on W for there to be an equilibrium position satisfying $0 < \theta < 180°$. Show that if the equilibrium position exists it is stable.

7. Discuss the stability of a half-cylinder of radius r balanced as shown on another half-cylinder of radius R against rolling disturbances without slip.

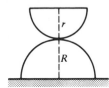

Exercise 7

8. The body shown, of uniform density, is formed of a right circular cone of height h joined to a hemisphere of radius R. Discuss the stability of the vertical equilibrium position on a rough horizontal plane.

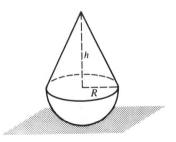

Exercise 8

9. The spring, unstretched when $\theta = 0$, is attached to the drum of radius R by a cord. The arm AB of length L and negligible weight is rigidly attached to the drum and rotates with it, lowering the weight W. (a) Show that if the position $\theta = 0$ is unstable, then there exists a second equilibrium position for $0 < \theta < 180°$. (b) By sketching graphs of $\sin \theta$ and $(kR^2/WL)\theta$, argue that the second equilibrium position is stable if it exists.

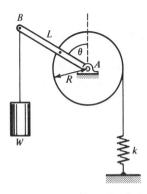

Exercise 9

10. If in Ex. 9, $L = 8$ in., $R = 3$ in., $k = 125$ lb/in., and $W = 200$ lb, determine by a numerical or trial-and-error solution the equilibrium position for $0 < \theta < 180°$, and show that it is stable.

6.3   VIRTUAL DISPLACEMENTS AND VIRTUAL WORK *

The principle of virtual displacements (sometimes called the principle of virtual work) is an alternative formulation of the conditions for equilibrium of a system that is at rest in the equilibrium configuration.† The method of virtual displacements may be used instead of writing the equilibrium equations for each free body to solve for unknown support reactions or to determine the equilibrium configuration of the system. The virtual displacement method is especially useful for systems of connected bodies moving subject to guides or constraints such that the system has only one or two degrees of freedom, although it can be applied to other cases. We will be mainly concerned with a mechanism having one degree of freedom or with a completely constrained structure that can be hypothetically transformed into a one-degree-of-freedom mechanism by replacing one constraint component by a force (equal to the constraint reaction replaced) that is imagined to remain constant as the system moves. The whole discussion is concerned with *small displacements from a possible equilibrium configuration.*

A *virtual displacement* $\delta\mathbf{r}$ is a kinematically admissible infinitesimal displacement of a material point whose position vector is $\mathbf{r}$. We use the notation $\delta\mathbf{r}$ for it instead of $d\mathbf{r}$ to distinguish it from an actual displacement $d\mathbf{r}$ that would occur under the given loads. Since we are talking about a condition of equilibrium at rest relative to an inertial reference, in general no actual displacements will occur under the given loads. For a free particle in space, any virtual displacement $\delta\mathbf{r}$ is kinematically admissible. The rectangular Cartesian components δx, δy, δz of $\delta\mathbf{r}$ may be independently prescribed. Although the principle includes this case, it offers no advantage for the treatment of an unconstrained particle or body. For a particle to move on a plane curve $y = f(x)$, we must have $\delta z = 0$ and $\delta y = f'(x)\,\delta x$ for the displacement to be kinematically admissible.

Kinematically admissible displacements for a mechanism are displacements that do not violate the constraint or interconnection conditions of the connected bodies forming the mechanism. We may also make a hypothetical relaxation of one of the constraint conditions at a time in order to determine the corresponding constraint reaction (see Example 6.3.1).

Virtual Work. The work δU that would be done by the forces acting in the initial rest configuration *if these forces remained constant* during a kinematically admissible infinitesimal virtual displacement from the initial configuration is called *virtual work.* We assume that the forces can be separated into two categories:

---

* This section may be omitted.

† The ideas contained in the principle of virtual displacements may go back to the school of Aristotle. They were used by Simon Stevinus (1548–1620), by John Bernoulli (1667–1748) and others. The principle was taken as the principal axiom of mechanics by Joseph-Louis Lagrange (1736–1813) in his *Mécanique Analytique* (1788).

Active forces that do work δU_a
(often called impressed forces)

(6.3.1)

Workless constraint and
connection forces, for which $\delta U_c = 0$

in any kinematically admissible displacement. Friction at a surface guiding the motion would have to be included among the active forces. The method is most successful where all constraint and interconnection forces are either workless or are linear elastic spring forces whose work δU_e can be accounted for by potential energy changes

$$\delta U_e = -\delta V_e = -\delta(\tfrac{1}{2}ke^2) = -ke\,\delta e \tag{6.3.2}$$

where the *spring elongation has been denoted by* e instead of δ to avoid confusion with the virtual change symbol δ.

Types of constraints; generalized coordinates. The configuration of a finite system is specified by prescribing N **generalized coordinates** $\theta_1, \theta_2, \ldots, \theta_N$, where N is the number of degrees of freedom of the system. We consider only cases where each θ_k can vary independently of the others, and we assume that the constraints can be modeled as fixed guides that the constrained point cannot leave. A block sliding on a table top, for example, will be assumed not to jump off the table.

The Principle of Virtual Displacements then asserts that a mechanical system at rest in an initial configuration where certain active forces are applied, and constrained only by "fixed-guide" constraints, is in equilibrium if, and only if, the total virtual work δU_a of all the active forces is zero for any arbitrary kinematically admissible virtual displacement from the initial configuration.

$$\delta U_a = 0$$

(6.3.3)

The derivation of this principle for the most general kind of system of rigid bodies is difficult because of the complicated kinematics associated with describing rotational motion of the rigid bodies under all possible kinds of constraints. For this reason the principle of virtual displacements is usually taken as a basic postulate for the theory of **variational mechanics**. Before proceeding to applications we will see how the principle can be derived from Newton's laws for the special case of a single particle moving on a smooth surface.

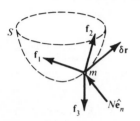

Figure 6-10 shows schematically a free-body diagram of a particle of mass m constrained to move on a smooth surface S under the action of forces $\mathbf{f}_1$, $\mathbf{f}_2$, ..., and the normal constraint force $N\hat{\mathbf{e}}_n$. (N may be negative.) If the particle is in equilibrium,

$$N\hat{\mathbf{e}}_n + \mathbf{f}_1 + \mathbf{f}_2 + \cdots = 0. \tag{6.3.4}$$

Fig. 6-10 Particle on Which Active Forces $\mathbf{f}_k$ and Smooth Surface Constraint Force $N\hat{\mathbf{e}}_n$ Act

Hence the total virtual work

$$\delta U = (N\hat{\mathbf{e}}_n + \mathbf{f}_1 + \mathbf{f}_1 + \cdots) \cdot \delta\mathbf{r} \tag{6.3.5}$$

will be zero in any virtual displacement. But $\delta U_c = N\hat{\mathbf{e}}_n \cdot \delta\mathbf{r} = 0$ for any kinematically admissible $\delta\mathbf{r}$, because $\delta\mathbf{r}$ must be perpendicular to $\hat{\mathbf{e}}_n$. Since $\delta U = 0$ and $\delta U_c = 0$, it follows that

$$\delta U_a = \delta U - \delta U_c = 0. \tag{6.3.6}$$

We have thus shown that equilibrium implies $\delta U_a = 0$ without using the hypothesis that the particle was at rest. That hypothesis will be needed for the converse, however.

For the converse proposition, we assume that $\delta U_a = 0$ in every kinematically admissible $\delta\mathbf{r}$. But since δU_c is also zero, it follows that the total $\delta U = 0$, whence the resultant force on the particle is either zero or perpendicular to all kinematically admissible $\delta\mathbf{r}$ [see Eq. (6.3.5)] and hence perpendicular to the surface S. But a nonzero resultant force in the direction of the normal would imply, by Newton's second law, a nonzero acceleration component a_n in the direction of the normal. In Volume II, however, it is shown that on a curved path the normal component of acceleration (centripetal acceleration) is zero when the speed is zero.* Therefore, for a particle at rest, the vanishing of δU_a for every kinematically admissible $\delta\mathbf{r}$ implies that the resultant force is zero, and hence that the particle is in equilibrium.

The method of virtual displacements for a connected system of rigid bodies consists in calculating the virtual work δU_a that would be done by the active forces and couples. We then set $\delta U_a = 0$ for certain arbitrarily chosen kinematically admissible virtual displacements from an initial configuration. By choosing various sets of admissible displacements we obtain enough equations to solve for the unknowns in a statically determinate system. The unknowns may be reaction forces in the equilibrium configuration, or we may seek to determine the generalized coordinates of possible equilibrium configurations. We shall illustrate the use of the method by examples.

* See Vol. II, Sec. 8.6.

EXAMPLE 6.3.1

The coplanar structure of Fig. 6-11(a) has two pin supports with four unknown reaction components. If you consider a free-body diagram (not shown) of the whole structure, you will be able to conclude that $A_y + C_y = Q$, but to determine the values of A_y and C_y you would have to dismember the structure and consider equilibrium of the parts as in Sec. 5.1. Here we use an alternative procedure, replacing pin A by an ideal frictionless roller in a vertical slot as in Fig. 6-11(b) acted upon by external force A_y.

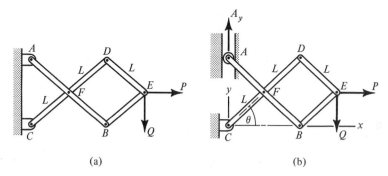

(a)   (b)

Fig. 6-11   Example 6.3.1

We have transformed the structure into a one-degree-of-freedom mechanism, whose configuration can be specified by the single generalized coordinate θ. For any virtual displacement of the system the total virtual work δU_a is

$$\delta U_a = A_y\, \delta y_A + P\, \delta x_E - Q\, \delta y_E. \tag{6.3.7}$$

The negative sign in the last term is needed because Q acts in the negative y-direction. For any kinematically admissible displacements we set $\delta U_a = 0$. To ensure that the displacements are kinematically admissible, we require

$$x_E = 3L \cos \theta, \qquad y_E = L \sin \theta, \qquad y_A = 2L \sin \theta$$
$$\delta x_E = -3L \sin \theta\, \delta\theta, \qquad \delta y_E = L \cos \theta\, \delta\theta, \qquad \delta y_A = 2L \cos \theta\, \delta\theta. \tag{6.3.8}$$

The virtual displacements δx_E, δy_E, and δy_A in Eq. (6.3.8) were obtained as the differentials of the expressions for x_E, y_E, and y_A. For these kinematically admissible displacements, we set

$$0 = \delta U_a = 2A_y L \cos \theta\, \delta\theta - 3PL \sin \theta\, \delta\theta - QL \cos \theta\, \delta\theta,$$

whence division by $L\delta\theta$ gives

$$2A_y \cos \theta = 3P \sin \theta + Q \cos \theta. \tag{6.3.9}$$

This equation gives the equilibrium value of the unknown A_y in terms of the equilibrium value of θ and the assigned loads P and Q. If we were dealing with an actual mechanism such as that shown in Fig. 6-11(b), then (neglecting friction) we could use this procedure to determine θ in the unknown equilibrium configuration in terms of assigned loads A_y, P, and Q. For the structure of Fig. 6-11(a) the equilibrium value of θ is known, and we are able to solve for A_y without detailed free-body analysis of the parts of the structure.

Note that Fig. 6-11(b) is not a free-body diagram, since the supports have not been completely removed. It is called **an active-force diagram**, since only those forces that do work are included in it. The reaction A_x is perpendicular to any kinematically admissible displacement of A. Since A_x does no work, it is not shown. The internal interconnection forces do no net work in this case, since they occur in equal and opposite pairs of forces whose points of application have the same displacement.

We shall also find the horizontal support reaction A_x of the structure of Fig. 6-11 by the method of virtual displacement. Because A_x can be determined more easily from a free-body analysis of the whole structure, the virtual displacement method is not a good choice of method for finding A_x. But it will illustrate kinematic ideas that may be useful in other problems. When we sought A_y, we freed the point A to displace vertically, causing the mechanism to "scissor." When we permit a horizontal velocity at A but no vertical velocity, the structure will be instantaneously rotating about pin C as in the active-force diagram of Fig. 6-12, where A_y is omitted because it does no work.

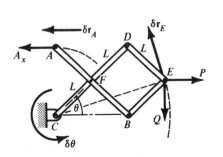

Point A is moving on a circle of radius $CA = 2L \sin\theta$. Point E is moving on a circle with radius $CE = R = [(3L \cos\theta)^2 + (L \sin\theta)^2]^{1/2}$ in direction $\hat{\mathbf{e}}_E = (1/R)[(-L \sin\theta)\mathbf{i} + (3L \cos\theta)\mathbf{j}]$. The kinematically admissible (tangential) displacements are each perpendicular to the radius R drawn from C to the point and have magnitude $R\,\delta\theta$ for rotation through angle $\delta\theta$ radians. Thus

$$\delta\mathbf{r}_A = -(2L \sin\theta\,\delta\theta)\mathbf{i} \quad \text{or} \quad \delta x_A = -2L\,\delta\theta \sin\theta,$$
$$(6.3.10)$$

and

Fig. 6-12 Active-Force Diagram for Horizontal Motion of Point A in Example 6.3.1

$$\delta\mathbf{r}_E = (-L\,\delta\theta \sin\theta)\mathbf{i} + (3L\,\delta\theta \cos\theta)\mathbf{j}. \qquad (6.3.11)$$

For these kinematically admissible virtual displacements (for A_x positive to the left),

$$0 = \delta U_a = A_x(2L\,\delta\theta \sin\theta) + (P\mathbf{i} - Q\mathbf{j})\cdot\delta\mathbf{r}_E$$
$$= 2A_xL\,\delta\theta \sin\theta - PL\,\delta\theta \sin\theta - 3QL\,\delta\theta \cos\theta. \qquad (6.3.12)$$

Hence

$$A_x = \frac{P \sin \theta + 3Q \cos \theta}{2 \sin \theta},$$

as could more easily have been found by taking moments about C on a free body of the whole structure.

EXAMPLE 6.3.2

As an illustration of the procedure in systems with more than one degree of freedom, consider the coplanar system shown in Fig. 6-13, consisting of two identical links, each of weight W and length $2b$, hinged together and to the support at O and acted upon at B by the horizontal force P. The configuration can be completely defined by the two generalized coordinates θ_1 and θ_2, the angles the members make with the horizontal, as shown. To determine the equilibrium configuration values of θ_1 and θ_2 we apply the method of virtual displacements. The virtual work is

$$\delta U_a = W\,\delta d_1 + W\,\delta d_2 + P\delta x_B.$$

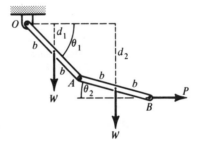

Fig. 6-13   Two-Degree-of-Freedom Example 6.3.2

For kinematically admissible displacements,

$$d_1 = b \sin \theta_1, \qquad d_2 = 2b \sin \theta_1 + b \sin \theta_2$$

$$x_B = 2b \cos \theta_1 + 2b \cos \theta_2. \qquad \delta d_1 = b\,\delta\theta_1 \cos \theta_1$$

$$\delta d_2 = 2b\,\delta\theta_1 \cos \theta_1 + b\,\delta\theta_2 \cos \theta_2$$

$$\delta x_B = -2b\,\delta\theta_1 \sin \theta_1 - 2b\,\delta\theta_2 \sin \theta_2.$$

Since θ_1 and θ_2 are independent generalized coordinates, we can choose $\delta\theta_1$ and $\delta\theta_2$, independently. It simplifies matters to make two different choices of $(\delta\theta_1\ \delta\theta_2)$, choosing one of them to be zero and the other nonzero in each case. We must have $\delta U_a = 0$ for each of these choices, and we therefore obtain two equations to solve for θ_1 and θ_2, as follows:

Case 1:

$$\delta\theta_1 = 0, \qquad \delta\theta_2 \neq 0$$

$$0 = \delta U_a = Wb\,\delta\theta_2 \cos\theta_2 - 2Pb\,\delta\theta_2 \sin\theta_2,$$

whence

$$\tan\theta_2 = \frac{W}{2P}.$$

Case 2:

$$\delta\theta_1 \neq 0, \qquad \delta\theta_2 = 0$$

$$0 = \delta U_a = Wb\,\delta\theta_1 \cos\theta_1 + 2Wb\,\delta\theta_1 \cos\theta_1 - 2Pb\,\delta\theta_1 \sin\theta_1,$$

whence

$$\tan\theta_1 = \frac{3W}{2P}.$$

Note that the principle of virtual displacements requires $\delta U_a = 0$ for every arbitrary kinematically admissible virtual displacement from the equilibrium configuration. In applying the principle, we are free to choose combinations of $\delta\theta_1$ and $\delta\theta_2$ that will simplify the computations.

Potential Energy Variation and Virtual Displacements

When an internal connection force is exerted by a linear elastic spring, this is an active force whose virtual work δU_e is equal to the negative of the change δV_e in its potential energy; see Eq. (6.3.2).

When all the external and internal active forces are conservative and the constraints are workless, the total

$$\delta U_a = -\delta V, \qquad (6.3.13)$$

where V is the total potential energy and δV is the *first variation* of V (a differential formed with the virtual displacements δr instead of actual displacements dr). *For kinematically admissible virtual displacements of a system under the action only of conservative active forces and workless constraints, the requirement $U_a = 0$ is equivalent to the condition that the first variation of the potential energy vanish at the equilibrium configuration.*

For a one-degree-of-freedom system with $V = V(\theta)$

$$\delta V = V'(\theta)\,\delta\theta, \tag{6.3.14}$$

While for a two-degree-of-freedom system with $V = V(\theta_1, \theta_2)$

$$\delta V = \frac{\partial V}{\partial \theta_1}\delta\theta_1 + \frac{\partial V}{\partial \theta_2}\delta\theta_2. \tag{6.3.15}$$

The vanishing of the first variation δV for **arbitrary variations** $\delta\theta$ (or $\delta\theta_1$ and $\delta\theta_2$) thus implies the equations that we used to determine the equilibrium configuration by potential energy methods in Sec. 6.2, namely,

$$V'(\theta_0) = 0 \tag{6.3.16}$$

in the equilibrium configuration θ_0 of a one-degree-of-freedom system or

$$\frac{\partial V}{\delta\theta_1} = 0 \quad \text{and} \quad \frac{\partial V}{\partial \theta_2} = 0 \qquad \text{at } (\theta_1^0, \theta_2^0) \tag{6.3.17}$$

for a two-degree-of-freedom system. The virtual displacement method, however, furnishes no information about stability. Since the virtual work is computed with the active forces assumed to remain constant during the virtual displacement, it gives only the first-order change $\delta V = -\delta U_a$ in the potential energy and furnishes no information about the second-order terms in ΔV needed to discuss stability.

Statically Indeterminate Problems

The principle of virtual displacements is an alternative formulation of the equilibrium conditions. In statically determinate problems, when one equation is obtained for each degree of freedom as in our solution of Example 6.3.2, there will be enough equations to solve for the unknowns. In statically indeterminate problems we know that the equilibrium equations are not sufficient in number but must be supplemented with force-deformation equations. It should come as no surprise therefore that the equations obtained from the virtual displacement method are also not sufficient to solve for all the unknowns in a statically indeterminate problem, since they are equivalent only to the usual equilibrium conditions.

Work of a Couple

In Sec. 12.2 of Vol. II we shall see that the power $P = dU/dt$ of a couple **c** acting on a body rotating at angular velocity $\boldsymbol{\omega}$ is $P = \mathbf{c}\cdot\boldsymbol{\omega}$. For the special cases of coplanar motion with the couple vector perpendicular to the plane and of rotation about a fixed axis with the couple vector parallel to the axis, $|P| = c|\dot{\theta}|$, where c and $|\dot{\theta}|$ are the magnitudes of $\mathbf{c}$ and $\boldsymbol{\omega}$. In an infinitesimal displacement

$$dU = P\,dt = c\,d\theta.$$

For a virtual displacement $\delta\theta$,

Virtual Work of Couple

(Coplanar Case)

$$\delta U = c\,\delta\theta. \qquad\qquad (6.3.18)$$

Nonconservative forces can be included in the principle of virtual displacements, but friction forces and nonconservative interaction forces such as those involved in plastic deformation of connecting members must be included among the active forces, and this complicates the problem. A known constant friction force can be included easily, as also can a known constant frictional couple in a hinge, but when the friction force is variable and must be related to the variable normal force, the simplifying features of the method disappear. This book considers only cases where friction work is negligible, and where all constraint and interconnection forces are workless, except for linear spring connections that can be treated as active forces whose work is accounted for by potential energy changes as in Sample Problem 6.3.1.

Workless constraints include

1. Reactions at fixed perfectly smooth surfaces.
2. Reactions at fixed perfectly rough surfaces (no relative motion).
3. Reactions at smooth pins, hinges, or ball-and-socket joints.
4. Inextensible cable connections.
5. Pin-connected rods of negligible weight.
6. Track reactions on a rigid wheel rolling without slipping.

As the examples so far presented have indicated, successful use of the method of virtual displacements depends mainly on establishing the geometric and kinematic relationships that express the displacement of all points where active forces are applied in terms of the variations of the few generalized coordinates that define the configuration. The methods of rotational rigid-body kinematics to be presented in Chapter 10 of Vol. II will be helpful in applying the principle to complicated mechanisms, including three-dimensional mechanisms. In the following exercises we shall consider only coplanar cases that can be treated by the direct analysis of the kinematics, differentiating expressions for the position coordinates in terms of the generalized coordinates as in Examples 6.3.1 and 6.3.2 or by using the circular-motion kinematics $|\delta \mathbf{r}| = R \, \delta\theta$.

SAMPLE PROBLEM 6.3.1

The spring in the pin-connected coplanar mechanism of Fig. 6-14 is unstretched when $\theta = 60°$. The roller at A moves in a smooth vertical guide. The four members have equal lengths L and negligible weight. Determine the load P required for equilibrium in a configuration where $0 < \theta < 60°$.

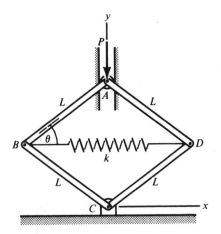

Fig. 6-14 Sample Problem 6.3.1

SOLUTION. The only active forces are the load P and the internal·spring force. Hence

$$\delta U_a = -P \, \delta y_A - \delta V_e$$

$$V_e = \tfrac{1}{2}ke^2, \qquad\qquad \delta V_e = ke \, \delta e$$

$$y_A = 2L \sin \theta, \qquad\qquad e = 2L \cos \theta - L$$

$$\delta y_A = 2L \, \delta\theta \cos \theta, \qquad \delta e = -2L \, \delta\theta \sin \theta,$$

where e is the elongation of the spring. For kinematically admissible virtual displacements,

$$0 = \delta U_a = -2PL\,\delta\theta\,\cos\theta + k(2L\cos\theta - L)(2L\,\delta\theta\,\sin\theta),$$

whence

$$P = kL(2\sin\theta - \tan\theta)\quad \textbf{\textit{Answer.}}$$

EXERCISES

(Additional practice may be gained by solving by virtual displacement methods that part of the exercises at the end of Sec. 6.2 requiring the determination of the equilibrium configuration or the forces acting in equilibrium. The virtual displacement methods gives no information about stability.)

1. Use the method of virtual displacements to determine the force in bar AB of the truss loaded as shown. (*Hint*: Transform it into a mechanism by removing the bar and replacing it by the forces it exerts at A and B.)

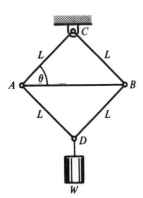

Exercise 1

2. Solve Ex. 16, Sec. 5.1, by virtual displacements.
3. Solve Ex. 18, Sec. 5.1, by virtual displacements.

4. Determine the support reactions for the pin-connected frame loaded as shown. Use a judicious combination of methods.

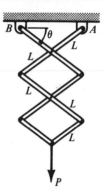

Exercise 4

5. Add a third link of weight W and length $2b$ in Fig. 6-13, and solve for the three angles for equilibrium with a horizontal force P at the lower end.
6. Determine the couple c required to balance the weight W. Neglect member weights.
7. The "black box" shown has a horizontal shaft sticking out through a hole at A, a vertical shaft at B, and an unknown mechanism inside. Experimentation shows that when the shaft at A is rotated by 10° the shaft at B comes out $\frac{1}{2}$ in. How much couple must be applied to the shaft at A to balance a 1000-lb weight on top of the shaft at B. Comment on the assumptions implicit in your solution.

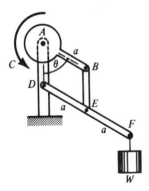

Exercise 6

9. The three unequal uniform links of weight
   w lb-ft are pin-connected and supported by
   rollers at A and D. Determine the angles of
   inclination for equilibrium in terms of w and
   the three lengths. Assume that $|L_3 - L_1| <$
   L_2. Note that the three angles are not
   independent. (*Hint*: Eliminate θ_2.)

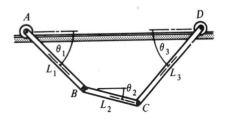

Exercise 9

10. If the spring is unstretched when $\theta = \theta_0$,
    determine the relation between θ and W for
    equilibrium of the coplanar mechanism.

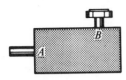

Exercise 7

8. Use virtual displacements to determine the
   force in bar ED of the loaded truss
   shown. All bar lengths are equal.

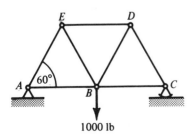

Exercise 8

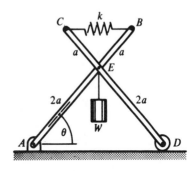

Exercise 10

6.4 SUMMARY

The work U done by a variable force $\mathbf{f}$ acting on a particle that moves along a path C
is given by the *line integral*.

$$U = \int_C \mathbf{f} \cdot d\mathbf{r} = \int_C f \cos \theta \, ds, \qquad (6.1.2)$$

where $f_t = f \cos \theta$ is the **working component** of the force. The dimensions of work are $[U] = [FL]$, but only when the force is constant and acts parallel to a straight-line motion is it possible to calculate U simply as force times distance.

A **conservative force field**

1. Is one that depends only on position, and
2. Is such that the work done by it on a particle that moves between any two positions is independent of the path followed.

When the force field $\mathbf{f}$ is conservative, the **potential energy** $V(P) = V(x, y, z)$ is given by

$$V(P) = -\int_{P_0}^{P} \mathbf{f} \cdot d\mathbf{r}, \qquad (6.1.14)$$

defined only up to an arbitrary additive constant depending on the choice of P_0. The work $U_{1 \to 2}$ done, by the conservative field whose potential energy is V, between P_1 and P_2 is

$$U_{1 \to 2} = -\Delta V = -(V_2 - V_1). \qquad (6.1.15)$$

Potential energy formulas were obtained for

$$\text{uniform gravitational field:} \quad V_g = Wh \qquad (6.1.17)$$

$$\text{linear elastic spring with datum at } \delta = 0: \quad V_e = \tfrac{1}{2}k\delta^2 \qquad (6.1.21)$$

$$\text{inverse square gravitational field:} \quad V(r) = -\frac{GMm}{r} + \text{const.} \qquad (6.1.26)$$

Potential energy methods were used in Sec. 6.2 to examine the stability of equilibrium of a conservative system. At equilibrium, the potential energy has a stationary value.

Equilibrium Position Is	For Potential Energy	
Stable	Relative minimum	
Unstable	Relative maximum or minimax	(6.2.3)
Neutral	Locally constant	

The **principle of virtual displacements** of Sec. 6.3 is an alternative formulation of the equilibrium conditions, sometimes called the principle of **virtual work**. It is especially useful for determining support reactions of a structural element or for the equilibrium configuration of a mechanism with negligible friction. Subject to certain conditions stated in connection with Eq. (6.3.3), the principle states that the system is in equilibrium in a certain initial rest configuration if, and only if, the total *virtual work* δU_a of all the active forces is zero for every kinematically admissible (infinitesimal) *virtual displacement* from the initial configuration. The active forces are assumed to remain constant during the virtual displacement in calculating δU_a.

The work and energy concepts of Sec. 6.1 will be important in dynamics as well as in statics. We shall now conclude Vol. I with a chapter on computation of centroids, second moments of area, and mass moments of inertia.

CHAPTER SEVEN

Properties of Geometric Shapes and Masses— Fluid Statics

7.1 INTRODUCTION

In Sec. 7.2 we shall review the definitions of centroid and center of mass of Sec. 4.2. In Sec. 7.3 methods are given for locating the centroid of a composite body when the centroids of the parts are known. In Sec. 7.4 single-integral methods, and in Sec. 7.5, multiple-integral methods for locating the centroid of a body are given.

The second moments of area treated in Secs. 7.6 and 7.7 are important for calculating the bending stiffness of an elastic beam in the mechanics of materials and structures. They are also useful for finding the mass moments of inertia of a flat plate.

In Sec. 7.8 on fluid statics we shall make use of the centroid and also the second moment of an area to locate the *center of pressure* on a submerged area. The mass moments of inertia and products of inertia of Sec. 7.9 are concepts of the dynamics of rotation. They will be extensively used in Chapter 12 of Vol. II. Although they are not used in statics, they are usually taught in a statics course as preparation for dynamics courses.

Centroids and moments of inertia of common geometric shapes are summarized in Tables A3 and A4 of Appendix A.

7.2 CENTER OF MASS

The mass center of a collection of particles $m_1, \ldots, m_N$ was defined in Eqs. (4.2.2) as the point whose coordinates $(\bar{x}, \bar{y}, \bar{z})$ were given by

$$m\bar{x} = \sum_{k=1}^{N} m_k x_k \qquad (7.2.1)$$

and two similar equations for $\bar{y}$ and $\bar{z}$, where m is the total mass of the system. This was extended in Eqs. (4.2.9) to define the mass center $(\bar{x}, \bar{y}, \bar{z})$ for a body with distributed mass by

$$m\bar{x} = \int_V x \, dm, \qquad \text{where} \qquad m = \int_V dm, \qquad (7.2.2)$$

or

$$\bar{x} = \frac{\int_V x\rho \, dV}{\int_V \rho \, dV}, \qquad \text{where} \qquad dm = \rho \, dV, \qquad (7.2.3)$$

with two similar formulas for $\bar{y}$ and $\bar{z}$. Here ρ is the density or mass per unit volume. In general ρ is a variable function of position (and, in a deformable body, of time). For most of the cases we shall consider, however, ρ will be uniform throughout the body, or else the body can be subdivided into parts with uniform density in each part.

When ρ is uniform, it can be canceled from the numerator and denominator of Eq. (7.2.3) and the similar equations for $\bar{y}$ and $\bar{z}$. Then the center of mass is located at the *centroid*. When ρ varies from point to point (or from part to part of a composite body) the center of mass and centroid will in general be different points.

The mass center of a collection of two or more finite bodies (or of a composite body made of two or more parts) can be found when the center of mass of each part is known. The total mass of N bodies is [see Eq. (4.2.8)]

$$m = m_1 + m_2 + \cdots + m_N, \qquad (7.2.4)$$

and the *first moment of the mass with respect to the* yz-*plane is*

$$\int_{V_1} x \, dm + \int_{V_2} x \, dm + \cdots + \int_{V_N} x \, dm = m_1 \bar{x}_1 + m_2 \bar{x}_2 + \cdots + m_N \bar{x}_N. \quad (7.2.5)$$

Thus we obtain

Composite Body

$$m\bar{x} = \sum_{k=1}^{N} m_k \bar{x}_k \qquad (7.2.6)$$

---

with two similar formulas for $\bar{y}$ and $\bar{z}$. Note the similarity to Eq. (7.2.1). The only difference is the missing overbar in Eq. (7.2.1) on x_k, which is the coordinate of the mass-point particle.

EXAMPLE 7.2.1

A uniform sphere of radius R and density ρ_2 is mounted on top of a circular cylinder of length L, radius a, and density ρ_1. Determine the center of mass of the composite body.

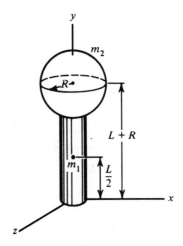

Fig. 7-1 Example 7.2.1

SOLUTION.   Choose axes as shown in Figure 7-1. Then, by symmetry $\bar{x} = \bar{z} = 0$. And by Eq. (7.2.6),

$$(m_1 + m_2)\bar{y} = m_1 \bar{y}_1 + m_2 \bar{y}_2.$$

Hence $\bar{y}$ is given by

$$\left[\rho_1(\pi a^2 L) + \rho_2\!\left(\frac{4}{3}\pi R^3\right)\right]\bar{y} = \rho_1(\pi a^2 L)\frac{L}{2} + \rho_2\!\left(\frac{4}{3}\pi R^3\right)(L + R).$$

In Sec. 7.3 we shall consider additional examples of the location of the centroid of a composite body, which is also its center of mass if the density is uniform. Exercises on center of mass are given with the centroid exercises at the end of Sec. 7.3.

7.3 CENTROIDS OF VOLUME, AREA, ARC LENGTH; COMPOSITE BODIES

We saw in Secs. 4.2 and 7.2 that when the density ρ is uniform throughout a volume V the density factor can be canceled from the integrals of Eqs. (4.2.10) and (7.2.3) for locating the coordinates $(\bar{x}, \bar{y}, \bar{z})$ of the center of mass. The point defined by omitting the density factors is called the **centroid** of the volume. For example, Eq. (7.2.3), written as

$$\bar{x}\int_V \rho \, dV = \int_V x\rho \, dV,$$

leads to the first equation in Eq. (7.3.1) below when the constant ρ is divided out. The centroid of an area or of an arc length is defined in a similar manner. Figure 7-2 illustrates the concept of centroid C for a volume V, for a plane area A, and for a pla..e arc L. Equation (7.3.1) gives the formula for the x-coordinate of the three centroids:

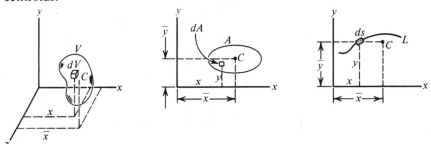

Fig. 7-2 Centroid of Volume V, Area A, Arc Length L

Centroids

$$\bar{x}V = \int_V x \, dV, \qquad \bar{x}A = \int_A x \, dA, \qquad \bar{x}L = \int_L x \, ds. \qquad (7.3.1)$$

These equations should be easy to remember if you note that there is really just one kind of equation to learn (noting that $ds = dL$). The elements of volume dV, area dA, or arc length ds in the definitions are infinitesimal, so that the integration variable x-coordinate of the element can be considered as the coordinate to a point. (The method of composite bodies, however, will permit us to evaluate the area and volume integrals in most cases as single integrals with an element of finite dimensions in all but one direction; see Sec. 7.4.) Formulas for the y-coordinate of the centroid can be written by replacing x by y and $\bar{x}$ by $\bar{y}$. And for a three-dimensional body a similar formula gives $\bar{z}$. (The area or the arc length could be on a curved surface in space or on a space curve.)

The centroid of area has applications in beam theory. It also is the center of mass of a uniform thin plate, just as the centroid of a volume is the center of mass of a body of uniform density. The centroid of an arc is the center of mass for a uniform thin bar or wire. When the density is not uniform, the mass center and the centroid do not coincide. In that case, if both points are to be considered, some different notations should be used, e.g., (x^*, y^*, z^*) for centroid and $(\bar{x}, \bar{y}, \bar{z})$ for mass center. The necessity for the distinctive notation seldom occurs, however, since when ρ is not uniform we usually are not interested in both the center of mass and the centroid of the same body. Notice that, in the case illustrated in Fig. 7-2, the centroid is not a point of the arc. This is typical for a curved arc. The centroid of an area or volume can also occur at a point not in the volume, as some of our examples will show.

Application to Elastic Beam Theory

It is shown in books on mechanics of deformable solids that the normal stress acting on the cross section of a bent *elastic* beam is proportional to the distance from the neutral plane separating the tension side of the beam from the compression side. (Stress is internal distributed force per unit area.) Figure 7-3(a) illustrates a distribution of bending stress proportional to distance y from the neutral plane N. The normal stress σ acts on the cross section marked A-A in Fig. 7-3(a) and marked A in the cross-sectional view of Fig. 7-3(b). The tension side ($\sigma > 0$) is above the neutral plane, while the compressive side is below (where $\sigma < 0$, is indicated by

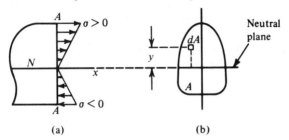

(a)   (b)

Fig. 7-3 Bending Stress in Elastic Beam: (a) Stress Distribution; (b) Cross Section

the arrows pointing to the left in Fig. 7-3). For an elastic beam σ is proportional to y. This means that $\sigma = ky$, where k is some constant. We assume that there is no net axial force on the beam. Hence

$$0 = \sum f_x = \int_A \sigma \, dA = k \int_A y \, dA,$$

whence

$$\bar{y} = 0, \qquad (7.3.2)$$

since

$$\bar{y}A = \int_A y \, dA.$$

This shows that in the elastic bending of a beam under no net axial force the neutral plane separating the tension region of the beam from the compression region must contain the centroid of the area of the cross section. This is the reason for our interest in being able to locate the centroid of the area of a beam cross section.

Composite Bodies

A derivation quite similar to that of Eq. (7.2.6) shows that in each case (volume, area, or arc) the centroid of a total body composed of N parts may be located by formulas such as

---

Composite Body

$$\bar{x}A = \bar{x}_1 A_1 + \bar{x}_2 A_2 + \cdots + \bar{x}_N A_N, \qquad (7.3.3)$$

---

where A is the total area and each $\bar{x}_k$ is the x-coordinate of the centroid of the area A_k of the part. You should have no difficulty writing similar formulas for volume or arc length and for $\bar{y}$ or $\bar{z}$.

EXAMPLE 7.3.1

The centroidal coordinate $\bar{x}$ for the L-section area A of Fig. 7-4 can be located as follows. We divide A into two rectangles A_1 and A_2 as shown.

$$A = A_1 + A_2 = 12 + 8 = 20 \text{ in.}^2$$

$$\bar{x}A = \bar{x}_1 A_1 + \bar{x}_2 A_2$$

$$20\bar{x} = (1)(12) + (4)(8) \qquad \bar{x} = 2.2 \text{ in.}$$

$$\bar{y}A = \bar{y}_1 A_1 + \bar{y}_2 A_2$$

$$20\bar{y} = 3(12) + (1)(8) \qquad \bar{y} = 2.2 \text{ in.}$$

We could have foreseen that $\bar{y} = \bar{x}$, since the centroid must lie on the 45° line of symmetry of the area, but calculating them both provides a check.

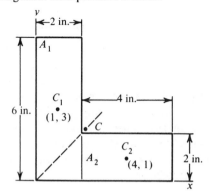

Fig. 7-4   L-Section Composite Area

EXAMPLE 7.3.2

The 8 by 16 in. rectangle shown in Fig. 7-5 has a circular area of radius 3 in. cut out. The centroid of the remaining area A may be located as follows. By symmetry $\bar{y} = 4$ in. Let A_1 denote the rectangle and A_2 the circle. Then

$$A_1 = 128 \text{ in.}^2 \qquad A_2 = 9\pi = 28.27 \text{ in.}^2$$

$$A_1 = A + A_2 \qquad A = 128 - 28.27 = 99.7 \text{ in.}^2$$

$$\bar{x}_1 A_1 = \bar{x}A + \bar{x}_2 A_2 \qquad 99.7\bar{x} = 8(128) - 12(28.27)$$

$$= 1024 - 339.2 = 684.8.$$

$$\bar{x} = 6.87 \text{ in.}$$

[This solution can be regarded as the solution of

$$A = A_1 + A_2$$

$$\bar{x}A = \bar{x}_1 A_1 + \bar{x}_2 A_2$$

interpreted as Eq. (4.3.3) written for "negative area" $A_2 = -28.27 \text{ in.}^2$]

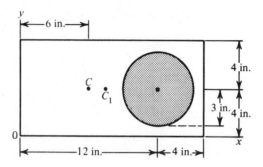

Fig. 7-5   Rectangle with Circular Cutout

Centroids of common arcs, areas, and volumes are given in Tables A3 and A4 of Appendix A. These tabulated values can be used to determine the centroids of composite bodies. The procedure with composite arcs or composite volumes is quite similar to that illustrated for composite areas in Examples 7.3.1 and 7.3.2. *In all cases be careful to correct for the fact that the origin of coordinates you are using for the composite body may be different from the origin used in the table.*

Single-integral methods of determining centroids will be given in Sec. 7.4 and multiple-integral methods in Sec. 7.5.

EXERCISES

1–8. Locate the centroids of the plane areas shown in the figures for Exs. 1–8.

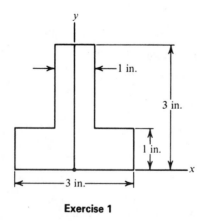

Exercise 1

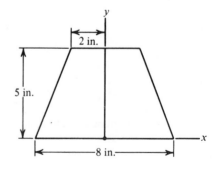

Exercise 2

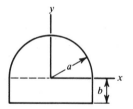

Exercise 3

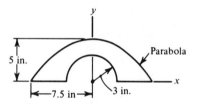

Exercise 8

9–10. Locate the centroids of the composite arcs shown in the figures for Exs. 9 and 10.

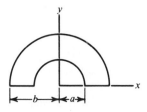

Exercise 4

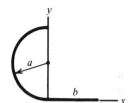

Exercise 9

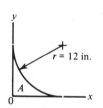

Exercise 5

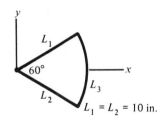

Exercise 10

11–16. Locate the volume centroids as illustrated in the figures for Exs. 11–16.

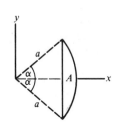

Exercise 6

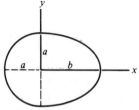

Half ellipse and
half circle

Exercise 7

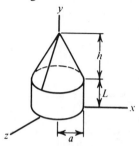

Cone and Cylinder

Exercise 11

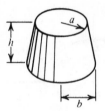

Frustum

Exercise 12

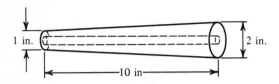

Tapered tube with
uniform 1/2-in.
diameter hole

Exercise 13

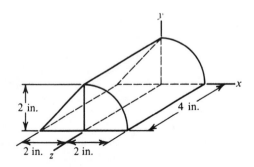

Exercise 14

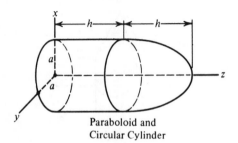

Paraboloid and
Circular Cylinder

Exercise 15

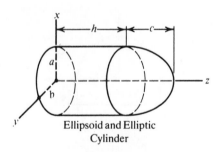

Ellipsoid and Elliptic
Cylinder

Exercise 16

17. If the area of Ex. 3 is formed from two thin plates—a semicircular plate of aluminum weighing w lb/in.^{2} and a rectangular plate of steel weighing $3w$ lb/in.^{2}—locate its center of mass.

18. If the composite arc is formed from two wires—a semicircular aluminum wire weighing w lb/in. and straight steel wire weighing $3w$ lb/in.—locate its center of mass. See the figure for Ex. 9.

19. If the quarter-circular cylinder in Ex. 14 is made of aluminum (170 lb/ft^3) and the triangular prism is of steel (490 lb/ft^3), locate the center of mass of the composite body.

20. Locate the center of mass of the composite body shown, consisting of a 4-in.-diameter steel shaft 16 in. long (weighing 0.283 lb/in.^{3}) and a thin sheet metal plate weighing 5 lb/ft^2.

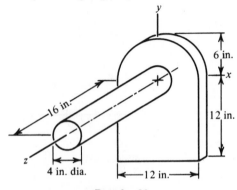

Exercise 20

21. A uniform rod of length 80 cm and mass 4 kg is welded to the end of another uniform rod of length 60 cm and mass 3 kg to form a right-angled L-shaped body. Locate the center of mass of the combination, taking the two legs as coordinate axes.

---

*7.4 CENTROID BY SINGLE INTEGRALS

We shall illustrate the method by examples. In rectangular coordinates, the element of area must be chosen so that it is a rectangle parallel to the y-axis, of infinitesimal width dx, whose length $L(x)$ is known as a function of x, or else a rectangle parallel to the x-axis, of width dy, whose length $L(y)$ is known as a function of y.

In each case we **denote the centroid of the element of area dA by** $(\bar{x}_{el}, \bar{y}_{el})$. Then, for the whole area the method of composite bodies [see Eq. (7.3.3)] gives

$$\bar{x}A = \sum (\bar{x}_{el}\, dA), \qquad \bar{y}A = \sum (\bar{y}_{el}\, dA). \tag{7.4.1}$$

EXAMPLE 7.4.1

We shall determine the centroid of a triangle ABC, shown in Fig. 7-6, as follows. We choose x,y-coordinate axes with the x-axis parallel to the base AB (of length b). The element of area is

$$dA = L(y)\, dy,$$

where, by similar triangles,

$$\frac{L(y)}{b} = \frac{y}{h} \quad \text{or} \quad L(y) = \frac{b}{h}y.$$

$$A = \sum dA = \int_0^h \frac{b}{h} y\, dy = \frac{1}{2}bh,$$

as is well known. Also

$$\bar{y}A = \sum (\bar{y}_{el})\, dA = \int_0^h \frac{b}{h} y^2\, dy = \frac{1}{3}bh^2,$$

whence

$$\bar{y} = \frac{2}{3}h. \tag{7.4.2}$$

* This section may be omitted.

For the element parallel to the x-axis, of infinitesimal width dy, we took the $\bar{y}$ of the element, $\bar{y}_{el}$, to be the integration variable y. Equation (7.4.2) shows that the centroid of any triangle is located at distance from the base equal to one-third the altitude, as shown in Fig. 7-7. Since any side can be chosen as base, the centroid is completely determined. It is located at the intersection of the three medians of the triangle, since the medians intersect at a point two-thirds of the distance from any vertex to the midpoint of the opposite side; see Fig. 7-7.

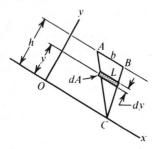

Fig. 7-6 Element of Area $dA = L\, dy$

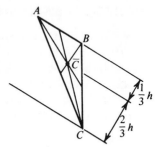

Fig. 7-7 Centroid $\bar{C}$ at Intersection of Medians of Triangle

EXAMPLE 7.4.2

We shall determine the centroid of the area in the first quadrant bounded by the x-axis, the line $x = a$, and the parabola shown in Fig. 7-8(a). We choose an element of area of width dy parallel to the x-axis as shown. An alternative possible choice of dA is shown in Fig. 7-8(b).

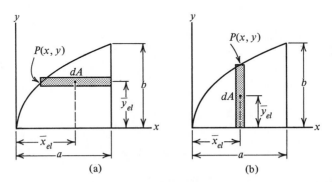

(a)   (b)

Fig. 7-8 Parabolic Quadrant

In general,

$$\bar{x}A = \int \bar{x}_{el}\, dA \qquad \bar{y}A = \int \bar{y}_{el}\, dA.$$

The parabola has an equation of the form $x = ky^2$. We determine k by requiring the parabola to pass through (a, b). Then $kb^2 = a$, and

$$x = \frac{a}{b^2} y^2 \quad \text{or} \quad y = b\sqrt{\frac{x}{a}}$$

is satisfied by any $P(x, y)$ on the curve. For the horizontal dA of Fig. 7-8(a).

$$dA = (a - x)\, dy = \left(a - \frac{a}{b^2} y^2\right) dy$$

$$\bar{x}_{el} = \frac{1}{2}(a + x) = \frac{1}{2}\left(a + \frac{a}{b^2} y^2\right) \qquad \bar{y}_{el} = y$$

$$A = \int dA = \int_0^b \left(a - \frac{a}{b^2} y^2\right) dy = \frac{2}{3} ab$$

$$\bar{x}A = \int \bar{x}_{el}\, dA = \int_0^b \frac{1}{2}\left(a^2 - \frac{a^2}{b^4} y^4\right) dy = \frac{2}{5} a^2 b$$

$$\bar{y}A = \int \bar{y}_{el}\, dA = \int_0^b \left(ay - \frac{a}{b^2} y^3\right) dy = \frac{1}{4} ab^2.$$

Hence

$$\frac{2}{3} ab\bar{x} = \frac{2}{5} a^2 b, \qquad \frac{2}{3} ab\bar{y} = \frac{1}{4} ab^2$$

$$\bar{x} = \frac{3}{5} a, \qquad\qquad \bar{y} = \frac{3}{8} b. \tag{7.4.3}$$

Polar Coordinates

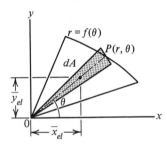

Fig. 7-9   Polar Coordinates

An area bounded by two radii from the origin and an arc given as $r = f(\theta)$, as shown schematically in Fig. 7-9, is conveniently handled by polar coordinates. The triangular polar-coordinate element of area of infinitesimal vertex angle $d\theta$ has its centroid at $2r/3$ as shown. Hence, by the results of Example 7.4.1,

$$dA = \tfrac{1}{2} r^2\, d\theta, \qquad \bar{x}_{el} = \tfrac{2}{3} r \cos\theta, \qquad \bar{y}_{el} = \tfrac{2}{3} r \sin\theta. \tag{7.4.4}$$

EXAMPLE 7.4.3

We shall use the polar-coordinate formulas of Eqs. (7.4.4) to find the centroid of area of the circular sector of radius a and central angle 2α in Fig. 7-10. By symmetry, $\bar{y} = 0$. Also $r = \text{const.} = a$.

$$\bar{x}A = \int \bar{x}_{el}\, dA = \int_{-\alpha}^{\alpha} (\tfrac{2}{3}a \cos \theta)(\tfrac{1}{2}a^2\, d\theta) = \tfrac{2}{3}a^3 \sin \alpha.$$

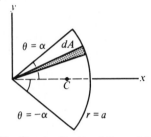

Fig. 7-10   Circular Sector of Central Angle 2α

Also, $A = a^2\alpha$, whence

$$a^2\alpha\bar{x} = \frac{2}{3}a^3 \sin \alpha$$

$$\bar{x} = \frac{2a}{3\alpha}\sin \alpha, \tag{7.4.5}$$

as given in Appendix A, Table A3.

   Arc length integrals are easy to formulate but frequently difficult to evaluate. For a plane curve, we have

$$(ds)^2 = (dx)^2 + (dy)^2 \quad \text{or} \quad (ds)^2 = (dr)^2 + (r\, d\theta)^2. \tag{7.4.6}$$

Hence, we take

$$ds = \sqrt{1 + \left(\frac{dy}{dx}\right)^2}\, dx \qquad \text{when the curve is } y = f(x).$$

$$ds = \sqrt{\left(\frac{dx}{dy}\right)^2 + 1}\, dy \qquad \text{when the curve is } x = f(y) \tag{7.4.7}$$

$$ds = \sqrt{r^2 + \left(\frac{dr}{d\theta}\right)^2}\, d\theta \qquad \text{when the curve is } r = f(\theta).$$

Since these are all infinitesimal elements, we take $\bar{x}_{el} = x$ and $\bar{y}_{el} = y$ for the element at $P(x, y)$ on the curve. In three dimensions

$$(ds)^2 = (dx)^2 + (dy)^2 + (dz)^2. \tag{7.4.8}$$

If the curve is specified by giving the three coordinates in terms of a single variable parameter, say u,

$$x = x(u), \qquad y = y(u), \qquad z = z(u),$$

then

$$dx = x'(u)\, du, \qquad \text{etc.,}$$

and

$$ds = [x'(u)^2 + y'(u)^2 + z'(u)^2]^{1/2}\, du. \tag{7.4.9}$$

Volume Centroids by Single Integrals

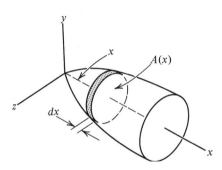

Fig. 7-11   Volume of Revolution

For volumes where a slab element of volume perpendicular to a coordinate axis can be suitably defined, the centroid may be located by a single integral. The most common case is a volume of revolution like that shown in Fig. 7-11, where the volume element dV is a circular disk of height dx and area $A(x) = \pi r^2$.

By symmetry $\bar{y} = \bar{z} = 0$, while

$$\bar{x}_{el} = x \quad \text{and} \quad dV = \pi r^2\, dx, \tag{7.4.10}$$

whence

$$\bar{x} V = \int x_{el}\, dV$$

can be evaluated if r is known as a function of x. A pyramid can also be treated by single integrals if its cross-sectional area can be expressed in terms of the distance from the apex and the centroid of the cross-sectional area is known.

EXAMPLE 7.4.4

We can locate the centroid of the volume of the half right circular cone shown in Fig. 7-12 as follows. Since the yz-plane is a plane of symmetry, the centroid lies in this plane ($\bar{x} = 0$). The slab element of volume dV of thickness dz is a half-disk.

$$dV = \frac{1}{2} \pi r^2\, dz$$

$$\bar{z}_{el} = z, \qquad \text{while} \qquad \bar{y}_{el} = \frac{4r}{3\pi}.$$

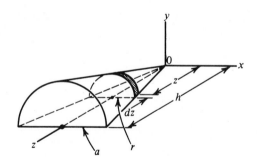

Fig. 7-12 Half-cone with Disk Element

See Appendix A, Table A3 (semicircular area). Also

$$\frac{r}{z} = \frac{a}{h} \quad \text{or} \quad r = \frac{a}{h}z.$$

$$V = \int dV = \int_0^h \frac{1}{2}\pi r^2 \, dz = \int_0^h \frac{1}{2}\pi \frac{a^2}{h^2}z^2 \, dz = \frac{1}{6}\pi a^2 h$$

$$\bar{y}V = \int \bar{y}_{el} \, dV = \int_0^h \frac{4r}{3\pi}\left(\frac{1}{2}\pi r^2\right) dz = \frac{2}{3}\int_0^h \frac{a^3}{h^3}z^3 \, dz = \frac{1}{6}a^3 h$$

$$\bar{z}V = \int \bar{z}_{el} \, dV = \int_0^h z\left(\frac{1}{2}\pi r^2\right) dz = \frac{1}{2}\pi \int_0^h \frac{a^2}{h^2}z^3 \, dz = \frac{1}{8}\pi a^2 h^2.$$

Hence

$$\frac{1}{6}\pi a^2 h\bar{y} = \frac{1}{6}a^3 h, \qquad \frac{1}{6}\pi a^2 h\bar{z} = \frac{1}{8}\pi a^2 h^2$$

$$\bar{x} = 0, \qquad \bar{y} = \frac{a}{\pi}, \qquad \bar{z} = \frac{3}{4}h.$$

Cylindrical Shell Element of Volume

A volume of revolution can also be evaluated by a single integral using a cylindrical shell element of volume. We shall illustrate the procedure as an alternative solution for the half-cone of Example 7.4.4. Figure 7-13 shows the half cylindrical shell element [note that $z_1(r) = hr/a$]

$$dV = \pi r[h - z_1(r)] \, dr = \pi hr\left(1 - \frac{r}{a}\right) dr$$

with

$$\bar{x}_{el} = 0, \qquad \bar{y}_{el} = \frac{2r}{\pi}, \qquad \bar{z}_{el} = \frac{1}{2}(z_1 + h) = \frac{h}{2}\left(1 + \frac{r}{a}\right).$$

(The $\bar{y}_{el}$ is obtained for a semicircular arc from Table A3 of Appendix A.) Then

$$V = \frac{\pi h}{a}\int_0^a (ar - r^2)\, dr = \frac{1}{6}\pi a^2 h$$

$$\bar{y}V = \int \bar{y}_{el}\, dV = \frac{2h}{a}\int_0^a (ar^2 - r^3)\, dr = \frac{1}{6}a^3 h$$

$$\bar{z}V = \int \bar{z}_{el}\, dV = \frac{\pi h^2}{2}\int_0^a \left(r - \frac{r^3}{a^2}\right)\, dr = \frac{1}{8}\pi a^2 h^2$$

in agreement with our previous solution for Example 7.4.4 using a slab element. In Sec. 7.5 we shall illustrate the use of multiple integrals for centroids.

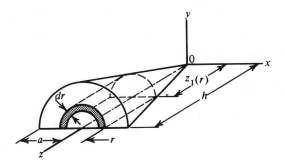

Fig. 7-13   Half-cone with Shell Element

EXERCISES

1. Derive by single integration the centroid formulas for the following figures given in Appendix A, Table A3:
   (a) The quarter-circle area.
   (b) The quarter-ellipse area.
   (c) The semicircle area.
   (d) The circular arc of central angle 2α.

2. Derive by single integration the volume centroid formulas given in Appendix A, Table A4, for the following figures:
   (a) Hemisphere.
   (b) Cone.
   (c) Pyramid.
   (d) Paraboloid of revolution ($a = b$).

3. Determine the centroid of the area bounded by the two curves $y = x^2$ and $x = y^2$.

4. Determine the centroid of the area below one half-wave of the sine curve $y = A \sin qx$.

5. Determine by integration the centroid of the area below the line $y = 2x$ and above the curve $y = x^2$. Length units are inches.

6. Determine the centroid of the parabolic arc $y = 0.04x^2$ in the first quadrant between $y = 0$ and $y = 10$ in.

7. Use polar coordinates to locate the centroid of the area bounded by the cardioid $r = a(1 + \cos \theta)$.

8. Find the center of mass of a solid hemisphere of radius R if its density is proportional to the distance from its circular base.

9. (a) Determine by integration the centroid of the area above the parabola of Fig. 7-8 and below a line $y = b$ (not shown). (b) Check your results by showing that they give the correct location for the centroid of the rectangular composite area of side lengths a and b.

---

*7.5 MULTIPLE-INTEGRAL EVALUATION

We shall consider a few examples of multiple-integral evaluation. For additional information, consult a calculus textbook. Our first example illustrates the procedure for a plane area whose boundary can be separated into an upper boundary $y = y_2(x)$ and a lower boundary $y = y_1(x)$ and possibly one or two vertical segments on the sides and such that a vertical line through the interior cuts the lower boundary only once and the upper boundary only once. We then integrate first in the y-direction. The procedure can be modified in an obvious way to integrate first in the x-direction if it is simpler to describe the right and left boundaries as $x = x_2(y)$ and $x = x_1(y)$. More complicated boundary shapes can be treated by subdividing the area into parts each of which is one of the types described above.

EXAMPLE 7.5.1

We shall find the centroid of the area OAB bounded above by the parabola $y = 4 - \frac{1}{4}x^2$, below by the line $y = \frac{3}{2}x$, and on the left by the y-axis, as shown in Fig. 7-14. The upper and lower boundaries intersect at $A(2, 3)$. The dashed arrow represents the direction of the first integration. Then $dA = dx\, dy$.

$$A = \int dA = \int_0^2 dx \int_{(3/2)x}^{4-(1/4)x^2} dy$$

$$= \int_0^2 [4 - \tfrac{1}{4}x^2 - \tfrac{3}{2}x]\, dx = 4\tfrac{1}{3} \text{ in.}^2$$

$$\bar{x}A = \int x\, dA = \int_0^2 x\, dx \int_{(3/2)x}^{4-(1/4)x^2} dy = 3 \text{ in.}^3$$

* This section may be omitted.

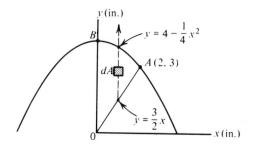

Fig. 7-14 Double-Integral Formulation for Example 7.5.1

Hence

$$(4\tfrac{1}{3})\bar{x} = 3 \qquad \bar{x} = \tfrac{9}{13} = 0.692 \text{ in.}$$

$$\bar{y}A = \int y \, dA = \int_0^2 dx \int_{(3/2)x}^{4-(1/4)x^2} y \, dy$$

$$= \tfrac{1}{2} \int_0^2 \left[(4 - \tfrac{1}{4}x^2)^2 - (\tfrac{3}{2}x)^2 \right] dx = 10.53 \text{ in.}^3$$

Hence

$$(4\tfrac{1}{3})\bar{y} = 10.53 \text{ in.}^3, \qquad \bar{y} = 2.43 \text{ in.}$$

This example could of course have been evaluated by a single integral with respect to *x*, with

$$dA = \left[(4 - \tfrac{1}{4}x^2) - \tfrac{3}{2}x \right] dx.$$

Polar Coordinates

The polar coordinate element of area is

$$dA = r \, d\theta \, dr. \tag{7.5.1}$$

The following example will illustrate its use in a simple problem that could also easily be solved by the method of composite areas. Notice that we seek $\bar{x}$ and $\bar{y}$ even though we use polar coordinates.

EXAMPLE 7.5.2

We seek the centroid of the area of the segment of a circle shown in Fig. 7-15. The area is bounded on the left by the line $x = $ const. on which

$$r \cos \theta = a \cos \alpha$$

and on the right by the circle $r = a$.

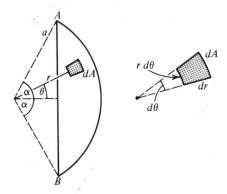

Fig. 7-15   Circular Segment Bounded by Chord *AB* and Arc *AB*

Hence the limits on r are

$$r_1 = \frac{a \cos \alpha}{\cos \theta} \quad \text{and} \quad r_2 = a.$$

$$A = \int_{-\alpha}^{\alpha} d\theta \int_{r_1}^{r_2} r \, dr$$

$$= \frac{1}{2} \int_{-\alpha}^{\alpha} [a^2 - a^2 \cos^2 \alpha \sec^2 \theta] \, d\theta = a^2(\alpha - \frac{1}{2} \sin 2\alpha)$$

$$\bar{x} A = \int x \, dA = \int_{-\alpha}^{\alpha} d\theta \int_{r_1}^{r_2} r^2 \cos \theta \, dr$$

$$= \frac{1}{3} \int_{-\alpha}^{\alpha} [a^3 \cos \theta - a^3 \cos^3 \alpha \sec^2 \theta] \, d\theta$$

$$= \frac{2}{3} a^3 \sin^3 \alpha.$$

Hence

$$\bar{x} = \frac{4a}{3}\frac{\sin^3 \alpha}{2\alpha - \sin 2\alpha}, \qquad \bar{y} = 0.$$

Volume by Triple or Double Integral

We shall consider one simple example first as a triple integral and then the same problem as a double integral. Notice that it is sometimes useful to mix cylindrical and rectangular coordinates in the same problem. The cylindrical coordinate element of volume is

$$dV = R \, d\theta \, dR \, dz. \qquad (7.5.2)$$

EXAMPLE 7.5.3

We seek the centroid of the volume bounded by the cylinder $R = a$, above the xy-plane and below the plane $z = (h/a)y$ shown in Fig. 7-16.

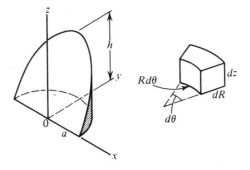

Fig. 7-16   Volume for Example 7.5.3

The upper limit on z is

$$z_2 = \frac{h}{a}y = \frac{h}{a}R \sin \theta$$

$$V = \int_0^{\pi} d\theta \int_0^a R \, dR \int_0^{z_2} dz$$

$$= \frac{h}{a}\int_0^{\pi} \sin \theta \, d\theta \int_0^a R^2 \, dR = \frac{2}{3}ha^2.$$

By symmetry, $\bar{x} = 0$.

$$\bar{y}V = \int y\, dV = \int_0^\pi \sin\theta\, d\theta \int_0^a R^2\, dr \int_0^{z_2} dz$$

$$= \frac{h}{a} \int_0^\pi \sin^2\theta\, d\theta \int_0^a R^3\, dR = \frac{1}{8}\pi h a^3.$$

Hence

$$\bar{y} = \frac{3}{16}\pi a.$$

$$\bar{z}V = \int z\, dV = \int_0^\pi d\theta \int_0^a R\, dR \int_0^{z_2} z\, dz$$

$$= \frac{h^2}{2a^2} \int_0^\pi \sin^2\theta\, d\theta \int_0^a R^3\, dR = \frac{1}{16}\pi h^2 a^2.$$

Hence

$$\bar{z} = \frac{3}{32}\pi h.$$

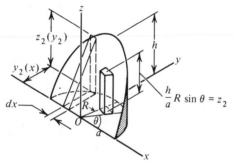

Fig. 7-17 Single-Integral and Double-Integral Volume Elements for Example 7.5.3

We shall now show how the same example can be solved by a double integral in cylindrical coordinates or by a single integral in rectangular coordinates. Figure 7-17 shows both the double-integral element of volume

$$dV = z_2 R\, d\theta\, dR$$

in cylindrical coordinates and the single-integral element

$$dV = A(x)\, dx,$$

where $A(x)$ is the triangular area of the element whose thickness is dx.

$$A(x) = \frac{1}{2}y_2(x)z_2[y_2(x)]$$

$$= \frac{1}{2}(a^2 - x^2)^{1/2}\left[\frac{h}{a}(a^2 - x^2)^{1/2}\right]$$

$$= \frac{h}{2a}(a^2 - x^2).$$

$$V = 2\int_0^a A(x)\, dx,$$

since $A(x)$ is an even function. Thus

$$V = \frac{h}{a} \int_0^a (a^2 - x^2)\, dx = \frac{2}{3} h a^2.$$

For the single-integral element

$$\bar{x}_{el} = x \qquad \bar{y}_{el} = \frac{2}{3} y_2(x) = \frac{2}{3}(a^2 - x^2)^{1/2}$$

$$\bar{z}_{el} = \frac{1}{3} z_2[y_2(x)] = \frac{1}{3}\frac{h}{a}(a^2 - x^2)^{1/2},$$

where $\bar{y}_{el}$ and $\bar{z}_{el}$ are known, since the centroid of a triangle is known to be at a distance from the base equal to one-third the altitude.

For the double-integral element, recall that, by Eq. (7.5.3),

$$dV = z_2 R\, d\theta\, dR.$$

Hence, for the example of Fig. 7-17,

$$dV = \frac{h}{a} R^2 \sin\theta\, dR\, d\theta$$

$$\bar{x}_{el} = x = R \cos\theta \qquad \bar{y}_{el} = y = R \sin\theta \qquad \bar{z}_{el} = \frac{1}{2} z_2 = \frac{h}{2a} R \sin\theta.$$

We need only last two, since for the whole volume, $\bar{x} = 0$, by symmetry. For the double-integral evaluations,

$$V = \frac{h}{a} \int_0^\pi \sin\theta\, d\theta \int_0^a R^2\, dR$$

$$\bar{y}V = \int \bar{y}_{el}\, dV = \frac{h}{a} \int_0^\pi \sin^2\theta \int_0^a R^3\, dR$$

$$\bar{z}V = \int \bar{z}_{el}\, dV = \frac{h^2}{2a^2} \int_0^\pi \sin^2\theta \int_0^a R^3\, dR.$$

Comparing this with the second line of each of the triple-integral evaluations, we see that after the first integration with respect to z has been made, the triple integral has reduced to the double-integral form. The remainder of the double-integral solution is therefore identical to that previously given for the triple integrals.

As an independent check on the results, you can evaluate the single integrals in rectangular coordinates. They lead to the same final values of $\bar{x}$ and $\bar{z}$.

In Sec. 7.6 we shall consider second moments of plane areas.

EXERCISES

1. Use double integrals to derive the following centroid formulas given in Appendix A, Table A3:
   (a) Quarter circle.
   (b) Quarter ellipse.
   (c) Semicircle.
2. Use a double integral to determine the centroid of the area bounded by the cardioid $r = a(1 + \cos \theta)$.
3. Use double integrals to derive the following centroid formulas given in Appendix A, Table A4.
   (a) Hemisphere.
   (b) Cone.

   (c) Half ellipsoid.
   (d) Elliptic paraboloid.
4. Use triple integrals to derive the following centroid formulas given in Appendix A, Table A4.
   (a) Elliptic paraboloid.
   (b) Half Ellipsoid.
5. Find by triple integration the centroid of the volume bounded by the sphere $x^2 + y^2 + z^2 = a^2$ and inside the paraboloid of revolution $(a/\sqrt{2})z = x^2 + y^2$.
6. Locate the centroid of the volume of the part of the sphere given in cylindrical coordinates by $R^2 + z^2 = a^2$ for $-(\pi/4) < \theta < \pi/4$. (*Note*: $x = R \cos \theta$, $y = R \sin \theta$.)

---

*7.6 SECOND MOMENTS OF AREA

The second moment of a plane area about an axis, often called the ***moment of inertia of the area*** despite the fact that area as such has no inertia, is important in mechanics because the second moment of a beam cross section about an axis through its centroid is a measure of the elastic stiffness of the beam against bending about the axis. For example, an area in the xy-plane has the second moment I_x with respect to the x-axis, defined by

$$I_x = \int_A y^2 \, dA. \qquad (7.6.1)$$

For a given cross-sectional area A, ***the second moment measures the extent to which the area is distributed away from the x-axis.*** A larger I_x is obtained with the

---

* This section may be omitted or postponed.

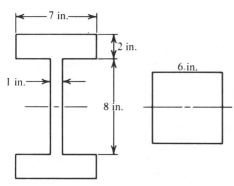

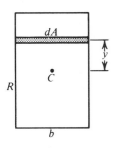

Fig. 7-18 I-Beam Section and Square with Same Area

Fig. 7-19 Rectangle with $dA = b\,dy$

I-beam cross section of Fig. 7-18, for example, than with the square section shown in the same figure, although both have the same area. An I-beam of the same material and cross-sectional area as a square beam has a correspondingly greater elastic bending stiffness.

Second moments of area for common geometrical figures can be evaluated by double integrals, but they are usually easy to evaluate by single integration. If the strip element of area is

$$dA = L(y)\,dy \qquad (7.6.2)$$

perpendicular to the y-axis, then since dy is infinitesimal, the second moment dI_x of the element dA ***about the axis parallel to the element*** is

$$dI_x = y^2\,dA = y^2 L(y)\,dy. \qquad (7.6.3)$$

For the rectangle of Fig. 7-19 the length $L(y) = b$ is constant. Hence $dI_x = by^2\,dy$, and integrations from 0 to h and from $-h/2$ to $h/2$, respectively, give

Second Moments of Rectangular Area		
about the base:	$I_{\text{base}} = \frac{1}{3}bh^3$	(7.6.4)
about the centroidal axis parallel to the base:	$\bar{I} = \frac{1}{12}bh^3$	(7.6.5)

The last formula is the one needed for the stiffness of an elastic beam with rectangular cross section. The second moment with respect to the base may be useful for computing the mass moment of inertia of a thin plate, as will be illustrated in Sec. 7.9. It is also useful for single-integral evaluation of a second moment of area using a strip element of area perpendicular to the axis about which the second moment is computed when that axis is a boundary of the area; see Sample Problem 7.6.1.

Composite Areas

Since the area integrals are additive, the total second moment of a composite area about an axis is the sum of the second moments of its parts *about the same axis*. The second moments for a few common areas are tabulated in Appendix A, Table A3. Formulas for standard structural shapes (I-beams, channels, and angle sections) may be found in structural engineering handbooks. Since the centroidal axis for the composite section is usually different from the axis listed in Appendix A, Table A3, it is necessary to use parallel axis theorems. [See Eq. (7.6.10).]

Before stating the parallel-axis theorems, we shall summarize the definitions of four kinds of second moments of area: I_x, I_y, the *area product of inertia* P_{xy}, and the *polar second moment* J_O (often called the polar moment of inertia) of the area.

Second Moments of Area in xy-Plane

$$I_x = \int_A y^2 \, dA, \qquad I_y = \int_A x^2 \, dA$$

$$P_{xy} = \int_A xy \, dA \qquad\qquad (7.6.6)$$

$$J_O = \int_A r^2 \, dA = I_x + I_y.$$

The second moment I_x was defined in Eq. (7.6.1). I_y is the same kind of second moment defined with respect to the y-axis (with moment arm x measured from the y-axis and squared to give the *second moment*). I_x or I_y measures the "awayness" of the area from the axis in question.

The *area product* P_{xy} measures the asymmetry of the area. It has applications to nonsymmetrical beams, which may twist as well as bend under a vertical load. Note that $P_{xy}=0$ whenever either the x-axis or y-axis is a line of symmetry of the area. The area product P_{xy} is also needed for computing $I_{x'}$ and $I_{y'}$ with respect to x',y'-axes rotated with respect to the x,y-axes; see Sec. 7.7.

The *polar second moment* $J_O=I_x+I_y$ measures the awayness of the area from the z-axis. It is proportional to the twisting stiffness of a circular elastic torsion bar of a given material; see Exs. 18 and 19 of Sec. 3.4.

All the second moments defined in Eq. (7.6.6) depend on the total area involved as well as on the shape of the area. Quantities explicitly characterizing the shape can be obtained by dividing each second moment by the area and taking the square root. For reasons related to the mass moments of inertia to be defined in Sec. 7.9, the number obtained is called a radius of gyration, denoted by k.

The **radius of gyration of an area** is defined as the distance k such that

$$I = k^2 A \quad \text{or} \quad k = \sqrt{\frac{I}{A}}. \tag{7.6.7}$$

In particular, with respect to the x- and y-axes

$$I_x = k_x^2 A, \qquad k_x = \sqrt{\frac{I_x}{A}}$$

$$\tag{7.6.8}$$

$$I_y = k_y^2 A, \qquad k_y = \sqrt{\frac{I_y}{A}},$$

and for the polar second moment

$$J_O = k_O^2 A, \qquad k_O = \sqrt{\frac{J_O}{A}}$$

$$k_O^2 = k_x^2 + k_y^2. \tag{7.6.9}$$

The radius of gyration might be interpreted as the distance from the axis at which all the area would have to be concentrated in order to have the same second moment about the axis. In statistics the **standard deviation** is the radius of gyration k_y of the area under a frequency distribution curve with the y-axis at the mean, a normalized measure of the awayness of the distribution from the mean.

Parallel-Axis Theorem

If AA is an arbitrary axis (which may or may not intersect the area A) and $A'A'$ is a **parallel axis through the centroid** of A (see Fig. 7-20), then

---

Parallel-Axis Theorem

$$I = \bar{I} + d^2 A, \tag{7.6.10}$$

---

where I denotes I_{AA}, $\bar{I}$ denotes $I_{A'A'}$, A is the total area, and d is the perpendicular distance from AA to $A'A'$. Note that $\bar{I}$ is the smallest possible second moment about any axis parallel to $A'A'$, since the term $d^2 A$ is nonnegative.

The proof of the theorem is easily made by choosing a coordinate axis to coincide with AA. In Fig. 7-21, for example, if we choose the x-axis to be AA and the x'-axis to be $A'A'$, then $d^2 = \bar{y}^2$, and we have to prove that $I_x = \bar{I}_x + \bar{y}^2 A$. From the figure it is clear that the ordinates to dA satisfy $y = y' + \bar{y}$. Hence

$$I_x = \int_A y^2 \, dA = \int (y')^2 \, dA + 2\bar{y} \int y' \, dA + \bar{y}^2 \int dA. \qquad (7.6.11)$$

Since $\int_A y' \, dA = 0$, because the x'-axis passes through the centroid, Eq. (7.6.11) shows that

$$I_x = \bar{I}_x + \bar{y}^2 A, \qquad (7.6.12)$$

as was to be proved.

You should have no difficulty in also using Fig. 7-21 to prove the parallel-axis theorems stated below for P_{xy} and for J_O with $d^2 = \bar{r}^2 = \bar{x}^2 + \bar{y}^2$ in the case of J_o.

Parallel-Axis Theorems for P_{xy} and J_O

$$P_{xy} = \bar{P}_{xy} + \bar{x}\bar{y} \, A \qquad (7.6.13)$$

$$J_O = \bar{J}_C + \bar{r}^2 A. \qquad (7.6.14)$$

By dividing Eqs. (7.6.10), (7.6.12), and (7.6.14) through by A, we obtain ***parallel-axis theorems for the radius of gyration:***

$$k^2 = \bar{k}^2 + d^2$$

$$k_x^2 = \bar{k}_x^2 + \bar{y}^2 \qquad k_O^2 = \bar{k}_C^2 + \bar{r}^2. \qquad (7.6.15)$$

The last two equations are just special cases of the first equation in Eq. (7.6.15).

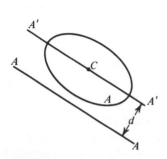

Fig. 7-20 Parallel Axis for I

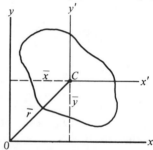

Fig. 7-21 Parallel Axes through Centroid C

Rotation of the coordinate axes leaves J_O unchanged but changes I_x, I_y, and P_{xy}, as will be considered in Sec. 7.7. Those special axes for which $P_{xy} = 0$ are called **principal axes**. It will be shown in Sec. 7.7 that the I_x and I_y for two perpendicular principal x,y-axes are the maximum and minimum values of the area second moment for all axes through the same origin.

SAMPLE PROBLEM 7.6.1

Use the same vertical strip single-integral element of area dA to evaluate I_x, I_y, and P_{xy} for the parabolic quadrant of Fig. 7-22.

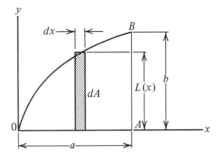

Fig. 7-22 Parabolic Quadrant with Vertical dA

SOLUTION. The equation for the parabola with vertex at O, opening to the right, is $y^2 = cx$. Since it passes through $B(a, b)$, $c = b^2/a$. Hence the equation is $y^2 = (b^2/a)x$ or $y = (b/a^{1/2})x^{1/2}$.

$$dA = L(x) \, dx = \left(\frac{b}{a^{1/2}}\right)x^{1/2} \, dx$$

$$dI_y = x^2 \, dA.$$

$$I_y = \frac{b}{a^{1/2}} \int_0^a x^{5/2} \, dx = \frac{2}{7}ba^3 \quad \textbf{\textit{Answer.}}$$

For dI_x we use the known result for a rectangular area, Eq. (7.6.4):

$$I_{\text{base}} = \frac{1}{3}(\text{base})(\text{altitude})^3$$

$$dI_x = \frac{1}{3}(dx)[L(x)]^3$$

$$I_x = \frac{1}{3}\frac{b^3}{a^{3/2}} \int_0^a x^{3/2} \, dx = \frac{2}{15}ab^3.$$

For dP_{xy} we note that the rectangular element has zero area product of inertia about its centroid. Hence the parallel-axis theorem, Eq. (7.6.13), gives

$$dP_{xy} = 0 + \bar{x}_{el}\bar{y}_{el}\, dA = x\frac{1}{2}\left(\frac{b}{a^{1/2}}\right)x^{1/2}\left(\frac{b}{a^{1/2}}\right)x^{1/2}\, dx.$$

$$P_{xy} = \frac{1}{2}\frac{b^2}{a}\int_0^a x^2\, dx = \frac{1}{6}a^2b^2 \quad \textbf{Answer.}$$

Remark. A variation of this last procedure could be used to compute I_x for an area not bounded by the x-axis by using the parallel-axis theorem to give

$$dI_x = \tfrac{1}{12}L(x)^3\, dx + \bar{y}^2{}_{el}\, dA,$$

but it is probably easier to use a double integral.

SAMPLE PROBLEM 7.6.2

Determine the second moment, about the x-axis, of the area of the fillet shown in Fig. 7-23.

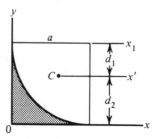

Fig. 7-23   Fillet of Radius a

SOLUTION.   The second moment can be found by subtracting the quarter-circle from the square.

For the quarter-circle, according to Appendix A, Table A3,

$$I_{x_1} = \frac{1}{16}\pi a^4 \qquad d_1 = \frac{4a}{3\pi}$$

$$I_{x_1} = \bar{I}_{x'} + d_1^2 A$$

$$\bar{I}_{x'} = \frac{1}{16}\pi a^4 - \left(\frac{4a}{3\pi}\right)^2\left(\frac{1}{4}\pi a^2\right)$$

$$I_x = \bar{I}_{x'} + d_2^2 A$$

$$I_x = \frac{1}{16}\pi a^4 - \left(\frac{4a}{3\pi}\right)^2 \left(\frac{1}{4}\pi a^2\right) + \left(a - \frac{4a}{3\pi}\right)^2 \left(\frac{1}{4}\pi a^2\right)$$

$$= \left(\frac{5}{16}\pi - \frac{2}{3}\right)a^4.$$

For the square,

$$I_x = \tfrac{1}{3}bh^3 = \tfrac{1}{3}a^4.$$

For the fillet,

$$I_x = \tfrac{1}{3}a^4 - (\tfrac{5}{16}\pi - \tfrac{2}{3})a^4$$
$$= a^4(1 - \tfrac{5}{16}\pi) = 0.01825a^4 \quad \textbf{Answer.}$$

EXERCISES

(Figures for Exs. 4–14 are on the following page.)

1. Verify by integration, as indicated below, the following formulas listed in Appendix A, Table A3:
   (a) I_x for the triangle by single integral.
   (b) I_y for the elliptic quadrant by single integral.
   (c) I_x and P_{xy} for the elliptic quadrant by double integrals.
   (d) I_x and P_{xy} for the quarter-circle by double integrals in polar coordinates.
   (e) I_y for the circular sector by a polar-coordinate double integral.

2. Use parallel-axis theorems to determine the following second moments of the areas shown in Appendix A, Table A3:
   (a) $\bar{I}_x$ for the triangle, given I_x.
   (b) $\bar{I}_x$, $\bar{I}_y$, and $\bar{P}_{xy}$ for the elliptic quadrant.
   (c) $\bar{I}_x$, $\bar{I}_y$, and $\bar{P}_{xy}$ for the parabolic quadrant.
   (d) I_{x_1}, I_{y_1}, and $P_{x_1 y_1}$ for the circle about x_1, y_1-axes through point $x = a$, $y = b$.
   (e) I_{x_1} for the triangle about an axis through its upper vertex.

3. In Appendix A, Table A3, find I_{x_1} and $P_{x_1 y_1}$ for x_1, y_1-axes both tangent to the circle, so that the circle is in the positive quadrant.

4. (a) Determine $\bar{I}_x$ and $\bar{I}_y$ for the I-beam shown in Fig. 7-24(a). (b) Evaluate for $a = 1$ in., $t = 0.5$ in., $h = 10$ in., $b = 4$ in.

5. Determine $\bar{I}_x$, $\bar{I}_y$, and $\bar{P}_{xy}$ for the angle section of Fig. 7-24(e) if $t = 2$ in., $b = 6$ in., and $h = 8$ in.

6. Determine $\bar{I}_x$ and $\bar{I}_y$ for the hollow box beam section of Fig. 7-24(f).

7. Determine $\bar{I}_x$ and $\bar{I}_y$ for the T-section shown in Fig. 7-24(b) if $a = t = 1$ in., $b = 9$ in., and $h = 8$ in.

8. If the channel section of Fig. 7-24(c) has $a = t = 0.5$ in., $b = 3$ in., and $h = 7$ in., determine $\bar{I}_x$.

9. Determine $\bar{I}_y$ for the channel of Ex. 8.

10. Evaluate $\bar{P}_{xy}$ for the channel of Ex. 8.

11. Determine $\bar{I}_x$ for the Z-section of Fig. 7-24(d). (Hint: Would $\bar{I}_x$ change if one flange were reversed in direction?)

12. For $t = 0.5$ in., $b = 3$ in., and $h = 4$ in., determine $\bar{I}_x$, $\bar{I}_y$, and $\bar{P}_{xy}$ for the Z-section of Fig. 7-24(d).

13. Determine I_x, I_y, and P_{xy} for the axes shown with the angle section of Fig. 7-24(e).

14. If $a = t = 0.5$ in., $b = 8$ in., and $h = 16$ in., determine $\bar{I}_x$ and $\bar{I}_y$ for the hollow box beam of Fig. 7-24(f).

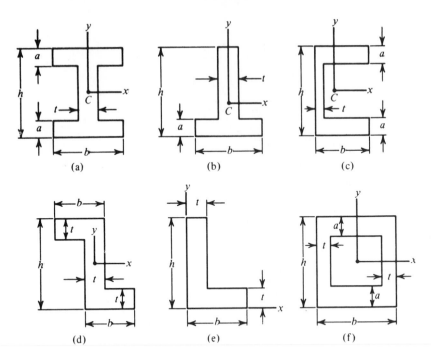

Fig. 7-24   Figures for Exs. 4–14

---

*7.7   ROTATION OF AXES FOR SECOND MOMENTS OF AREA; MOHR'S CIRCLE; PRINCIPAL AXES

In Fig. 7-25 the x',y'-axes are rotated through counterclockwise angle θ from the x,y-axes. Hence for the element of area dA the coordinates are related, according to Eq. (3.2.15), by

$$x' = x \cos \theta + y \sin \theta$$

$$y' = -x \sin \theta + y \cos \theta. \tag{7.7.1}$$

By definitions of Eqs. (7.6.6) the second moments and products are

$$I_{x'} = \int_A y'^2 \, dA, \quad I_{y'} = \int_A x'^2 \, dA, \quad \text{and} \quad P_{x'y'} = \int_A x'y' \, dA,$$

which, by use of Eq. (7.7.1), become

$$I_{x'} = I_x \cos^2 \theta - 2P_{xy} \sin \theta \cos \theta + I_y \sin^2 \theta$$

$$I_{y'} = I_x \sin^2 \theta + 2P_{xy} \sin \theta \cos \theta + I_y \cos^2 \theta \tag{7.7.2}$$

$$P_{x'y'} = I_x \sin \theta \cos \theta + P_{xy}(\cos^2 \theta - \sin^2 \theta) - I_y \sin \theta \cos \theta.$$

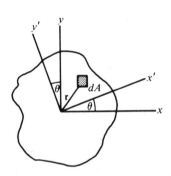

Fig. 7·25   Rotation of Axes

* This section may be omitted.

These results may be given a convenient form by using the trigonometric identities

$$\cos^2\theta = \tfrac{1}{2}(1+\cos 2\theta), \qquad \sin^2\theta = \tfrac{1}{2}(1-\cos 2\theta)$$

$$\cos^2\theta - \sin^2\theta = \cos 2\theta, \qquad 2\sin\theta\cos\theta = \sin 2\theta. \tag{7.7.3}$$

We obtain

$$I_{x'} = \tfrac{1}{2}(I_x+I_y)+\tfrac{1}{2}(I_x-I_y)\cos 2\theta - P_{xy}\sin 2\theta$$

$$I_{y'} = \tfrac{1}{2}(I_x+I_y)-\tfrac{1}{2}(I_x-I_y)\cos 2\theta + P_{xy}\sin 2\theta \tag{7.7.4}$$

$$P_{x'y'} = +\tfrac{1}{2}(I_x-I_y)\sin 2\theta + P_{xy}\cos 2\theta.$$

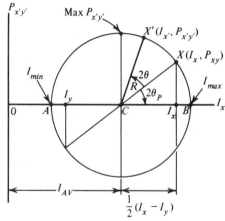

Fig. 7-26 Mohr's Circle for Second Moments of Area

The first and last of these equations are parametric equations for the graph of $I_{x'}$ versus $P_{x'y'}$ in a representation plane where the point corresponding to any value of θ has coordinates $(I_{x'}, P_{x'y'})$. The graph is a circle, as shown in Fig. 7-26, called Mohr's circle. The point marked $X(I_x, P_{xy})$ is the point corresponding to $\theta = 0$, while $X'(I_{x'}, P_{x'y'})$ represents a typical value of θ. The Cartesian equation of the circle can be obtained by transposing $\tfrac{1}{2}(I_x+I_y)$ in the first equation and then squaring and adding the first and third equations to eliminate θ. The result is

$$\left[I_{x'}-\frac{1}{2}(I_x+I_y)\right]^2 + P_{x'y'}^2 = \left[\frac{I_x-I_y}{2}\right]^2 + P_{xy}^2. \tag{7.7.5}$$

For given I_x, I_y, P_{xy} we let

$$I_{av} = \tfrac{1}{2}(I_x+I_y) \quad \text{and} \quad R = \sqrt{[\tfrac{1}{2}(I_x-I_y)]^2 + P_{xy}^2}$$

to obtain the *equation of the circle*

$$(I_{x'}-I_{av})^2 + P_{x'y'}^2 = R^2 \tag{7.7.6}$$

with radius R and center at $(I_{av}, 0)$ in the $I_{x'}$, $I_{x'y'}$ representation plane of Fig. 7-26.

The figure shows that where $P_{x'y'}=0$ the second moment $I_{x'}$ is maximum or minimum. These special directions for the x'-axis are called *principal axes*, and the values I_{max} and I_{min} are the corresponding *principal second moments of inertia*. The two special values of θ (in the physical xy-plane of Fig. 7-25) corresponding to the principal directions are denoted by θ_P (θ_{P_1} and θ_{P_2}, or sometimes by θ_1 and θ_2). They are found by setting $P_{x'y'} = 0$ in the last equation in Eqs. (7.7.4) to obtain

$$\tan 2\theta_P = \frac{-P_{xy}}{\frac{1}{2}(I_x - I_y)}. \tag{7.7.7}$$

This defines two values of $2\theta_P$, differing by 180°. Hence the two angles θ_P differ by 90°, so that the two principal directions are perpendicular. It is usually easy to determine by inspection which of the two principal directions determined by the formula for $\tan 2\theta_P$ corresponds to I_{max} and which to I_{min}, since for I_{max} more of the area will be far from the axis.

Principal Second Moments of Area

$$\left.\begin{array}{c} I_{max} \\ I_{min} \end{array}\right\} = I_{av} \pm R.$$

$$I_{av} = \tfrac{1}{2}(I_x + I_y) = \tfrac{1}{2}(I_{max} + I_{min}) \tag{7.7.8}$$

$$R = \sqrt{[\tfrac{1}{2}(I_x - I_y)]^2 + P_{xy}^2} = \tfrac{1}{2}(I_{max} - I_{min})$$

$$\tan 2\theta_P = \frac{-P_{xy}}{\frac{1}{2}(I_x - I_y)}.$$

I_{max} is of course only the maximum value for axes lying in the xy-plane. The second moment about the z-axis is

$$J_O = I_x + I_y = I_{max} + I_{min}, \tag{7.7.9}$$

which is obviously greater than I_{max}.

The Mohr's circle also shows that the maximum product of inertia *in the plane*, i.e., max $P_{x'y'}$, is equal to the radius of the circle,

$$\max P_{x'y'} = R = \tfrac{1}{2}(I_{max} - I_{min}). \tag{7.7.10}$$

EXERCISES

1. Determine the principal axes and principal second moments for the quarter-circle area in Appendix A, Table A3, for the origin shown.

2. For the rectangle of Appendix A, Table A3, if $b = 3$ in. and $h = 2$ in., determine the principal axes and principal second moments of area for the origin shown.

3. For the rectangle of Ex. 2, determine $I_{x'}$ and $P_{x'y'}$ for x' rotated 60° from x.
4. For the quarter-circle of Ex. 1, determine $I_{x'}$ and $P_{x'y'}$ for x' rotated 30° from x.
5. For the I-beam of Ex. 4(b) in Sec. 7.6, determine $\bar{I}_{x'}$, $\bar{I}_{y'}$, and $\bar{P}_{x'y'}$ for x' rotated 45° from x.
6. For the T-section of Ex. 7 in Sec. 7.6, determine the angle θ such that $\bar{I}_{x'} = \bar{I}_{y'}$, and compute the corresponding $\bar{P}_{x'y'}$.
7. Determine the centroidal principal axes and principal second moments for the Z-section of Ex. 12 in Sec. 7.6.
8. For the Z-section of Ex. 12 in Sec. 7.6, determine $\bar{I}_{x'}$, and $\bar{P}_{x'y'}$ for x'-axis rotated 60° from the x-axis.

*7.8 FLUID STATICS

In this section we shall consider one case of a deformable continuous medium that can be treated very simply, namely, a fluid that is either at rest or in unaccelerated motion with a uniform velocity so that it is not deforming. In such a nondeforming fluid there can be no tangential force across any internal surface, since the characteristic that distinguishes a fluid (liquid or gas) from a solid is that it will continue to deform as long as there is any tangential force applied. It follows that the internal forces (stresses) across an internal surface must all act in a direction normal to the internal surface. Moreover, the pressure is the same on all planes through a given point.

Fluid Statics

1. In a fluid at rest, or with a uniform velocity field, the normal stress at a point (pressure) is the same in all directions.
2. If in addition the fluid is unaccelerated and the only body force is gravity acting vertically downward (negative z-direction), then the pressure p in the fluid varies only with depth according to

$$\frac{dp}{dz} = -\gamma, \tag{7.8.1}$$

where

$$\gamma = \rho g \tag{7.8.2}$$

is the *specific weight* (weight per unit volume).

---

* This section may be omitted.

To derive Eq. (7.8.1) consider the free-body diagram of Fig. 7-27. Normal pressure p distributed over the lateral surface of the cylinder does not affect the vertical force summation. Equilibrium of the vertical cylinder requires

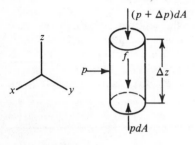

$$\left.\begin{aligned} \sum f_z &= 0 \\ p\,dA - (p+\Delta p)\,dA - \gamma^*(\Delta z)\,dA &= 0 \\ \frac{\Delta p}{\Delta z} &= -\gamma^*, \end{aligned}\right\} \quad (7.8.3)$$

whence

Fig. 7-27 Cylindrical Free-Body Diagram of Cross-Section Area dA

where γ^* is the average specific weight in the cylinder of finite height Δz. We now pass to the limit as $\Delta z \to 0$ (and $\gamma^* \to \gamma$) to obtain Eq. (7.8.1).

Hydrostatics of a Uniform Liquid

The density of a liquid varies so little with position that for most engineering purposes it may be considered constant, whence $\gamma = \rho g$ is also constant. Equation (7.8.1) can then be integrated to yield

---

Hydrostatics

$$p_2 - p_1 = -\gamma(h_2 - h_1)$$

or

$$\Delta p = -\gamma\,\Delta h, \qquad (7.8.4)$$

---

where h *is the elevation above an arbitrary datum level. Thus the pressure in a liquid increases linearly with the depth.*

Gage Pressure and Absolute Pressure

The pressure in the foregoing discussion was the ***absolute pressure***, the normal force per unit area acting across the surface. Pressure is often measured and quoted as the pressure difference from the local atmospheric pressure; this is called gage pressure.

---

absolute pressure = gage pressure

+ local atmospheric pressure.   (7.8.5)

The atmospheric pressure at sea level under "standard conditions" is $14.70 \, \text{lb/in.}^2$ (at a barometer reading of 29.92 in. of mercury). The atmospheric pressure for any other barometer reading can be calculated by proportion. The unit of pressure in the SI system is the newton per square meter, also called the pascal (Pa). A more convenient unit is the kilonewton per square meter ($1000 \, \text{N/m}^2$).

$$1 \, \text{Pa} = 1 \, \text{N/m}^2$$

$$1 \text{ standard atmosphere} = 101.325 \, \text{kN/m}^2 = 101.325 \, \text{kPa}$$

at a barometric pressure of 760 mm of mercury.

If point 1 in Eq. (7.8.4) is the free surface of a liquid (exposed to atmospheric pressure), then $-(h_2 - h_1)$ is the depth to point 2, and $p_2 - p_1$ is the gage pressure at point 2. Thus *in a liquid with free surface* (*exposed to atmospheric pressure*)

$$\text{gage pressure} = \gamma \, (\text{depth}).$$ (7.8.6)

Resultant Force on Submerged Plane Surface; Center of Pressure

A plane area submerged in a liquid (for example, a dam gate) is subjected to a parallel force distribution because of the fluid pressure. We saw in Secs. 4.1 and 4.2 that a parallel force system is equipollent to a single force. The point where the resultant single force acts on the plane area is called the *center of pressure*, a point in general below the centroid of the area (except, of course, when the plane area is horizontal), as the following example for a rectangular gate illustrates.

EXAMPLE 7.8.1

A rectangular gate has edge lengths b parallel to the liquid surface and a at angle α to the vertical, as shown in Fig. 7-28, where O is at the level of the free surface.

Let s denote the distance to a strip element of area dA with edge parallel to the free surface ($dA = b \, ds$ for a rectangle). Since the atmospheric pressure acts on the right of the gate, the net force f on the gate is found by integrating the gage pressure p_g

$$p_g = \gamma(\text{depth}) = \gamma s \cos \alpha$$

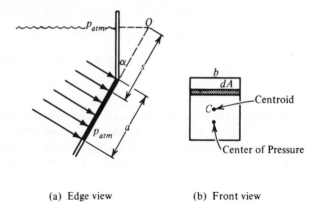

(a) Edge view   (b) Front view

Fig. 7-28   Rectangular Dam Gate

over the area A.

$$f = \int p_g \, dA$$
$$= \gamma \cos \alpha \int s \, dA \qquad (7.8.7)$$
$$= (\gamma \cos \alpha)\bar{s}A,$$

where $\bar{s}$ is the s-coordinate of the centroid, since $\bar{s}A = \int s \, dA$. Now $\gamma(\bar{s} \cos \alpha)$ is the gage pressure at the centroid of the area. This result (not limited to the rectangular gate case) shows the following fact:

The total force on plane area A equals the area A multiplied by the pressure at the centroid of the area. ***But the resultant force acts at a point below the centroid.***

$$(7.8.8)$$

---

as we now demonstrate for the rectangular example. Let s_p be the s-coordinate of the center of pressure. We determine s_p, the s-coordinate of the line of action of the resultant force, by taking moments about an axis through O lying in the free surface and also in the plane of the gate:

$$f s_p = \int sp \, dA$$
$$= \gamma \cos \alpha \int s^2 \, dA. \qquad (7.8.9)$$

For the rectangle

$$\int s^2 \, dA = b \int_{\bar{s}-(1/2)a}^{\bar{s}+(1/2)a} s^2 \, ds$$

$$= \tfrac{1}{3}b[(\bar{s}+\tfrac{1}{2}a)^3 - (\bar{s}-\tfrac{1}{2}a)^3]$$

$$= b\bar{s}^2 a + \tfrac{1}{12}ba^3,$$

where $\bar{s}$ is the s-coordinate to the centroid. Hence, after use of Eq. (7.8.7) with $A = ab$, Eq. (7.8.9) takes the form

$$\gamma ab\bar{s}s_p \cos \alpha = \gamma \cos \alpha (b\bar{s}^2 a + \tfrac{1}{12}ba^3),$$

whence

$$s_p = \bar{s} + \frac{a^2}{12\bar{s}}, \tag{7.8.10}$$

showing that the center of pressure is below the centroid.

Remark. The integral $\int s^2 \, dA$ in Eq. (7.8.9) is the **second moment of the area** about the axis through O. See Sec. 7.6.

Force on a curved surface can also be calculated by an integral, but this is more complicated than the plane area case, since with a curved surface the pressure forces do not form a parallel force system. It is usually possible to avoid the complication by a trick illustrated in the following example.

EXAMPLE 7.8.2

Figure 7-29(a) shows the cross-sectional view of a straight dam of length 40 ft perpendicular to the cross section. The part of the cross section in contact with the water is a quarter-circle of radius 60 ft, as shown.

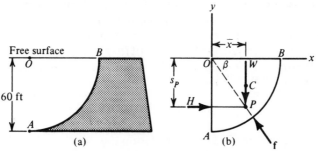

Fig. 7-29 (a) Cross Section of Dam; (b) Free Body of Fluid

To find the resultant force exerted against the dam by the water, we can consider the free body of fluid whose cross section is the quarter-circle shown in Fig. 7-29(b). If **f** is the resultant force exerted on the fluid by the dam, then the free body must be in equilibrium under the action of its weight *W*, the force **f**, and the resultant horizontal force of magnitude *H* exerted by the fluid to the left of *OA* acting on the free body. The horizontal force and its line of action can be determined as in Example 7.8.1. The rectangular face *OA* has an area of 2400 ft^2 and a centroid at a depth of 30 ft. For fresh water,

$$\gamma = 62.4 \text{ lb/ft}^3.$$

Then

$$H = (62.4)(30)(2400) = 4,490,000 \text{ lb}$$

acts at depth s_p given by Eq. (7.8.10) as

$$s_p = 30 + \frac{(60)^2}{12(30)} = 40 \text{ ft.}$$

The total weight of the free body is

$$W = \tfrac{1}{4}\pi(60^2)(40)(62.4) = 7,060,000 \text{ lb,}$$

and its line of action goes through the center of gravity *C* at $\bar{x} = 4a/3\pi$:

$$\bar{x} = \frac{(4)(60)}{3\pi} = 25.5 \text{ ft.}$$

Since all the pressure forces exerted by the dam on the free body are perpendicular to the cylindrical surface *AB*, they all intersect the cylinder axis *OZ* through *O* and therefore have no moment about *OZ*. Hence their resultant **f** can have no moment about *OZ*. Thus **f** intersects the axis *OZ*. By symmetry, all three forces *W*, *H*, and **f** act in the middle cross section of the free body, the cross section containing *C*. *W* and *H* intersect at *P*(25.5, −40) ft. For equilibrium **f** must also pass through *P* in order that the total moment about *P* will be zero. Hence the line of action is along *PO* and

$$\mathbf{f} = -H\mathbf{i} + W\mathbf{j},$$

while the force exerted by the water on the dam is

$$-\mathbf{f} = H\mathbf{i} - W\mathbf{j}$$

of magnitude

$$f = 8,370,000 \text{ lb}$$

at angle

$$\beta = 57.5°,$$

as shown, determined by

$$\tan \beta = \frac{W}{H}.$$

EXERCISES

1. Show that the quantity $z + (p/\gamma)$ is independent of position in a static liquid of uniform density.

2. A closed tank is partly filled with carbon tetrachloride of specific gravity 1.59 (Specific gravity is the ratio of its density to that of water.) The pressure at the liquid surface is 15 lb/in.² Calculate the pressure at a depth of 15 ft below the liquid surface.

3. How many inches of mercury (barometer reading) are equivalent to a pressure of 30 lb/in.²? How many feet of water.

4. The gas equation $pv = RT$ (where $v = 1/\rho$ is specific volume) implies that in an isothermal gas ($T = $ const.) we have (for negligible variation in g) $p/\gamma = C$, where C is a constant. Hence show that in an isothermal atmosphere

$$p = p_1 e^{-z/C},$$

where z is elevation above the surface where $p = p_1$. (Note that $C = p_1/\gamma_1$.)

5. In the U.S. "standard atmosphere" (below the stratosphere) the temperature variation is assumed to be linear with elevation $T = T_1 - kz$. Neglecting the variation in g, the gas equation $pv = RT$ may be written as $p/(\gamma T) = C$, where C is a constant. Show that this

implies that

$$C \frac{dp}{p} = -\frac{dz}{T_1 - kz}$$

and hence show that

$$p = p_1 \left(1 - \frac{kz}{T_1}\right)^{1/(Ck)}$$

[Note that $C = p_1/(\gamma_1 T_1)$.]

6. For the U.S. standard atmosphere of Ex. 5, the constants are $T_1 = 519°F$ (absolute), $\gamma_1 = 0.0765 \text{ lb/ft}^3$, $p_1 = (14.7)(144) \text{ lb/ft}^2$, and $k = 0.00356°F/ft$. Calculate the pressure and specific weight at an altitude of 20,000 ft.

7. A vertical rectangular dam gate 8 ft wide and 6 ft high has its center 5 ft below a water surface. Calculate the resultant force on the gate, and locate its line of action.

8. A circular dam gate 6 ft in diameter lies in a plane inclined at 45° to the vertical with its center located at a depth of 8 ft below the water surface. Determine the resultant force on the gate, and locate its line of action.

9. A triangular dam gate lies in a plane making a 30° angle with the vertical. The triangle has a horizontal lower 6-ft base at a depth of 12 ft

below the water surface and an altitude of 5 ft. Calculate the resultant force on the gate, and locate its line of action.

10. A vertical square gate has 8-ft edges. Calculate the distance between the center of pressure and the centroid when the top of the gate is (a) in the water surface and (b) 40 ft below the water surface.

11. A hemispherical shell 4 ft in diameter is used as a gate to close a circular hole in a vertical dam (with the shell extending into the water). If the center of the shell is located 6 ft below the water surface, what are the vertical and horizontal components of the resultant fluid force on (a) the hemisphere and (b) the bottom half of the hemisphere.

---

*7.9 MASS MOMENTS OF INERTIA AND PRODUCTS OF INERTIA

For translational motion without rotation the inertia of a rigid body is characterized by a single scalar constant, the mass of the body. For rotation, however, six additional constants are needed to characterize the **rotational inertia**. These six rotational **inertia components** are three **moments of inertia** and three **products of inertia**. These components are defined and methods for calculating them are presented in this section.

The **moment of inertia** I_{AA} of a mass-point particle of mass m with respect to an axis AA is defined in elementary physics by $I_{AA} = mR^2$, where R is the distance to the particle from the axis AA. For a given mass, I_{AA} is a positive number proportional to the square of the distance from the axis to the particle. For a collection of particles

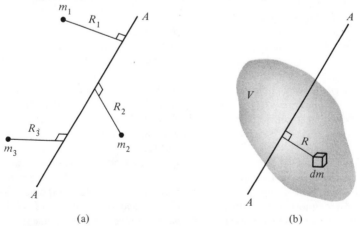

Fig. 7-30 Illustrations for Moment of Inertia Definitions
(a) Mass-Point Particles; (b) Element of Mass *dm*

* This section may be omitted.

$m_1, m_2, m_3, \ldots$, as in Fig. 7-30(a), the total moment of inertia with respect to AA is

$$I_{AA} = m_1 R_1^2 + m_2 R_2^2 + m_3 R_3^2 + \cdots. \tag{7.9.1}$$

For given masses, the magnitude of I_{AA} characterizes what might be called the "awayness" of the masses from AA. For a rigid body occupying volume V, as in Fig. 7-30(b),

$$I_{AA} = \int_V R^2 \, dm = \int_V R^2 \rho \, dV$$

$$[I] = [ML^2], \tag{7.9.2}$$

where ρ is density and again (for given total mass) I_{AA} measures the awayness of the mass distribution from AA.

A flywheel with most of its mass in the rim evidently has a bigger I_{AA} than a uniform disk of the same radius and same mass.

The **product of inertia** is defined with respect to two perpendicular reference planes. For example, if the two reference planes are the two coordinate planes $x = 0$ and $y = 0$, the product of inertia for a single particle m is $P_{xy} = mxy$, involving m times the product of the two coordinates x and y, measured from the reference planes. For the system of particles in Fig. 7-30(a)

$$P_{xy} = m_1 x_1 y_1 + m_2 x_2 y_2 + m_3 x_3 y_3 + \cdots, \tag{7.9.3}$$

where x_k and y_k are the x- and y-coordinates of m_k. For a rigid body occupying volume V

$$P_{xy} = \int_V xy \, dm = \int_V xy\rho \, dV$$

$$[P_{xy}] = [ML^2], \tag{7.9.4}$$

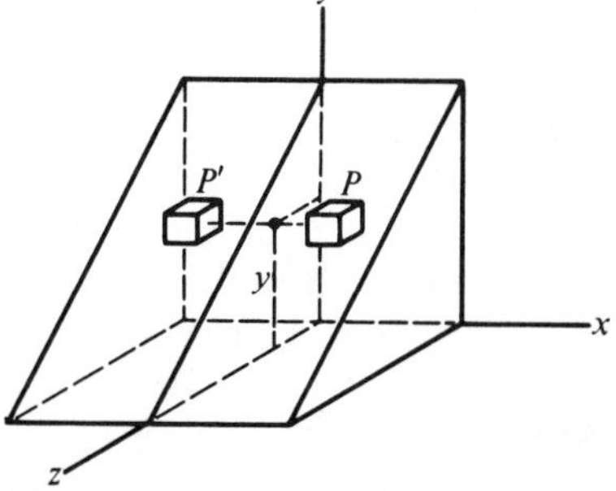

Fig. 7-31 Example Where $x = 0$ Is Plane of Symmetry, So That $P_{xy} = 0$ and $P_{xz} = 0$

where x and y are coordinates of the volume element dV. The product of inertia is a measure of the skewness or asymmetry of the mass distribution with respect to the two planes. If either of the planes $x = 0$ or $y = 0$ is a plane of symmetry of a body of uniform density, then $P_{xy} = 0$, because the integrand is odd in x or y. See, for example, Fig. 7-31 where the plane $x = 0$ is a plane of symmetry. For each mass element dm at a point P with coordinates (x, y, z) there is another equal dm at

$P'(-x, y, z)$, such that $xy\,dm - xy\,dm = 0$. When the integral is extended over the whole body we obtain $P_{xy} = 0$. (In this case $P_{xz} = 0$ also.)

Actually P_{xy} can be zero when neither the plane $x = 0$ nor the plane $y = 0$ is a plane of symmetry. In fact, no matter what the shape of the body, for any choice of reference point A it is always possible to find three mutually perpendicular axes through A such that all the products of inertia are zero referred to these axes. These special axes are called ***principal axes of inertia*** of the body with respect to the point A. Methods for determining the principal axes for an arbitrary body will be given in Vol. II, Sec. 11.3.

Inertia components with respect to x,y,z-***axes*** are given by the following integrals. Note that $y^2 \times z^2$ is the square of the distance from the x-axis to the element of mass $dm = \rho\,dV$, etc.

Mass Moments of Inertia

$$I_{xx} = \int_V \rho(y^2 + z^2)\,dV, \qquad I_{yy} = \int_V \rho(x^2 + z^2)\,dV$$

$$I_{zz} = \int_V \rho(x^2 + y^2)\,dV. \tag{7.9.5}$$

Mass Products of Inertia

$$P_{xy} = \int_V \rho xy\,dV, \qquad P_{yx} = \int_V \rho yx\,dV, \qquad \text{etc.} \tag{7.9.6}$$

Note that there are only three independent products of inertia, since

$$P_{xy} = P_{yx}, \qquad P_{yz} = P_{zy}, \qquad P_{zx} = P_{xz}. \tag{7.9.7}$$

The defining volume integrals must be thought of as triple integrals with dV infinitesimal in all directions, so that the variables x, y, z, and ρ appearing in the integrands can be unambiguously associated with the position of the volume element. We shall see, however, that the inertia components (that is, the moments and products of inertia with respect to the chosen axes) can often be calculated by single integrals. For many common shapes of uniform density the moments of inertia are tabulated in Table A4 of Appendix A with respect to principal axes (usually with the origin at the center of mass).

For other bodies whose boundaries can be readily defined it is a straightforward process to evaluate the defining integrals of Eqs. (7.9.5) and (7.9.6) as triple integrals, choosing limits on the interior integrals as illustrated in Sec. 7.5, either explicitly if the geometry is simple enough or numerically using multiple-integral routines available in most computer centers. We shall not give any examples of such multiple-integral evaluations here. A few exercises of this type are given at the end of this section. The rectangular block and the circular cylinder are especially easy to calculate in Cartesian coordinates and cylindrical coordinates, respectively. We shall see at the end of this section that many cases of interest can be calculated by single integrals by making use of the method of composite bodies.

The ***method of composite bodies*** is used to compute the moments and products of inertia for bodies made of several parts when the moments and products of the parts are known ***with respect to the axes of interest***. Since the axes of interest for a composite body are not usually the same as the reference axes in the tabulation of Appendix A, Table A4, we shall first have to make use of the parallel-axis theorems to be presented in this section to obtain the inertia properties of each part of the composite body with respect to the axes of interest. Then the method of composite bodies consists of simply adding the values for the separate parts, since the defining integrals are additive, that is; $\int_{V_1+V_2} = \int_{V_1} + \int_{V_2}$. By this procedure the rotational inertia properties of many bodies of engineering interest can be computed in terms of known simple geometric bodies, neglecting small irregularities such as fillets and small holes.

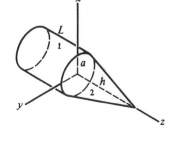

Fig. 7-32 Composite Body

For example, Fig. 7-32 shows a cone mounted on the end of a cylinder. The total I_{zz} is given by

$$I_{zz} = I_{zz}^{(1)} + I_{zz}^{(2)}$$
$$= \tfrac{1}{2}m_1 a^2 + \tfrac{3}{10}m_2 a^2, \qquad (7.9.8)$$

where m_1 and m_2 are the masses of the cylinder and cone, respectively, since the z-axis is the same for both bodies in Appendix A, Table A4. But we cannot simply add the values of I_{xx} given in Table A4 for the two bodies, since the x-axis for the cylinder in Table A4 does not coincide with either the x- or the x_1-axis given for the cone. Exercise 7 at the end of this section will ask for the computation of I_{xx} for this composite body. It will be necessary to refer both bodies to the same x-axis, by using the following parallel-axis theorem, before the two values of $I_{xx}^{(1)}$ and $I_{xx}^{(2)}$ can be added.

Parallel-Axis Theorem for Moment of Inertia

If AA is an arbitrary axis (which may or may not intersect the body) and $A'A'$ is ***a parallel axis through the center of mass***, then

---

Parallel-Axis Theorem

$$I = \bar{I} + d^2 m, \qquad\qquad (7.9.9)$$

---

where I denotes I_{AA}, $\bar{I}$ denotes $I_{A'A'}$, m is the mass of the body, and d is the perpendicular distance from AA to $A'A'$.

Note that $\bar{I}$ is the smallest possible moment of inertia about any axis parallel to $A'A'$, since the term $d^2 m$ is nonnegative.

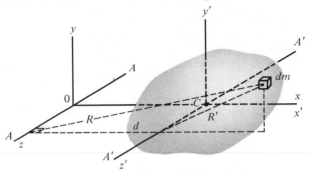

Fig. 7-33   Axes for Parallel-Axis Theorem Proof

For the proof refer to Fig. 7-33. We drop a perpendicular to AA from the mass center C to locate O and then choose axes as shown with z along AA and z' along $A'A'$. By the definition of Eq. (7.9.2), I_{AA} and $I_{A'A'}$ are given by

$$I = \int_V R^2\,dm \quad \text{and} \quad \bar{I} = \int_V R'^2\,dm,$$

where R and R' are the distances from the z- and z'-axes, respectively, to dm.

$$R^2 = x^2 + y^2 = (x' + d)^2 + y'^2 = x'^2 + y'^2 + 2dx' + d^2$$
$$= R'^2 + 2dx' + d^2.$$

Hence

$$I = \int_V R'^2\,dm + 2d \int x'\,dm + d^2 \int dm.$$

Since $\int x'\,dm = m\bar{x}' = o$, because C is the mass center, this gives $I = \bar{I} + d^2 m$, as asserted in Eq. (7.9.9).

This important result permits us to extend the tabulated information of Appendix A, Table A4, to arbitrary parallel axes. Note that a few of the cases in Table A4 do not give $\bar{I}$. For them $\bar{I}$ must be first be calculated from Eq. (7.9.9) as $\bar{I} = I_1 - d_1^2 m$, where I_1 is the tabulated value and d_1 is the distance from C to the tabulated axis. Then I can be calculated for an arbitrary axis. For uniform bodies the mass center coincides with the centroid. See the tabulated centroid locations of Appendix A, Table A4. The quantity $\bar{I}$ is usually in fact called the **centroidal moment of inertia**, even for nonuniform bodies, but it really means with respect to the mass center and not the centroid.

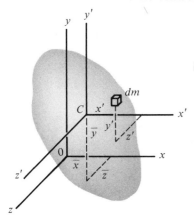

Fig. 7-34 Parallel Axes with x', y', z' Through Mass Center

Parallel-Axis Theorem for Products of Inertia

Let x, y, z be an arbitrary set of rectangular Cartesian axes and $x'y'z'$ a parallel set through the mass center C whose coordinates are $(\bar{x}, \bar{y}, \bar{z})$ referred to the arbitrary x,y,z-axes; see Fig. 7-34.

Then we can prove the

Parallel-Axis Theorem for Products

$$P_{xy} = \bar{P}_{xy} + \bar{x}\bar{y}m, \qquad (7.9.10)$$

---

where

$$P_{xy} = \int xy \, dm \quad \text{and} \quad \bar{P}_{xy} = \int x'y' \, dm,$$

according to the definition of Eq. (7.9.6). The proof, which is quite similar to that of Eq. (7.9.9), is left as an exercise; see Ex. 5 at the end of this section. Theorems for the other products can be written down by obvious substitutions.

Since one or more of the coordinates $\bar{x}$, $\bar{y}$, $\bar{z}$ may be negative, P_{xy} may be smaller than $\bar{P}_{xy}$. Also P_{xy} can be negative, while for a real body I_{xx} is positive. (The slender rod approximation of Appendix A, Table A4, lists $\bar{I}_{zz} = 0$, but this merely means that it is negligible in comparison to $\bar{I}_{xx}$ and $\bar{I}_{yy}$.)

The products of inertia are all zero for most of the axes shown in Table A4, since they are principal axes, but the parallel axes will not as a rule be principal axes. For those bodies $P_{xy} = \bar{x}\bar{y}m$, since $\bar{P}_{xy} = 0$ for the axis directions shown.

Radius of Gyration

Since the moment of inertia about an axis depends on the total mass as well as on the way it is distributed, the **nature** of the mass distribution is characterized better by a quantity independent of the total mass. If I_{AA} is denoted by

$$I = \int_V r^2 \, dm,$$

then the **radius of gyration** k of the body about AA is defined by

$$I = mk^2 \quad \text{or} \quad k = \sqrt{\frac{I}{m}}$$

$$[k] = [L]. \tag{7.9.11}$$

It is equal to the distance from the axis that a concentrated mass particle, of the same total mass m as the actual distributed mass, would have to be located in order that the particle would have the same moment of inertia. Two different uniform bodies of the same size and shape will have the same radius of gyration when they are in the same position relative to the axis, but they will have different moments of inertia if they have different densities. A parallel-axis theorem for the radius of gyration follows immediately from Eq. (7.9.9) by dividing through by m, namely,

$$k^2 = \bar{k}^2 + d^2, \tag{7.9.12}$$

since $I = mk^2$ and $\bar{I} = m\bar{k}^2$, where k and $\bar{k}$ are the radii of gyration of the body with respect to axes AA and $A'A'$, respectively, in Fig. 7-33.

Thin plate approximations for flat plates of various shapes may be readily derived from the **second moments of area**

$$I_x = \int_A y^2 \, dA, \qquad I_y = \int_A x^2 \, dA, \qquad P_{xy} = \int_A xy \, dA$$

$$J_O = \int_A r^2 \, dA \equiv \int_A (x^2 + y^2) \, dA = I_x + I_y \tag{7.9.13}$$

discussed in Sec. 7.6. For a plate of uniform thickness t and density ρ we obtain approximations to the mass moments of inertia I_{xx}, I_{yy}, I_{zz}, and P_{xy} of these plates by multiplying through by ρt. We demonstrate here that the formula for I_{xx} can be approximated in this manner.

$$I_{xx} = \int_V \rho(y^2 + z^2)\, dV = \rho \int_A \left[\int_{-t/2}^{t/2} (y^2 + z^2)\, dz \right] dA$$

$$= \rho \int_A (y^2 t + \tfrac{1}{12}t^3)\, dA \tag{7.9.14}$$

$$= \rho t [I_x + \tfrac{1}{12}t^2 A].$$

We assume that t^2 is so small that $\tfrac{1}{12}t^2 A$ is negligible in comparison to the second moment of area I_x. Then, for the thin plate,

$$I_{xx}\ (\text{mass}) = \rho t I_x\ (\text{area})$$

and similarly $\left.\rule{0pt}{48pt}\right\}$ (7.9.15)

$$I_{yy}\ (\text{mass}) = \rho t I_y\ (\text{area}),$$

while the formulas

$$I_{zz}\ (\text{mass}) = \rho t J_O\ (\text{area})$$

$$P_{xy}\ (\text{mass}) = \rho t P_{xy}\ (\text{area}) \tag{7.9.16}$$

are exact for a uniform plate, where J_O is the ***polar second moment of area***.

   Some second moments of area are tabulated in Appendix A, Table A3. Their use to compute an approximate mass moment of inertia for a triangular plate is illustrated in Sample Problem 7.9.5.

   Single-integral evaluation of moments and products of inertia is possible when the inertia properties of a slab element of volume or a cylindrical shell element of volume are known as a function of a single parameter specifying the position of the element of volume. (See the discussion of centroids by single integration in Sec. 7.4.) The most obvious cases to treat this way are solids of revolution. The total moment of inertia about the axis of revolution is simply the sum of the moments of inertia of the disk or shell elements of volume, which are known from Appendix A, Table A4. For any other axis we must apply the parallel-axis theorems to the elements of volume before summing, as in Sample Problem 7.9.1.

   Other sample problems and exercises then illustrate the use of the parallel-axis theorems and the composite body method for calculating the inertia components at an arbitrary point. See Appendix A, Table A4, for centroid and volume formulas.

Units of Moments and Products of Inertia
Dimensions $[ML^2] = [FLT^2]$

SI system	kg-m^2
British system	slug-ft^2 = lb-ft-sec^2

$$(7.9.17)$$

SAMPLE PROBLEM 7.9.1

Compute the principal moments of inertia of a uniform cone with respect to the apex.

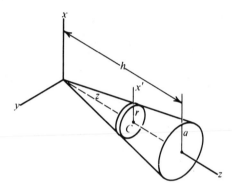

Fig. 7-35   Disk Element of Volume for Cone

SOLUTION.   The element of volume is

$$dV = \pi r(z)^2 \, dz,$$

where $r(z)$ is the variable radius of the element; see Fig. 7-35. Its mass is

$$\rho \, dV = \rho \pi r^2 \, dz,$$

and its moments of inertia with respect to C are, by the thin disk formulas of Appendix A, Table A4

$$(\bar{I}_{zz} = \tfrac{1}{2} mr^2 \quad \text{and} \quad \bar{I}_{xx} = \tfrac{1}{4} mr^2)$$

$$dI_{zz} = \tfrac{1}{2} \rho \pi r^4 \, dz$$

$$dI_{x'x'} = \tfrac{1}{4} \rho \pi r^4 \, dz.$$

By the parallel-axis theorem

$$dI_{xx} = dI_{x'x'} + z^2(\rho\pi r^2\,dz)$$
$$= \rho\pi(\tfrac{1}{4}r^4 + r^2 z^2)\,dz.$$

Hence

$$I_{zz} = \int_0^h \tfrac{1}{2}\rho\pi r^4\,dz$$

$$I_{xx} = \int_0^h \rho\pi(\tfrac{1}{4}r^4 + r^2 z^2)\,dz.$$

To evaluate these equations, we must substitute $r = (z/h)a$, by similar triangles. Hence

$$I_{zz} = \frac{1}{2}\rho\pi\frac{a^4}{h^4}\int_0^h z^4\,dz = \frac{1}{10}\rho\pi a^4 h$$

$$I_{xx} = \rho\pi\int_0^h\left(\frac{1}{4}\frac{a^4}{h^4}z^4 + \frac{a^2}{h^2}z^4\right)dz$$

$$= \frac{1}{5}\rho\pi\left[\frac{1}{4}a^4 h + a^2 h^3\right].$$

Since $m = \rho V = \tfrac{1}{3}\rho\pi a^2 h$, this gives

$$I_{zz} = \frac{3}{10}ma^2 \quad \text{and} \quad I_{xx} = \frac{3}{20}ma^2 + \frac{3}{5}mh^2,$$

in agreement with the values I_{zz} and $I_{x_1 x_1}$ in Appendix A, Table A4, for principal axes through the apex.

SAMPLE PROBLEM 7.9.2

Use the composite-body method to derive the cylindrical thin-shell approximate formulas of Appendix A, Table A4, from the formulas for a solid cylinder.

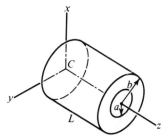

Fig. 7-36 Hollow Cylinder

SOLUTION. For the hollow cylinder of Fig. 7-36, with outer radius b and inner radius a, we obtain $\bar{I}_{xx}$ by subtracting the value for a solid cylinder of radius a from that for a solid cylinder of radius b, according to the formulas of Appendix A, Table A4. Thus

$$\bar{I}_{xx} = \tfrac{1}{12}m_b(3b^2 + L^2) - \tfrac{1}{12}m_a(3a^2 + L^2)$$

where

$$m_b = \rho\pi b^2 L \quad \text{and} \quad m_a = \rho\pi a^2 L,$$

so that

$$\bar{I}_{xx} = \tfrac{1}{12}(m_b - m_a)L^2 + \tfrac{3}{12}\rho\pi L(b^4 - a^4)$$

$$= \tfrac{1}{12}m[L^2 + 3(a^2 + b^2)],$$

where $m = m_b - m_a$. This result for the hollow cylinder could also be calculated easily by a multiple integral in cylindrical coordinates.

We now introduce the approximation by letting r denote the mean radius and t the thickness. Thus

$$r = \tfrac{1}{2}(a + b), \qquad t = b - a, \qquad b = r + \tfrac{1}{2}t, \qquad a = r - \tfrac{1}{2}t,$$

whence, for the hollow cylinder, we have exactly

$$\bar{I}_{xx} = \tfrac{1}{12}m[L^2 + 6r^2 + \tfrac{3}{2}t^2].$$

For the thin shell we neglect $\tfrac{3}{2}t^2$ in comparison to $L^2 + 6r^2$ to obtain the approximate formula

$$\bar{I}_{xx} = \tfrac{1}{2}mr^2 + \tfrac{1}{12}mL^2 \quad \textbf{Answer,}$$

listed in Table A4. We could obtain $\bar{I}_{zz} = mr^2$ similarly, but this is an obvious result by assuming all the mass to be at distance r from the axis.

SAMPLE PROBLEM 7.9.3

The thin rod shown in Fig. 7-37 weighs w lb/ft of length. If it is bent into the L-shape shown with lengths L_1 ft on the x-axis and L_2 ft parallel to the y-axis, determine the inertia components with respect to the x,y,z-axes at O.

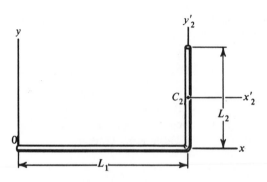

Fig. 7-37   Bent Rod

SOLUTION. The formulas of Appendix A, Table A4, give

for L_1: $I_{yy}^{(1)} = I_{zz}^{(1)} = \frac{1}{3}m_1 L_1^2$,   others zero.

for L_2: $\bar{I}_{xx} = \bar{I}_{zz} = \frac{1}{12}m_2 L_2^2$,   others zero at C_2.

Parallel-axis theorems give, for L_2,

$$I_{xx}^{(2)} = \tfrac{1}{12}m_2 L_2^2 + (\tfrac{1}{2}L_2)^2 m_2 = \tfrac{1}{3}m_2 L_2^2$$

$$I_{yy}^{(2)} = 0 + L_1^2 m_2 = m_2 L_1^2$$

$$I_{zz}^{(2)} = \frac{1}{12}m_2 L_2^2 + \left[L_1^2 + \left(\frac{1}{2}L_2\right)^2\right]m_2 = m_2 L_1^2 + \frac{1}{3}m_2 L_2^2$$

$$P_{xy}^{(2)} = 0 + (L_1)(\tfrac{1}{2}L_2)m_2 = \tfrac{1}{2}m_2 L_1 L_2.$$

Hence, after substituting $m_1 = wL_1/g$ and $m_2 = wL_2/g$ and adding the two parts together, we obtain the inertia components

$$I_{xx} = \frac{1}{3}\frac{w}{g}L_2^3, \qquad I_{yy} = \frac{w}{g}\left[\frac{1}{3}L_1^3 + L_1^2 L_2\right]$$

$$I_{zz} = \frac{w}{g}\left[\frac{1}{3}L_1^3 + L_1^2 L_2 + \frac{1}{3}L_2^3\right]$$

$$P_{xy} = \frac{1}{2}\frac{w}{g}L_1 L_2^2, \qquad P_{yz} = P_{zx} = 0.$$

SAMPLE PROBLEM 7.9.4

Determine moments of inertia of the uniform hemisphere (Fig. 7-38) with respect to the x,y,z-axes shown. (The zx-plane is parallel to the bounding plane and tangent to the hemisphere at its top point.)

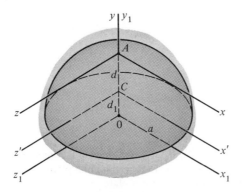

Fig. 7-38 Hemisphere

SOLUTION.   We use the tabulated formulas for a hemisphere, given in Appendix A, Table A4.

$$I_{x_1x_1} = I_{y_1y_1} = I_{z_1z_1} = \tfrac{2}{5}ma^2.$$

We cannot now apply the parallel-axis theorem to find I_{xx}, since neither of the axes x or x_1 passes through the mass center. We must first find $\bar{I}_{xx}$.

$$I_{x_1x_1} = \bar{I}_{xx} + d_1^2 m,$$

where $d_1 = OC = \tfrac{3}{8}a$ (see Table A4).   Then

$$I_{xx} = \bar{I}_{xx} + d^2 m,$$

where

$$d = \tfrac{5}{8}a.$$

Hence

$$I_{xx} = I_{x_1x_1} - d_1^2 m + d^2 m$$
$$= \tfrac{2}{5}ma^2 - \tfrac{9}{64}ma^2 + \tfrac{25}{64}ma^2$$
$$= \tfrac{2}{5}ma^2 + \tfrac{1}{4}ma^2 = \tfrac{13}{20}ma^2.$$

By symmetry

$$I_{zz} = I_{xx}, \qquad \text{while} \qquad I_{yy} = I_{y_1y_1}.$$

Hence

$$I_{xx} = I_{zz} = \tfrac{13}{20}ma^2, \qquad I_{yy} = \tfrac{2}{5}ma^2 \quad \textbf{\textit{Answer.}}$$

SAMPLE PROBLEM 7.9.5

Given that the second moment of area of a triangle is $\tfrac{1}{12}bh^3$ about the base and $\tfrac{1}{36}bh^3$ about a centroidal axis parallel to the base, determine the inertia components with respect to the x,y,z-axes shown for the isosceles triangular plate of mass m in Fig. 7-39. The xy-plane is the middle plane of the uniform plate.

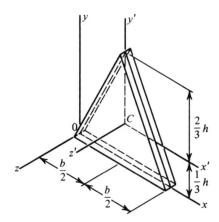

Fig. 7-39   Isosceles Triangular Plate

SOLUTION

$$\bar{I}_x = \frac{1}{36}bh^3$$

$$\bar{I}_y = \frac{1}{12}h\left(\frac{b}{2}\right)^3 + \frac{1}{12}h\left(\frac{b}{2}\right)^3 = \frac{1}{48}hb^3.$$

By Eqs. (7.9.13),

$$\bar{J}_C^{\cdot} \equiv \bar{I}_x + \bar{I}_y = \tfrac{1}{36}bh^3 + \tfrac{1}{48}hb^3.$$

By Eqs. (7.9.15) and (7.9.16) the centroidal mass moments of inertia are

$$\bar{I}_{xx} = \rho t \bar{I}_x, \qquad \bar{I}_{yy} = \rho t \bar{I}_y, \qquad \bar{I}_{zz} = \rho t \bar{J}_C,$$

where ρ is the density and t is the thickness. Hence

$$\bar{I}_{xx} = \tfrac{1}{36}\rho t b h^3, \qquad \bar{I}_{yy} = \tfrac{1}{48}\rho t h b^3$$

$$\bar{I}_{zz} = \tfrac{1}{2}\rho t b h (\tfrac{1}{18}h^2 + \tfrac{1}{24}b^2).$$

Since $m = \rho A t = \tfrac{1}{2}\rho t b h$, these equations become

$$\bar{I}_{xx} = \tfrac{1}{18}mh^2, \qquad \bar{I}_{yy} = \tfrac{1}{24}mb^2, \qquad \bar{I}_{zz} = m(\tfrac{1}{18}h^2 + \tfrac{1}{24}b^2),$$

while

$$\bar{P}_{xy} = \bar{P}_{yz} = \bar{P}_{zx} = 0,$$

since two centroidal coordinate planes are planes of symmetry.
We now use the parallel-axis theorems to obtain

$$I_{xx} = \bar{I}_{xx} + m\left(\frac{h}{3}\right)^2 = \frac{1}{6}mh^2$$

$$I_{yy} = \bar{I}_{yy} + m\left(\frac{b}{2}\right)^2 = \frac{7}{24}mb^2$$

$$I_{zz} = \bar{I}_{zz} + m\left[\left(\frac{b}{2}\right)^2 + \left(\frac{h}{3}\right)^2\right] = I_{xx} + I_{yy} = \frac{1}{6}m\left(h^2 + \frac{7}{4}b^2\right)$$

$$P_{xy} = 0 + \bar{x}\bar{y}m = \frac{1}{6}mbh, \qquad P_{yz} = P_{zx} = 0 \quad \textbf{Answer.}$$

EXERCISES

(Use the formulas of Appendix A, Table A4, except when asked to derive the formulas. Centroid and volume formulas are also given in Table A4.)

1. Use multiple integration to derive the formulas for moments of inertia listed in Table A4 for the
   (a) Rectangular block.
   (b) Solid cylinder.
   (c) Elliptical cylinder.
   (d) Ellipsoid.
   (e) Elliptic paraboloid.
2. Use a single integral to derive the formulas for moments of inertia listed in Table A4 for the (a) sphere, (b) paraboloid of revolution (elliptic paraboloid with $a = b$).
3. Use multiple integration to determine the products of inertia of one octant of an ellipsoid of uniform density ρ with respect to its semiaxes.
4. Use parallel-axis theorems to determine the inertia components of a rectangular block with respect to its edges. (Choose axes so that the block is in the positive octant.)
5. Prove the parallel-axis theorem for the product of inertia P_{xy}, Eq. (7.9.10).

6. Determine centroidal moments of inertia for a cone with respect to axes parallel to those shown in Appendix A, Table A4.
7. Determine I_{xx} for the composite body of Fig. 7-32.
8. A cubical block of edge length $4a$ has a concentric spherical hole of radius a. If the material is of uniform density ρ, what is the moment of inertia with respect to an axis through the center parallel to an edge of the block.
9. At what distance R from the axis AA may a uniform sphere of radius a be treated as a particle with an error of no more than 1% in computing I_{AA}?
10. The thin disk of mass m and radius r rolls as a wheel, supported by AB, around the circular path of radius R. In the position shown, what are the moments and products of inertia of the disk with respect to the XYZ-axes?
11. Determine the centroidal moments and products of inertia for a body consisting of the half of the solid cone of Appendix A, Table A4, for which $y > 0$, for axes parallel to x_1, y_1, z_1.
12. Determine the centroidal principal moments of inertia for a body consisting of the half torus of Appendix A, Table A4.

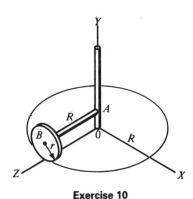

Exercise 10

13. Derive the formula in Appendix A, Table A4, for the thin spherical shell.
14. Determine the moment of inertia and the radius of gyration for the uniform frustrum of a cone with respect to its axis of symmetry.

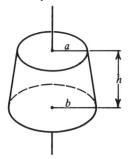

Exercise 14

15. (a) Use integration to determine the principal moments of inertia with respect to the apex

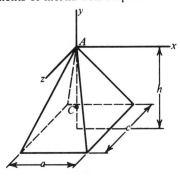

Exercise 15

for a uniform right rectangular pyramid with altitude h, mass m, and base dimensions a and

c parallel to the x- and z-axes as shown. (b) Now use parallel-axis theorems to determine centroidal principal moments of inertia.

16. Determine the inertia components with respect to the axes shown for the solid composite body consisting of a uniform pyramid with a 0.2×0.2 m square base, an altitude of 0.4 m, and a mass of 50 kg mounted on top of a 0.2-m cube of another material with a mass of 90 kg. [See the results of Ex.15(b).]

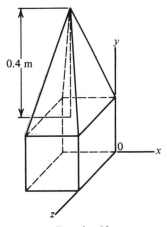

Exercise 16

17. Determine the inertia components for the flat steel plate with respect to the axes shown. The plate is 2 ft by 3 ft by 0.5 in. thick and has a circular hole in it of diameter 1 ft with center at $x = 1$ ft, $y = -1$ ft. Steel weighs 490 lb/ft^3.

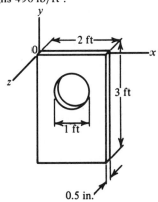

Exercise 17

18. Determine by arc length integrals the inertia components of a thin rod in the form of a circular arc of radius R ft and central angle α, with respect to the axes shown, if the rod weighs w lb/ft. (In the thin-rod approximation assume that $z = 0$ throughout the body.) Verify that the results for I_{zz} in any case and for I_{xx} and I_{yy} when α is a multiple of $\pi/2$ check with values you could determine from the complete ring formulas of Appendix A, Table A4, by symmetry arguments.

19. Determine the mass moment of inertia about a diameter for a hollow steel sphere with an inside radius of 10·cm and an outside radius of 12 cm. Density is 7880 kg/m³.

20. Two identical hollow steel balls like those of Ex. 19 are welded to the ends of a solid steel bar that has a length of 1 m and a diameter of 4 cm. Find I_{AA} about an axis perpendicular to the shaft at its center.

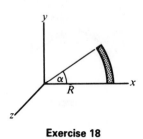

Exercise 18

7.10 SUMMARY

In this chapter we have considered several different special topics: centroids and mass centers in Secs. 7.2–7.5, second moments of area in Secs. 7.6 and 7.7, fluid statics in Sec. 7.8, and mass moments and products of inertia in Sec. 7.9. The last are to be used for rotational dynamics in Chapters 11 and 12 of Vol. II. This concludes Vol. I.

APPENDIX A

Some Useful Tables

Refer to Vol. I, Sec. 1.3, or Vol. II, Sec. 8.3, for explanations of Tables A1 and A2.

Table A1 Six Systems of Units (All Systems Use Time in Seconds)

	Mass	Force	Length	g_c
British:				
1.	Slug	lb(f)	ft	1
2.	lb (m)	Poundal	ft	1
3. Nonconsistent	lbm	lbf	ft	$32.174 \dfrac{\text{lbm-ft}}{\text{lbf-sec}^2}$
Metric:				
1. SI	kg	Newton (N)	m	1
2. CGS	g	Dyne	cm	1
3. Nonconsistent MKS	kgm	kgf	m	$9.81146 \dfrac{\text{kgm-m}}{\text{kgf-sec}^2}$

Table A2   Some Conversion Factors

Length
    1 meter = 100 cm = 39.37 in. = 3.280 ft
    1 foot = 0.3048 m = 30.48 cm
    1 inch = 0.0254 m = 2.540 cm

Mass
    1 lbm = 0.03108 slugs = 0.45359 kg
    1 slug = 32.174 lbm = 14.594 kg
    1 kg = 1000 g = 2.2046 lbm = 0.068521 slugs

Force
    1 lbf = 32.174 poundals = 4.450 N
    1 poundal = 0.03108 lbf = 0.1383 N
    1 newton = 10^5 dynes = 0.2247 lbf = 7.23 poundals

Work and energy
    1 joule (J) = 1 newton-meter (N-m)
              = 1 kg-m^2/s^2 = 10^7 ergs
    1 ft-lb = 1.356 J

Power
    1 watt (W) = 1 J/s = 1 N $-$ m/s = 1 kg-m^2/s^3
    1 hp = 550 ft-lb/sec = 746 W

Table A3 Properties of Plane Arcs and Areas

Centroids and Second Moments of Plane Areas

Rectangle

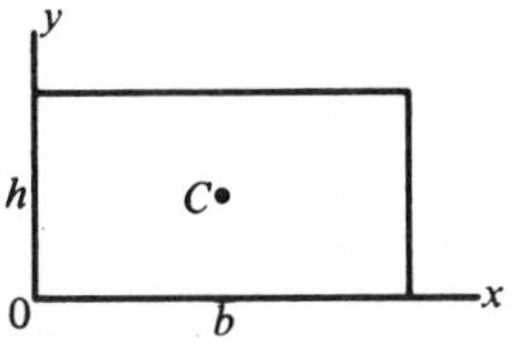

$I_x = \frac{1}{3} bh^3 \qquad I_y = \frac{1}{3} hb^3 \qquad \bar{x} = \frac{1}{2} b$

$\bar{I}_x = \frac{1}{12} bh^3 \qquad \bar{I}_y = \frac{1}{12} hb^3 \qquad \bar{y} = \frac{1}{2} h$

$P_{xy} = \frac{1}{4} b^2 h^2 \qquad A = bh$

Triangle

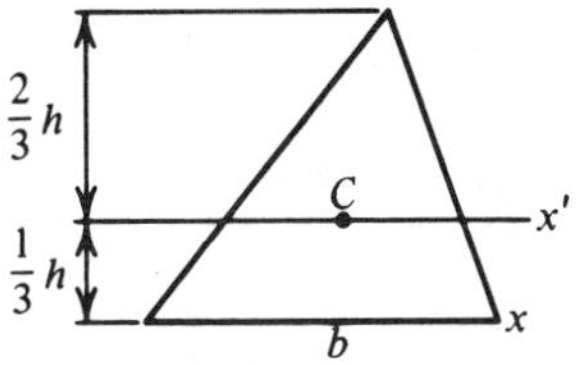

$\bar{I}_x = \frac{1}{36} bh^3 \qquad \bar{y} = \frac{1}{3} h$

$I_x = \frac{1}{12} bh^3 \qquad A = \frac{1}{2} bh$

Quarter Circle Area

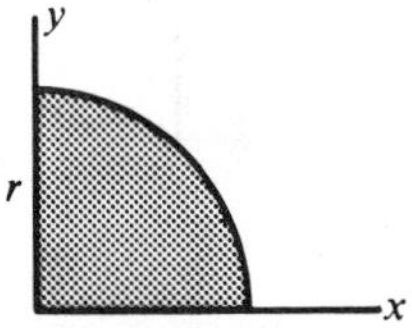

$I_x = I_y = \frac{1}{16} \pi r^4 \qquad \bar{x} = \bar{y} = \frac{4r}{3\pi}$

$P_{xy} = \frac{1}{8} r^4 \qquad A = \frac{1}{4} \pi r^2$

Semicircle

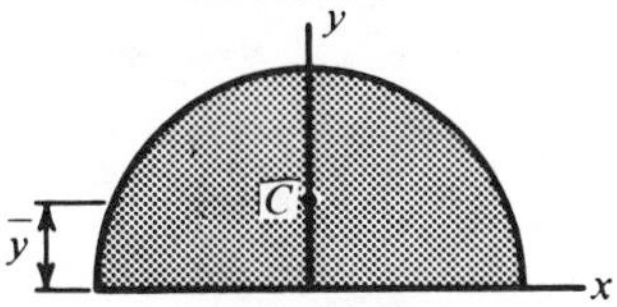

$I_x = I_y = \frac{1}{8} \pi r^4 \qquad \bar{y} = \frac{4r}{3\pi}$

$P_{xy} = 0 \qquad A = \frac{1}{2} \pi r^2$

Circular Sector Area

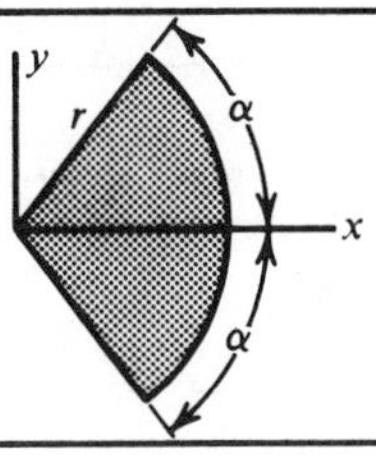

$I_x = \frac{1}{4} r^4 \left(\alpha - \frac{1}{2} \sin 2\alpha \right) \qquad \bar{x} = \frac{2}{3} r \frac{\sin \alpha}{\alpha}$

$I_y = \frac{1}{4} r^4 \left(\alpha + \frac{1}{2} \sin 2\alpha \right) \qquad A = r^2 \alpha$

Table A3   (continued)

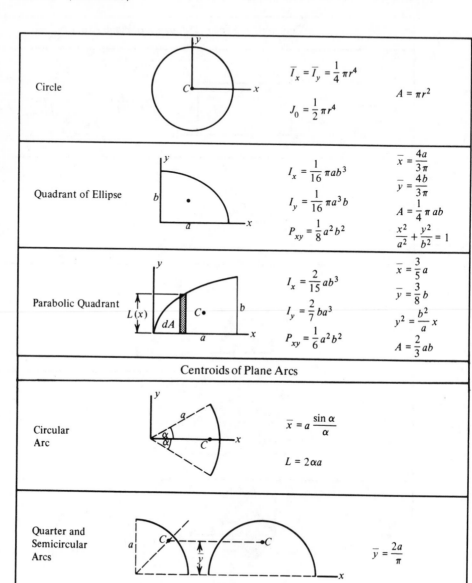

Circle	$\overline{I}_x = \overline{I}_y = \frac{1}{4}\pi r^4$ $J_0 = \frac{1}{2}\pi r^4$	$A = \pi r^2$
Quadrant of Ellipse	$I_x = \frac{1}{16}\pi ab^3$ $I_y = \frac{1}{16}\pi a^3 b$ $P_{xy} = \frac{1}{8}a^2 b^2$	$\overline{x} = \frac{4a}{3\pi}$ $\overline{y} = \frac{4b}{3\pi}$ $A = \frac{1}{4}\pi ab$ $\frac{x^2}{a^2} + \frac{y^2}{b^2} = 1$
Parabolic Quadrant	$I_x = \frac{2}{15}ab^3$ $I_y = \frac{2}{7}ba^3$ $P_{xy} = \frac{1}{6}a^2 b^2$	$\overline{x} = \frac{3}{5}a$ $\overline{y} = \frac{3}{8}b$ $y^2 = \frac{b^2}{a}x$ $A = \frac{2}{3}ab$

Centroids of Plane Arcs

Circular Arc	$\overline{x} = a\,\frac{\sin\alpha}{\alpha}$ $L = 2\alpha a$
Quarter and Semicircular Arcs	$\overline{y} = \frac{2a}{\pi}$

Table A4   Moments of Inertia and Center of Mass for Common Geometric Shapes of Uniform Density

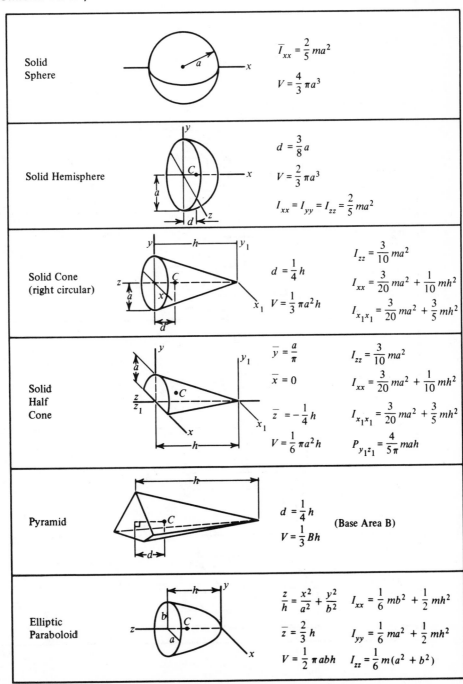

| Solid Sphere | $\bar{I}_{xx} = \frac{2}{5} ma^2$

 $V = \frac{4}{3}\pi a^3$ |

Solid Hemisphere
$d = \frac{3}{8} a$
$V = \frac{2}{3}\pi a^3$
$I_{xx} = I_{yy} = I_{zz} = \frac{2}{5} ma^2$

Solid Cone (right circular)
$d = \frac{1}{4} h$
$V = \frac{1}{3}\pi a^2 h$
$I_{zz} = \frac{3}{10} ma^2$
$I_{xx} = \frac{3}{20} ma^2 + \frac{1}{10} mh^2$
$I_{x_1 x_1} = \frac{3}{20} ma^2 + \frac{3}{5} mh^2$

Solid Half Cone
$\bar{y} = \frac{a}{\pi}$
$\bar{x} = 0$
$\bar{z} = -\frac{1}{4} h$
$V = \frac{1}{6}\pi a^2 h$
$I_{zz} = \frac{3}{10} ma^2$
$I_{xx} = \frac{3}{20} ma^2 + \frac{1}{10} mh^2$
$I_{x_1 x_1} = \frac{3}{20} ma^2 + \frac{3}{5} mh^2$
$P_{y_1 z_1} = \frac{4}{5\pi} mah$

Pyramid
$d = \frac{1}{4} h$ (Base Area B)
$V = \frac{1}{3} Bh$

Elliptic Paraboloid
$\frac{z}{h} = \frac{x^2}{a^2} + \frac{y^2}{b^2}$ $I_{xx} = \frac{1}{6} mb^2 + \frac{1}{2} mh^2$
$\bar{z} = \frac{2}{3} h$ $I_{yy} = \frac{1}{6} ma^2 + \frac{1}{2} mh^2$
$V = \frac{1}{2}\pi abh$ $I_{zz} = \frac{1}{6} m(a^2 + b^2)$

Table A4 (continued)

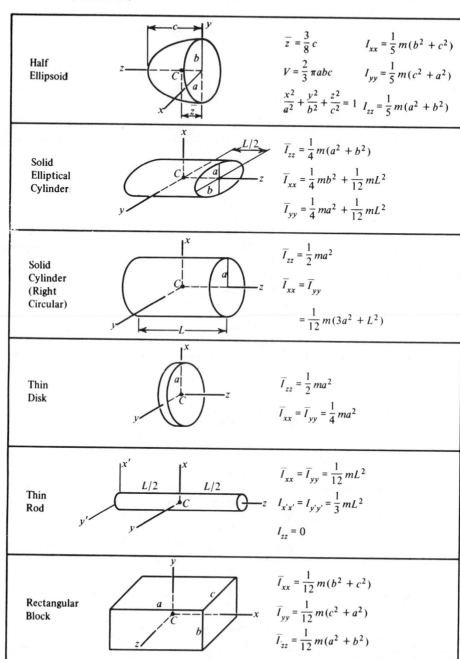

Half Ellipsoid

$$\bar{z} = \frac{3}{8}c \qquad I_{xx} = \frac{1}{5}m(b^2 + c^2)$$

$$V = \frac{2}{3}\pi abc \qquad I_{yy} = \frac{1}{5}m(c^2 + a^2)$$

$$\frac{x^2}{a^2} + \frac{y^2}{b^2} + \frac{z^2}{c^2} = 1 \quad I_{zz} = \frac{1}{5}m(a^2 + b^2)$$

Solid Elliptical Cylinder

$$\bar{I}_{zz} = \frac{1}{4}m(a^2 + b^2)$$

$$\bar{I}_{xx} = \frac{1}{4}mb^2 + \frac{1}{12}mL^2$$

$$\bar{I}_{yy} = \frac{1}{4}ma^2 + \frac{1}{12}mL^2$$

Solid Cylinder (Right Circular)

$$\bar{I}_{zz} = \frac{1}{2}ma^2$$

$$\bar{I}_{xx} = \bar{I}_{yy}$$

$$= \frac{1}{12}m(3a^2 + L^2)$$

Thin Disk

$$\bar{I}_{zz} = \frac{1}{2}ma^2$$

$$\bar{I}_{xx} = \bar{I}_{yy} = \frac{1}{4}ma^2$$

Thin Rod

$$\bar{I}_{xx} = \bar{I}_{yy} = \frac{1}{12}mL^2$$

$$I_{x'x'} = I_{y'y'} = \frac{1}{3}mL^2$$

$$I_{zz} = 0$$

Rectangular Block

$$\bar{I}_{xx} = \frac{1}{12}m(b^2 + c^2)$$

$$\bar{I}_{yy} = \frac{1}{12}m(c^2 + a^2)$$

$$\bar{I}_{zz} = \frac{1}{12}m(a^2 + b^2)$$

Table A4   (continued)

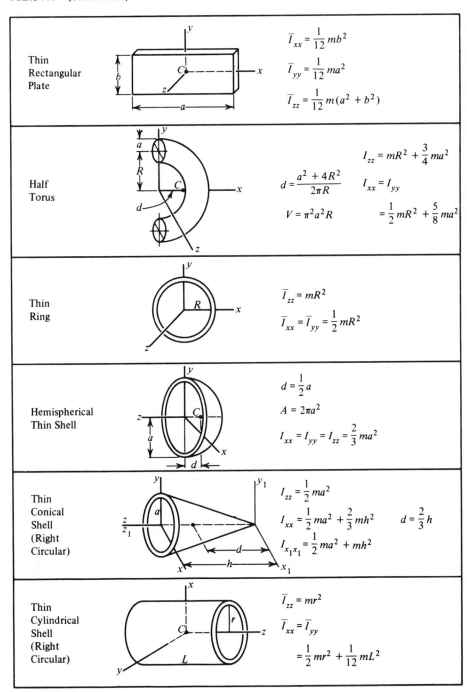

Thin Rectangular Plate

$$\bar{I}_{xx} = \frac{1}{12} mb^2$$

$$\bar{I}_{yy} = \frac{1}{12} ma^2$$

$$\bar{I}_{zz} = \frac{1}{12} m(a^2 + b^2)$$

Half Torus

$$I_{zz} = mR^2 + \frac{3}{4} ma^2$$

$$d = \frac{a^2 + 4R^2}{2\pi R}$$

$$I_{xx} = I_{yy}$$

$$V = \pi^2 a^2 R \qquad = \frac{1}{2} mR^2 + \frac{5}{8} ma^2$$

Thin Ring

$$\bar{I}_{zz} = mR^2$$

$$\bar{I}_{xx} = \bar{I}_{yy} = \frac{1}{2} mR^2$$

Hemispherical Thin Shell

$$d = \frac{1}{2} a$$

$$A = 2\pi a^2$$

$$I_{xx} = I_{yy} = I_{zz} = \frac{2}{3} ma^2$$

Thin Conical Shell (Right Circular)

$$I_{zz} = \frac{1}{2} ma^2$$

$$I_{xx} = \frac{1}{2} ma^2 + \frac{2}{3} mh^2 \qquad d = \frac{2}{3} h$$

$$I_{x_1 x_1} = \frac{1}{2} ma^2 + mh^2$$

Thin Cylindrical Shell (Right Circular)

$$\bar{I}_{zz} = mr^2$$

$$\bar{I}_{xx} = \bar{I}_{yy}$$

$$= \frac{1}{2} mr^2 + \frac{1}{12} mL^2$$

APPENDIX B

Answers to Exercises

CHAPTER 1

SECTION 1.2

1. 330 N $\measuredangle$ 46.5°.
2. 73.2 lb at +60°, 51.7 lb at −15°.
3. 47.1 lb $\measuredangle$ 45.9°.
4. $\alpha = 44.3°$, $\beta = 34.1°$.
5. (b) $\sum f_y = 0.3420A + 0.9848B - 0.8660C + 0.2488D$.
6. 340.3° for min., 160.3° for max.
7. (a) 940 N, (b) −866 N.
8. (b) $\mathbf{f}_1 = 200\mathbf{i} - 200\mathbf{j} - 100\mathbf{k}$ N.
9. (c) $\mathbf{f}_4 = 0.707f_4\mathbf{i} - 0.707f_4\mathbf{j}$.
10. (a) $\mathbf{f}^R = 330.9\mathbf{i} - 241.4\mathbf{j} - 194.8\mathbf{k}$ N, (b) $\hat{\mathbf{e}}_{R'} = 0.730\mathbf{i} - 0.533\mathbf{j} - 0.429\mathbf{k}$.
11. $\hat{\mathbf{e}}_a = 0.337\mathbf{i} - 0.421\mathbf{j} + 0.842\mathbf{k}$.
12. $-\hat{\mathbf{e}}_b = -0.706\mathbf{i} - 0.528\mathbf{j} - 0.471\mathbf{k}$.
13. $\mathbf{f} = -4.1\mathbf{i} + 20.4\mathbf{j} + 97.8\mathbf{k}$ N.
14. $\mathbf{f} = 700\mathbf{i} - 200\mathbf{j} + 685\mathbf{k}$ N.

SECTION 1.3

1. $m = (a_0/a)m_0$.
2. 58.9 lb $\measuredangle$ 81.9°.
3. (a) 331 lbf, (b) 2000 lbm, (c) 62.1 slugs.
4. $M = 7.35 \times 10^{22}$ kg.
5. (b) 400 mi.
6. (a) $GM = 3.978 \times 10^{14}$ m^3/sec^2, (b) $M = 5.96 \times 10^{14}$ kg.
7. (a) $GM = 1.408 \times 10^{16}$ ft^3/sec^2, (b) $M = 4.09 \times 10^{23}$ slugs.
8. $M_E/M_M = 81.3$.
9. 26.6×10^{-9} N, 5.98×10^{-9} lb.
10. Density = 62.4 lbm/ft^3 or 1.939 slugs/ft^3. Specific weight = 58.2 lbf/ft.^{3}
11. 101.4 kN/m^2.
12. 207,000 MN/m^2.

CHAPTER 2

SECTION 2.2
1. $T_{CA} = 87.9$ lb, $T_{CB} = 65.3$ lb.
2. $T_{BA} = 26.8$ lb, $T_{BC} = 30.0$ lb.
3. $T_{BC} = 1556$ lb, $f_A = 1192$ lb $\rightarrow$.
4. $f_A = 52.8$ lb, at $0.9°\triangle$.
5. $f_A = f_{AB} = 528$ lb at $\measuredangle\ 60°$.
6. $f_3 = 306$ lb, $f_4 = 543$ lb.
7. $f_1 = 2000$ lb, $f_3 = 866$ lb.
8. (b) $T_{CA} = 5760$ lb, $T_{AB} = 5670$ lb.
9. $\tan \theta = (500 - 0.866 T_{AC})/(T_{AB} + 0.5 T_{AC})$.
   $P^2 = (T_{AB} + 0.5 T_{AC})^2 + (500 - 0.866 T_{AC})^2$.
10. $T_{CE} = 1400$ lb.
11. 20 ft.
12. $P = 0.508 W$, $f_{AB} = 1.155 W$ compression,
    $f_{BC} = 0.577 W$ tension, $f_{AC} = 0.377\ W$
    compression.

SECTION 2.3
1. (a) $F_f = P - W \sin \alpha$ acts downhill on block,
   (b) $P_{max} = W(\mu_s \cos \alpha + \sin \alpha)$.
2. $P = W(\sin \alpha + \mu_s \cos \alpha)/(\cos \beta + \mu_s \sin \beta)$.
3. (a) 632 lb, (b) 506 lb, (c) 63.2 lb, (d) 79.0 lb.
4. 43 lb.
5. No, $F_A = 63.5$ lb.
6. (a) $P = 50.0 \cot(19.3° + \alpha) - 73.5$,
   (b) $\alpha = 14.9°$.
7. $P = W[2\mu \cos \alpha + (1 - \mu^2) \sin \alpha]/(\cos \alpha - \mu$
   $\sin \alpha)$.
8. $\alpha = \tan^{-1}[2\mu/(1 - \mu^2)]$.
9. $F = f_{BC} = 57.7$ lb, $f_{AB} = 115.4$ lb.
10. 323 lb.

SECTION 2.4
1. $\mathbf{f}_A = 600\mathbf{i} + 600\mathbf{j} + 300\mathbf{k}$ lb, $\mathbf{f}_B = 800\mathbf{j} -$
   $600\mathbf{k}$ lb, $\mathbf{f}_C = -600\mathbf{i} + 600\mathbf{j} + 300\mathbf{k}$ lb.
2. $T_{OA} = 850$ lb, $T_{OB} = 1950$ lb, $T_{OC} = 1400$ lb.
3. $f_{DA} = 21,000$ N, $f_{DB} = 12,000$ N
   (compression), $f_{DC} = 18,000$ N (tension).
4. $T_{AB} = 600$ lb, $T_{AC} = 450$ lb.
5. $T_{CA} = 1000$ lb, $T_{CB} = 848$ lb, $\mathbf{f}_o = 800\mathbf{i} +$
   $600\mathbf{k}$ lb.
6. $T_{DA} = 90$ lb, $T_{DB} = 120$ lb, magnitude $f_{CD} =$
   250 lb.
7. $T_{DC} = 250$ lb, $f_A = 210$ lb, $f_B = 160$ lb.
8. $\mathbf{f} = \mathbf{i} + 3.99\mathbf{j} - 3.99\mathbf{k}$ lb.

9. $T = 10.9$ lb.
10. (a) $T = W(\mu_k^2 \cos^2 \theta - \sin^2 \theta)^{1/2}$, (b) $\tan \beta =$
    $\sin \theta/(\mu_k^2 \cos^2 \theta - \sin^2 \theta)^{1/2}$.

CHAPTER 3

SECTION 3.2
1. $\mathbf{a} \cdot \mathbf{b} = 80$.
2. $\mathbf{u} \cdot \mathbf{v} = 5$.
3. (a) $+66$, (b) -66.
4. $AOB = 69.8°$.
5. $f_a = 59.7$ lb.
6. $\mathbf{a}(\mathbf{b} \cdot \mathbf{c}) = 100\mathbf{i} - 125\mathbf{j} - 75\mathbf{k}$.
7. $f_{AB} = 25$ lb.
8. $67.4°$.
9. $f_{AB} = -f(a^2 + c^2)^{1/2}/(a^2 + b^2 + c^2)^{1/2}$.
10. $\cos(\mathbf{f}, AB) = -(a^2 + c^2)^{1/2}/(a^2 + b^2 + c^2)^{1/2}$.
11. $f_{AC} = -83.3$ lb.
12. $BAC = 103.8°$.
13. $\mathbf{a} \perp \mathbf{b}$.
14. $x = 18$.
15. $a'_x = 18.66$, $a'_y = 12.32$.
17. $x + 4y - 2z + 10 = 0$.
18. $f_n = -700$ N.

SECTION 3.3
1. $\mathbf{a} \times \mathbf{b} = 30\mathbf{i} - 40\mathbf{k}$.
2. $\mathbf{u} \times \mathbf{v} = -9\mathbf{i} - \mathbf{j} - 7\mathbf{k}$.
3. (a) $-4\mathbf{j} + 8\mathbf{k}$, (b) $15\mathbf{i} - 16\mathbf{k}$, (c) $-10\mathbf{i} + 6\mathbf{j}$,
   (d) $-\frac{3}{5}bf\mathbf{i} - \frac{4}{5}bf\mathbf{j} + \frac{3}{5}af\mathbf{k}$.
4. 27.2 in.^{2}
5. $\hat{\mathbf{n}} = \pm(0.0368\mathbf{i} - 0.589\mathbf{j} + 0.810\mathbf{k})$.
6. $\hat{\mathbf{e}} = \pm(0.802\mathbf{i} + 0.535\mathbf{j} + 0.267\mathbf{k})$.
7. (a) $\mathbf{m}_A = 1800\mathbf{i} - 3000\mathbf{j} + 6600\mathbf{k}$ N-m,
   (b) $m_{OA} = 1400$ N-m.
8. (a) $\mathbf{m}_O = -10\mathbf{i} + 120\mathbf{j} + 90\mathbf{k}$ lb-ft,
   (b) $m_{OB} = 93.3$ lb-ft.
9. $m_{CD} = -144$ lb-ft.
10. (a) $\mathbf{m}_D = (10^3)\sqrt{2}\mathbf{i}$ lb-ft, (b) $m_{CD} = 471.3$ lb-ft.
11. $m_{AB} = -288$ lb-ft.
12. $m_{AB} = abcf/[(a^2 + c^2)(a^2 + b^2 + c^2)]^{1/2}$.
13. $m_{OA} = 993$ lb-ft.
14. $m_A = -640$ lb-in.
15. On line parallel to force shown, intersecting
    AB 6.4 in. above A.
16. $(24, 2, 0)$ ft.
17. $(2, 0, 0)$ m.

SECTION 3.4

1. $c = 12i + 6j - 6k$ lb-ft.
2. $c = 70i - 100j + 125k$ lb-ft.
3. $c = 57.7i + 57.7j + 57.7k$ N-m, $-c$.
4. $c = 150i + 200j$ lb-ft, $-c$.
5. $m_{OP} = -12$ lb-ft.
6. $m_O = 360i + 60j + 280k$ N-m.
7. $m_A = -40i + 70j + 420k$ lb-ft.
8. $m_{OA} = 297$ N-m.
9. $c = 100i + 150k$ lb-ft.
10. $c = fL \sin \theta$ ⤻.
11. f moved to intersect AB 28.3 in. to left of B.
12. (a) 100 lb down, 350 lb-in. ⤸.
13. (a) f, (b) $x = -2aP/(f \sin \theta)$.
14. $m_{OD} = 14$ lb-ft.
15. $c = -48i - 32j + 24k$ lb-ft.
16. $f = 60i - 10j + 40.7k$ lb.
    $c = -172.1i + 80j + 149.3k$ lb-ft.
17. $(6, 1, 0)$ ft.
19. $c = J_O G \, d\phi/dL$, where $J_O = \frac{1}{2}\pi(b^4 - a^4)$.
20. $c = ND\mu_k/3$.
21. $c = \frac{2}{3}N\mu_k(b^3 - a^3)/(b^2 - a^2)$.
23. f at $(5, -8, 0)$ m.
24. f at $(-1, 16, 0)$ in.
25. $(\frac{1}{3}a, -\frac{1}{2}a, 0)$ ft.

SECTION 3.5

1. $a \times (b \times c) = -8i - 106j - 28k$.
2. $(a \times b) \times c = -14i - 14j - 8k$.
4. $a \times b \cdot c =$ (a) 14, (b) 0.
5. $u \cdot v \times w =$ (a) -14, (b) 0.

CHAPTER 4

SECTION 4.1

1. (a) $f^R = 40i + 70j + 80k$ lb,
   $c_A^R = -270i + 440j + 250k$ lb-ft,
   (b) $m_{AD} = 480$ lb-ft.
2. $f_B^R = f_A^R$, $c_B^R = -240i + 880j - 150k$ lb-ft.
3. (a) $f^R = 100i + 50j - 80k$ lb,
   $c_A^R = -520i + 160j - 450k$ lb-ft,
   (b) $m_{AB} = 440$ lb-ft.
4. (a) $f^R = 400$ lb ↓, $c_A^R = 2800$ lb-ft ⤻,
   (b) $c_B^R = 2800$ lb-ft ⤸,
   (c) 400 lb ↓ at $x_R = 7$ ft.
5. $f^R = 0$, $c_A^R = c_B^R = 400$ lb-ft ⤻ .

6. (a) $f^R = -300i + 400j$ lb, $c_A^R = 1200$ lb-ft ⤻ ,
   (b) midpoint AB.
7. 100 N ↓ at $(11, 0, -4)$ m.
8. $f^R = 100i - 50j + 100k$ lb,
   $c_A^R = 100i + 140j + 120k$ lb-ft.
9. (a) $f^R = -30i + 40j - 20k$ N, $c_A^R = 160i + 180j + 40k$ N-m, (b) $m_{AB} = 64$ N-m.
10. (a) $f^R = 10i - 10j + 20k$ lb,
    $c_A^R = -100j + 50k$ lb-ft, $m_{OA} = -35.35$ lb-ft.
11. $f^R = 75.5$ lb, at $\theta_x = -148°$, intersects AB at $x = 12$ in.
12. $f^R = 0$, $c^R = -500i - 1000k$ N $- m$.
13. Wrench: $f^R = f$, $c^R = (392i + 224j - 56k)/33$ N-m acts on common line of $y + 4z = 70/33$, $x + 7z = -158/33$, $4x - 7y = -142/33$.
14. 23,000 lb ↓ at $(7.30, 15.83, 0)$ ft.

SECTION 4.2

1. $f^R = 2400$ kN ↓ at $x_R = 7$ m.
2. $f^R = \frac{1}{3}kL$ ↓ at $x_R = \frac{3}{4}L$.
3. $f^R = \frac{2}{3}w_0L$ at $x_R = \frac{3}{8}L$.
4. $x_R = 6.875$ ft.
5. $x_R = 8.2$ ft.

SECTION 4.3

1. $A_x = 86.6$ lb →, $A_y = 50$ lb ↑, $D = 100$ lb ⤢30°.
2. $F_A = \frac{3}{8}W$.
3. $A = 75$ lb →, $B = 125$ lb ∥ DB.
4. $T = 1312$ lb, $A_x = 1135$ lb →, $A_y = 444$ lb ↑.
5. $T = 50$ N.
6. $\mu_s = 0.24$.
7. $A_x = P$ →, $A_y = wL$ ↑, $c_A = \frac{1}{2}wL^2$ ⤻ .
8. $A_x = P$ →, $A_y = \frac{1}{2}w_BL$ ↑, $c_A = \frac{1}{3}w_BL^2$ ⤸.
9. $T_{AD} = 5380$ N, $B_x = 2690$ N →, $B_y = 9570$ N ↑.
10. $T = 0.217W$, $N_D = 1.217W$ ↑.
11. Slips, $P_{max} = 160.8$ lb.
12. (a) $A_x = \frac{5}{6}W$ →, $A_y = W$ ↑, $B = \frac{5}{6}W$ ←,
    (b) $T_D = 25W/24$, $A_x = \frac{5}{8}W$ →, $A_y = 11W/6$ ↑, (c) $A_x = 0$, $A_y = W$ ↑, $c_A = 5W$ ⤸.
13. (a) Completely constrained, determinate, $A_x = -4P/3$, $A_y = P$, $B = 4P/3$ →, (b) improperly constrained, indeterminate, (c)

partially constrained, determinate, $A = C = \frac{1}{2}P \uparrow$, (d) completely constrained, indeterminate, $A_x = -4P/3$, $B_x = 4P/3$.

14. (a) 202 lb, (b) 622 lb.

15. (a) $P = 2830$ N, (b) $h_{max} = 0.173$ m.

16. $A = 12$ kN $\uparrow$, $B = 18$ kN $\uparrow$.

17. $A_x = 3000$ lb $\leftarrow$, $A_y = 9170$ lb $\uparrow$, $B = 9830$ lb $\uparrow$.

18. $A = 30$ kN $\uparrow$, $c_A = 108$ kN-m $\searrow$.

19. $A_x = 3000$ lb $\leftarrow$, $A_y = 19{,}000$ lb $\uparrow$, $c_A = 118{,}000$ lb-ft $\searrow$.

20. $N_A = 869$ N, $N_B = 1073$ N, $F = 273$ N.

21. Yes, (a) $9.9°$, (b) $13.2°$.

22. $N_A = 0.812W$, $N_B = 0.1278W$, $F_A + F_B = 0.342W$.

23. (a) $\mu_s < 0.5$, (b) $\mu_s > 0.5$.

24. $A_x = 15$ lb, $A_y = -37$ lb, $B_x = 0$, $B_y = 63$ lb.

25. Front 5030 N/m, rear 2470 N/m, drag truss 1011 N/m (forward).

SECTION 4.4

1. $A_x = -45$ lb, $A_y = 60$ lb, $B_x = 45$ lb, $B_y = 60$ lb, $N_D = 90$ lb.

2. $A_x = -22.5$ lb, $A_y = 60$ lb, $B_x = 22.5$ lb, $B_y = 60$ lb, roller $N = 90$ lb.

3. $T_{CD} = 788$ lb, $T_{CE} = 1050$ lb, $A_x = 1138$ lb, $A_y = 444$ lb, $A_z = 0$.

4. $T_{CD} = 620$ lb, $T_{CE} = 908$ lb, $A_x = 888$ lb, $A_y = 212$ lb, $A_z = 0$.

5. $\mathbf{f}_A = 450\mathbf{j} - 600\mathbf{k}$ lb, $\mathbf{f}_B = 1350\mathbf{j} + 1800\mathbf{k}$ lb, $\mathbf{f}_D = -600\mathbf{j} - 400\mathbf{k}$ lb.

6. $\mathbf{f}_A = 24\mathbf{i} + 54\mathbf{j} + 48\mathbf{k}$ lb, $\mathbf{c}_A = -93\mathbf{i} + 264\mathbf{j} + 24\mathbf{k}$ lb-ft.

7. $\mathbf{f}_A = 153\mathbf{i} + 100\mathbf{j} + 125\mathbf{k}$ lb, $\mathbf{c}_A = 270\mathbf{i} - 1670\mathbf{j} + 382\mathbf{k}$ lb-ft.

8. $A_x = 15.3$ lb, $A_y = 140$ lb, $A_z = 15.3$ lb, $B_x = B_z = -15.3$ lb.

9. $A_x = 165.5$ lb, $A_y = 691$ lb, $A_z = 0$, $T_F = 276$ lb, $T_G = 125$ lb, $T_H = 75$ lb.

10. $\mathbf{f}_A = 270\mathbf{j}$ lb, $\mathbf{c}_A = 200\mathbf{i} + 825\mathbf{k}$ lb-ft.

11. $\mathbf{f}_A = -200\mathbf{i} + 300\mathbf{j} - 100\mathbf{k}$ lb, $\mathbf{c}_A = -420\mathbf{i} - 700\mathbf{j} + 160\mathbf{k}$ lb-in.

12. $R_x = 1150$ lb, $R_y = -428$ lb, $R_z = 1150$ lb, $f_C = 1435$ lb, $F_D = 1282$ lb.

13. $R_x = R_z = 761$ lb, $R_y = 714$ lb, $f_C = 631$ lb, $f_D = 612$ lb.

14. $A_x = 174.5$ lb, $A_y = 151$ lb, $B_x = 58$ lb, $B_y = 83.6$ lb.

15. $\mathbf{f}_B = 233\mathbf{i} + 235\mathbf{j}$ lb, $\mathbf{c}_B = -2415\mathbf{i} + 2800\mathbf{j}$ lb-in.

16. $T_{EF} = 1526$ N, $A_x + B_x = 654$ N, $A_y = 164$ N, $A_z = 436$ N, $B_y = 490$ N, $B_z = 0$.

17. $T_{DF} = 1722$ N, $A_x + B_x = 1309$ N, $A_y = 981$ N, $A_z = 544$ N, $B_y = B_z = 0$.

18. $A_y = B_y = 23$ lb, $A_z = -20$ lb, $B_x = -8$ lb, $C_x = 8$ lb, $C_y = 54$ lb, no.

19. (a) $A_x = \frac{1}{2}W \tan \theta \cos \phi$, $B_x = -A_x$, $A_z = \frac{1}{2}W \tan \theta \sin \phi$, $B_y = W$, $B_z = -Az$, (b) $\theta_{max} = \tan^{-1}(2\mu)$.

20. (a) $A_x = 0.265W$, $A_z = 0.0884W$, $B_x = -0.0884W$, $B_y = W$, $B_z = -0.265W$, (b) $\mu_{min} = 0.28$.

21. $\mathbf{f}_A = 100\mathbf{j} - 53.9\mathbf{k}$ lb, $\mathbf{c}_A = 68.8\mathbf{i} - 151.5\mathbf{j} + 119\mathbf{k}$ lb-in., $\mathbf{f}_C = 53.9\mathbf{k}$ lb.

22. $D = 40.8$ kN, $c_{Fz} = 2.91$ kN-m, $F_x = 18.26$ kN, $F_y = 11.75$ kN, $F_z = 62.3$ kN.

CHAPTER 5

SECTION 5.1

1. $A_x = 25$ lb $\rightarrow$, $A_y = 50$ lb $\uparrow$, $C_x = 25$ lb $\leftarrow$, $C_y = 100$ lb $\uparrow$.

3. $A_x = 0$, $A_y = 75$ lb $\uparrow$, $C_x \quad \therefore$, $C_y = 150$ lb $\downarrow$, $E = 75$ lb $\uparrow$.

5. $A_x = 1000$ N $\rightarrow$, $A_y = 5000$ N $\uparrow$, $C_x = 1000$ N $\leftarrow$, $C_y = 3000$ N $\uparrow$.

7. $B_x = (W/2b)[a + c \cot \theta + d \cos \theta] \rightarrow$, $B_y = W/2 \downarrow$, $C_x = (W/2b)[a + (b+c) \cot \theta + d \cos \theta] \leftarrow$, $C_y = 0$.

8. $A_x = 125$ lb $\rightarrow$, $A_y = 175$ lb $\uparrow$, $C_x = 200$ lb $\leftarrow$, $C_y = 100$ lb $\downarrow$, $E_x = 75$ lb $\rightarrow$, $E_y = 75$ lb $\downarrow$.

9. $\mu = 0.107$.

10. $A_x = 1.875$ kN $\leftarrow$, $A_y = 1.667$ kN $\uparrow$.

11. $A_x = 4.38$ kN $\leftarrow$, $A_y = 3.89$ kN $\uparrow$.

12. Magnitudes $A = 0.20W$, $C = 1.40W$.

13. $C = 833$ lb, $B_x = 437$ lb $\rightarrow$, $B_y = 583$ lb $\uparrow$.

14. $A_x = 2$ kN $\rightarrow$, $A_y = 0.5$ kN $\downarrow$, $B_x = 2$ kN $\leftarrow$, $B_y = 1.5$ kN $\downarrow$, $C_x = 0$, $C_y = 2$ kN $\uparrow$, $c = 6$ kN-m $\searrow$ at C.

15. $B_x = 600$ lb $\leftarrow$, $B_y = 300$ lb $\uparrow$, $C_x = 0$, $C_y = 400$ lb $\downarrow$, $D_x = 600$ lb $\rightarrow$, $D_y = 500$ lb $\uparrow$.

16. $M = aP \sin \theta[1 + (a \cos \theta)/(b^2 - a^2 \sin^2 \theta)^{1/2}]$.

17. $\theta_1 = 75°58'$, $\theta_2 = 255°58'$, $|A| = 0.103\,P$, $C_x = \frac{1}{4}P$, $C_y = P$, $D_x = \frac{3}{4}P$, $D_y = P$.

18. $H = \frac{1}{2}P \cot \theta$.

19. $M = 18P \sin \varphi \cos \varphi \cos(\theta - \varphi)/\sin \theta$ lb-in., where $\tan \varphi = 3 \sin \theta/(10 + 3 \cos \theta)$.

20. $A_y = B_y = 120$ lb., $D_x = 84.8$ lb, $D_y = D_z = 0$, $E_x = F_x = -42.2$ lb, $F_z = -E_z = 24.3$ lb.

21. $P = \frac{2}{3}W$, $f_{BC} = W/(3 \sin \theta)$, $D_x = \frac{1}{3}W \cot \theta$, $D_y = -W$.

22. $F = (L/b)P \sin \theta$, $\theta_{\max} = \tan^{-1} \mu$.

23. $A_x = 270$ lb, $A_y = 180$ lb, $A_z = 0$, $C_x = 0$, $C_y = 90$ lb, $C_z = 135$ lb.

24. $A_x = 540$ lb, $A_y = 180$ lb, $A_z = 0$, $C_x = 0$, $C_y = 90$ lb, $C_z = 270$ lb.

25. $D_x = E_x = 158.4$ lb, $D_y = E_y = 158.4$ lb, $D_z = -E_z = 25$ lb, $F_x = 183.3$ lb, $F_y = F_z = 0$.

SECTION 5.2

(*Note:* T or C after answer denotes tension or compression.)

1. $BD = AB = \frac{1}{2}P \cos \theta + \frac{1}{2}Q$, T; $AC = \frac{1}{2}P + \frac{1}{2}Q \tan \theta$, C; $BC = P$, T; $CD = (\frac{1}{2}P - \frac{1}{2}Q \tan \theta)/\sin \theta$, C.

2. $AH = 12.72^k$, C; $AB = 9^k$, T; $BH = 3^k$, C; $GH = 9^k$, C; $BG = 4.25^k$, T; $BC = 6^k$, T; $DF = 3^k$, T; $EF = 4.25^k$, C; $DE = 3^k$, T.

3. $CD = 10^k$, T; $DG = 10^k$, T.

4. $BC = 6^k$, T; $BG = 4.25^k$, T.

5. $AB = BC = CD = DE = 6^k$, T; $AH = EF = 7.5^k$, C; $BH = DF = 2^k$, T; $GH = FG = 5^k$, C; $CH = CF = 2.5^k$, C; $CG = 6^k$, T.

6. $DB = GE = EF = CE = 0$; $CG = GF = 400$ lb, C; $AD = CD = 800$ lb, C; $AB = 346.4$ lb, T; $BC = 692.8$ lb, T.

7. $BH = CG = FD = 0$; $AH = 1.46^k$, C; $AB = BC = CD = 1.168^k$, T; $DG = 2.4^k$, T; $DE = 2.5^k$, T; $EF = GF = 1.875^k$, C.

8. $AB = BC = CJ = EH = 0$; $DE = EF = 3.89^k$, T; $FG = GH = 3.33^k$, C; $CD = 3.33^k$, T; $DH = 1^k$, T; $CH = 1.3^k$, C; $AC = 3.9^k$, T; $AJ = JH = 2.5^k$, C.

9. $EF = ED = 0$; $DF = 424$ lb, C; $DC = 300$ lb, T; $AC = 894$ lb, C; $FC = 300$ lb, T; $AF = 300$ lb, C.

10. $CJ = DH = GE = 0$; $FG = AJ = 1200$ lb, T; $AC = CD = GH = HJ = 1697$ lb, T; $GD = JD = 1200$ lb, C.

11. $DJ = 0$; $DI = 3394$ lb, T; $KD = 2828$ lb, T; $CD = 5200$ lb, T.

12. $CG = 3^k$, T; $CF = 1.25^k$, C.

13. $AH = 1^k$, T; $JH = 8.75^k$, T.

14. $KJ = 6.07^k$, C; $KN = 2.02^k$, C

15. $JI = 8.42^k$, C; $IO = 3.47^k$, T.

SECTION 5.3

1. $AD = (2/3)^k$, T; $AE = (25/3)^k$, T; $AF = (\frac{2}{3}\sqrt{34})^k$, T; $AB = 2^k$, C; $BC = 0$, $BD = (10/3)^k$, T; $BF = (25/3)^k$, C.

2. $AD = BD = CD = 0.4082P$, C; $AB = BC = AC = 0.1361P$, T.

3. $AD = BD = CD = AE = BE = CE = 0.4082P$, T; $AB = BC = AC = 0.2722P$, C.

4. $EF = 715$ lb, T; $BF = CF = 429$ lb, C; $BE = CE = 509$ lb, C; $DE = AE = 274$ lb, T.

SECTION 5.4

1. $V = w(\frac{1}{2}L - x)$, $M = \frac{1}{2}wx(L - x)$, $M_{\max} = \frac{1}{8}wL^2$ at $x = \frac{1}{2}L$.

2. $V = -Q$, $M = Qx$.

3. $V = Q$, $M = Qx$ on $0 < x < \frac{1}{3}L$; $V = 0$, $M = -\frac{1}{3}QL$ on $\frac{1}{3}L < x < \frac{2}{3}L$; $V = -Q$, $M = Q(L - x)$ on $\frac{2}{3}L < x < L$.

4. $V = \frac{1}{2}(w_m/L)x^2$, $M = -(w_m/6L)x^3$.

5. $V = (w_m L/6) - (w_m/2L)x^2$, $M = (w_m L/6)x - (w_m/6L)x^3$, $M_{\max} = 0.385 w_m L^2$ at $x = 0.58L$.

6. $V = w_m[(x^2/2L) - x]$, $M = w_m[(x^3/6L) - \frac{1}{2}x^2]$, $M_{\max} = |M(L)| = \frac{1}{3}w_m L^2$.

7. $V = -(q_m L/\pi) \cos(\pi x/L)$, $M = -(q_m L^2/\pi^2) \sin(\pi x/L)$.

8. $V = -(L - a)Q/a$, $M = -(L - a)Qx/a$ on $0 < x < a$; $V = Q$, $M = -(L - x)Q$ for $a < x < L$; $M_{\max} = (L - a)Q = |M(a)|$.

9. $V = -wx - Q$, $M = -\frac{1}{2}wx^2 - Qx$ on $0 < x < \frac{1}{2}L$; $V = w(L - x) + Q$, $M = -\frac{1}{2}w(L - x)^2 -$

$Q(L-x)$ on $\frac{1}{2}L<x<1$; $M_{max}=|M(\frac{1}{2}L)|$
$=\frac{1}{8}wL^2+\frac{1}{2}QL$.

10. $V=6$kN, $M=6x-12$ kN-m on $0<x<$
1.5 m; $V=2$ kN, $M=2x-6$ kN-m on
1.5 m$<x<$ 4.5 m; $V=-2$kN,
$M=-2x+12$ kN-m on 4.5 m$<x<$6m.

11. $V=3600-300x$ lb, $M=-150x^2+3600x$
$-19,200$ lb-ft, $M_{max}=|M(0)|=19,200$ lb-ft.

13. $V=Q$, $M=Qx$ on $0<x<a$; $V=0$, $M=$
Qa on $a<x<L-a$; $V=-Q$, $M=$
$Q(L-x)$ on $L-a<x<L$.

14. $V=-2x$ k, $M=-x^2$ k-ft on $0<x<6$ ft;
$V=-12$ k, $M=36-12x$ k-ft on $6<x<$
9 ft; $V=-22$ k, $M=146-22x$ k-ft on $9<$
$x<18$ ft; $M_{max}=|M(18)|=250$ k-ft.

15. $V=10$ k, $M=10x-152$ k-ft on $0<x<8$ ft;
$V=10$ k, $M=10x-160$ k-ft on $8<x<$
16 ft; $M_{max}=152$ k-ft.

16. $V=-10$ k, $M=-10x$ k-ft on $0<x<5$ ft;
$V=15$ k, $M=15x-125$ k-ft on $5<x<$
10 ft; $V=-5$ k, $M=75-5x$ k-ft on $10<$
$x<15$ ft; $M_{max}=|M(5)|=50$ k-ft.

17. $V=8$ k, $M=8x$ k-ft on $0<x<6$ ft; $V=$
$20-2x$ k, $M=-x^2+20x-36$ k-ft; $M_{max}=$
$M(10)=64$ k-ft.

18. $V=16-4x$ k, $M=16x-2x^2$ k-ft on $0<x<$
4 ft; $V=0$, $M=32$ k-ft $=M_{max}$ on $4<x<$
11 ft; $V=44-4x$ k, $M=44x-2x^2-210$ k-
ft on $11<x<15$ ft.

19. $V=x^2-80x$ lb, $M=\frac{1}{3}x^3-40x^2$ lb-ft;
$M_{max}=13.33$ k-ft.

20. $V=-25x^2$ lb, $M=-25x^3/3$ lb-ft on $0<$
$x<8$ ft; $V=-1600$ lb, $M=-1600x+$
$(25,600/3)$ lb-ft on $8<x<12$ ft; $M_{max}=$
$|M(12)|=10,670$ lb-ft.

21. $V=880-40x$ lb, $M=-20x^2+880x-$
6480 lb-ft on $0<x<6$ ft; $V=680-40x$ lb,
$M=-20x^2+680x-5280$ lb-ft on $6<x<$
12 ft; $M_{max}=|M(0)|=6480$ lb-ft.

22. $V=1033$ lb, $M=1033x$ lb-ft on $0<x<$
4 ft; $V=-967$ lb, $M=8000-967x$ lb-ft on
$4<x<12$ ft; $V=200(18-x)$ lb, $M=$
$-100(18-x)^2$ on $12<x<18$ ft; $M_{max}=$
$M(4)=4132$ lb-ft.

23. $f_x=f_z=0$, $f_y=wR\theta$, $c_x=wR^2(\theta-\sin\theta)$,
$c_y=0$, $c_z=-wR^2(1-\cos\theta)$.

SECTION 5.5

1. $T_m=4540$ N.
2. $W=49.9$ lb.
3. 3%.
4. $T_{min}=9,000,000$ lb, $T_{max}=9,490,000$ lb.
5. 3050 ft.
6. 16.9 ft.
7. (a) 165 m, (b) 3780 N.
8. (a) 210 ft, 211w, (b) 215 ft, 186.4w,
   (c) 222 ft, 167.5w, (d) 236 ft, 153.9w.
10. (a) 6.22 m, (b) 243 N.
11. $T_{max}=2wx_m[1+(9x_m^2/16y_m^2)]^{1/2}$, $V=wx_m$,
    $H=\frac{3}{2}wx_m^2/y_m$.
12. (a) 1921 lb, (b) est. 1920 lb.

CHAPTER 6

SECTION 6.1

1. (a) $8k/3$ ft-lb, (b) $4k$ ft-lb, (c) $16k/3$ ft-lb,
   (d) $8k$ ft-lb, (e) $8k$ ft-lb.
2. $-2mgR$, same.
3. πkR^2.
4. $2\pi f$ N-m.

SECTION 6.2

2. $\theta=0$ (stable), $\theta=\pm60°$ (unstable).
3. $\theta=2$ radians (stable).
4. $\theta=0$ (stable) and $\theta=60°$ (unstable).
5. (a) $k=3.42W/b$ lb/ft, (b) yes.
6. (b) $W<\frac{1}{2}kb$.
7. Stable for $R>(\frac{3}{4}\pi-1)r$.
8. Stable for $h<R\sqrt{3}$.
10. 80.4°.

SECTION 6.3

1. $f_{AB}=W\cot\theta$ (compression).
4. $A_x=\frac{5}{2}P\cot\theta\rightarrow$, $B_x=\frac{5}{2}P\cot\theta\leftarrow$, $A_y=$
   $B_y=\frac{1}{2}P\uparrow$.
5. $\tan\theta_1=5W/2P$, $\tan\theta_2=3W/2P$, $\tan\theta_3=$
   $W/2P$.
6. $c=2aW\sin\theta$.

7. $c = 2860$ lb-in.
8. $f_{ED} = 2.89$ kN, compression.
9. $\sin \theta_2 = (L_3 - L_1)/L_2$, $\theta_1 = \theta_3 = 90°$.
10. $W = 2ka \tan \theta(\cos \theta - \cos \theta_0)$.

CHAPTER 7

Section 7.3

1. $\bar{y} = 1.1$ in.
2. $\bar{y} = 2.22$ in.
3. $\bar{y} = (\frac{2}{3}a^2 - b^2)/(\frac{1}{2}\pi a + 2b)$.
4. $\bar{y} = 4(b^3 - a^3)/3\pi(b^2 - a^2)$.
5. $\bar{x} = \bar{y} = 2.68$ in.
6. $\bar{x} = \frac{2}{3}a \sin^3 \alpha/(\alpha - \sin \alpha \cos \alpha)$.
7. $\bar{x} = 4(b - a)/3\pi$.
8. $\bar{y} = 2.29$ in.
9. $\bar{x} = (b^2 - 4a^2)/2(\pi a + b)$, $\bar{y} = \pi a^2/(\pi a + b)$.
10. $\bar{x} = 6.11$ in.
11. $y = [\frac{1}{3}h(L + \frac{1}{4}h) + \frac{1}{2}L^2]/(L + \frac{1}{3}h)$.
12. $\bar{y} = \frac{1}{4}h(b^4 - a^4)/(b - a)(b^3 - a^3)$.
13. $\bar{x} = 6.20$ in. from left end.
14. $\bar{x} = 0.259$, $\bar{y} = 0.778$, $\bar{z} = 2$ in.
15. $\bar{z} = 7h/9$.
16. $\bar{z} = [\frac{1}{2}h^2 + \frac{2}{3}c(h + \frac{3}{8}c)]/(h + \frac{2}{3}c)$.
17. $\bar{y} = -\frac{2}{3}(9b^2 - 2a^2)/(12b + \pi a)$.
18. $\bar{x} = \frac{1}{2}(3b^2 - 4a^2)/(\pi a + 3b)$, $\bar{y} = \pi a^2/(\pi a + 3b)$.
19. $\bar{x} = -0.132$, $\bar{y} = 0.730$, $\bar{z} = 2.00$ in.
20. $\bar{y} = -0.391$, $\bar{z} = 7.07$ in.
21. $\bar{x} = 22.9$, $\bar{y} = 12.85$ cm.

Section 7.4

3. $\bar{x} = \bar{y} = 0.45$.
4. $\bar{x} = \pi/2q$, $\bar{y} = \pi A/8$.
5. $\bar{x} = 1$, $\bar{y} = 1.6$ in.
6. $\bar{x} = 8.64$, $\bar{y} = 3.78$ in.
7. $\bar{x} = \frac{5}{8}a$.
8. $d = 8R/15$ from base.
9. $\bar{x} = 0.3a$, $\bar{y} = 0.75b$.

Section 7.5

2. $\bar{x} = \frac{5}{8}a$.
5. $\bar{z} = 0.574a$.
6. $\bar{x} = (3a\sqrt{2})/8$.

Section 7.6

2. (b) $\bar{I}_x = [(\pi/16) - (4/9\pi)]ab^3$, $\bar{I}_y = [(\pi/16) - (4/9\pi)]a^3b$, $\bar{P}_{xy} = [\frac{1}{8} - (4/9\pi)]a^2b^2$,
    (c) $\bar{I}_x = 19ab^3/480$, $\bar{I}_y = 8a^3b/175$, $\bar{P}_{xy} = a^2b^2/60$, (d) $I_{x_1} = \pi r^2[\frac{1}{4}r^2 + b^2]$, $I_{y_1} = \pi r^2[\frac{1}{4}r^2 + a^2]$, $P_{x_1y_1} = \pi r^2 ab$, (e) $I_{x_1} = \frac{1}{4}bh^3$.
3. $I_{x_1} = \frac{5}{4}\pi r^4$, $P_{x_1y_1} = \pi r^4$.
4. (a) $\bar{I}_x = \frac{1}{12}[bh^3 - (b - t)(h - 2a)^3]$, $\bar{I}_y = \frac{1}{6}ab^3 + \frac{1}{12}(h - 2a)t^3$, (b) $\bar{I}_x = 184$ in.^{4}, $\bar{I}_y = 10.75$ in.^{4}
5. $\bar{I}_x = 136$ in.^{4}, $\bar{I}_y = 64$ in.^{4}, $\bar{P}_{xy} = -48$ in.^{4}.
6. $\bar{I}_x = \frac{1}{6}[th^3 + (b - 2t)a^3 + 3a(b - 2t)(h - a)^2]$, $\bar{I}_y = \frac{1}{6}[ab^3 + (h - 2a)t^3 + 3t(h - 2a)(b - t)^2]$.
7. $\bar{I}_x = 92.5$ in.^{4}, $\bar{I}_y = 61.3$ in.^{4}
8. $\bar{I}_x = 40.8$ in.^{4}
9. $\bar{I}_y = 4.66$ in.^{4}
10. $\bar{P}_{xy} = 0$.
11. $\bar{I}_x = \frac{1}{12}[bh^3 - (b - t)(h - 2t)^3]$.
12. $\bar{I}_x = 10.38$ in.^{4}, $\bar{I}_y = 6.97$ in.^{4}, $\bar{P}_{xy} = -6.56$ in.^{4}
13. $P_{xy} = \frac{1}{4}t^2[b^2 + h^2 - t^2]$.
14. $\bar{I}_x = 762$ in.^{4}, $\bar{I}_y = 254$ in.^{4}

Section 7.7

1. $\theta_p = \pm 45°$ (or $45°$ and $135°$), $I_{min} = 0.0713r^4$, $I_{max} = 0.321r^4$.
2. $\theta_p = 30.5°$ and $120.5°$, $I_{max} = 23.3$ in.^{4}, $I_{min} = 2.7$ in.^{4}
3. $I_{x'} = 7.71$ in^4, $P_{x'y'} = -8.83$ in.^{4}
4. $I_{x'} = 0.881 r^4$, $P_{x'y'} = 0.625r^4$.
5. $\bar{I}_{x'} = 97.4$ in.^{4}, $\bar{I}_{y'} = 97.4$ in.^{4}, $\bar{P}_{x'y'} = 86.6$ in.^{4}
6. $\theta = \pm 45°$, $\bar{P}_{x'y'} = \pm 15.6$ in.^{4}
7. $\theta_p = 42.7°$ and $132.7°$, $I_{max} = 15.45$ in.^{4}, $I_{min} = 1.897$ in.^{4}
8. $\bar{I}_{x'} = 13.50$ in.^{4}, $\bar{P}_{x'y'} = 4.76$ in.^{4}

Section 7.8

2. 25.4 psi.
3. 61 in. Hg, 69.2 ft H_2O.
6. $p = 6.76$ psi, $\gamma = 0.0408$ lb/ft^3.
7. $f^R = 15,000$ lb, $s_p = 5.6$ ft.
8. $f^R = 14,100$ lb, $s_p = 11.5$ ft.
9. $f^R = 9890$ lb, $s_p = 12.27$ ft.
10. (a) 1.33 ft. (b) 0.12 ft.
11. (a) $H = 4700$ lb, $V = 1045$ lb $\uparrow$, (b) $H = 2690$ lb, $V = 2880$ lb $\uparrow$.

SECTION 7.9

3. $P_{xy} = 2mab/5\pi$, $P_{yz} = 2mbc/5\pi$, $P_{zx} = 2mac/5\pi$.

4. $I_{xx} = \frac{1}{3}m(b^2 + c^2)$, $I_{yy} = \frac{1}{3}m(a^2 + c^2)$, $I_{zz} = \frac{1}{3}m(a^2 + b^2)$, $P_{xy} = \frac{1}{4}mab$, $P_{yz} = \frac{1}{4}mbc$, $P_{zx} = \frac{1}{4}mac$.

6. $\bar{I}_{xx} = \bar{I}_{yy} = (3ma^2/20) + (3mh^2/80)$, $\bar{I}_{zz} = 3ma^2/10$.

7. $I_{xx} = m(\frac{1}{4}a^2 + \frac{1}{3}L^2) + m_2(0.15a^2 + 0.6h^2)$.

8. $153.9\,\rho a^5$.

9. $R \ge 6.29a$.

10. $I_{xx} = \frac{1}{4}m(4R^2 + 5r^2)$, $I_{yy} = m(R^2 + \frac{1}{4}r^2)$, $I_{zz} = \frac{3}{2}mr^2$, $P_{xy} = P_{yz} = 0$, $P_{zx} = rRm$.

11. $\bar{I}_{xx} = 0.0487\,ma^2 + 0.0375\,mh^2$, $\bar{I}_{yy} = 0.15ma^2 + 0.0375mh^2$, $\bar{I}_{zz} = 0.1987ma^2$, $\bar{P}_{yz} = 0.01592mah$, $\bar{P}_{xy} = \bar{P}_{xz} = 0$.

12. $\bar{I}_{yy} = \frac{1}{2}mR^2 + \frac{5}{8}ma^2 - m[(a^2 + 4R^2)/2\pi R]^2$, $\bar{I}_{xx} = \frac{1}{2}mR^2 - \frac{5}{8}ma^2$, $\bar{I}_{zz} = mR^2 + \frac{3}{4}ma^2 - $ $m[(a^2 + 4R^2)/2\pi R]^2$.

14. $I = 0.3m(b^5 - a^5)/(b^3 - a^3)$, $k = \sqrt{I/m}$.

15. (a) $I_{xx} = (m/20)[c^2 + 12h^2]$, $I_{yy} = (m/20)[a^2 + c^2]$, $I_{zz} = (m/20)[a^2 + 12h^2]$, (b) $\bar{I}_{xx} = (m/80)[4c^2 + 3h^2]$, $\bar{I}_{yy} = (m/20)[a^2 + c^2]$, $\bar{I}_{zz} = (m/80)[4a^2 + 3h^2]$.

16. $I_{xx} = 7.8$, $I_{yy} = 3.6$, $I_{zz} = 7.8$, $P_{xy} = -2.4$, $P_{yz} = 2.4$, $P_{zx} = -1.4$ kg-m^2.

17. $I_{xx} = 10.89$, $I_{yy} = 4.54$, $I_{zz} = 15.43$, $P_{xy} = -5.21$ slug-ft^2, $P_{yz} = P_{zx} = 0$.

18. $I_{xx} = \frac{1}{2}mR^2[1 - (\sin 2\alpha/2\alpha)]$, $I_{yy} = \frac{1}{2}mR^2[1 + (\sin 2\alpha/2\alpha)]$, $P_{xy} = \frac{1}{2}mR^2[(1 - \cos 2\alpha)/2\alpha]$, others zero. *Note:* $m = WR\alpha/g$

19. 0.191 kg-m^2.

20. 16.27 kg-m^2.

Index

Accelerated motion, 3
Acceleration, 3, 15
  external effect of force, 15
  of gravity (*see also* Gravity, acceleration of)
Accuracy, numerical, 27
Action and reaction, 19, 154–55
Active-force diagram, 234
Addition:
  of couples, 90
  of vectors, 4, 8
Admissible displacement, 230
d'Alembert, 119
Angle:
  of friction, 42
  between vectors, by scalar product, 63–64
Arc (*see* Centroid, of arc)
Arc length integral, 258–59
Area:
  product of inertia (*see* Product *and also* Second moments)
  properties of, Table A3, iii–iv
  second moments of (*see* Second moments)

Area, centroid of (*see* Centroid)
Area element:
  double integral, 263, 264
  polar coordinate, 257, 263, 264
  single integral, 255, 256, 257
Associative law of vector addition, 5
Atmosphere:
  isothermal, 285
  standard, 281, 285
Axes:
  coordinate, rotation of, 64–65, 276–77
  principal:
    of area, 277–78
    of inertia, 288

Ball-and-socket joint or support, 50, 51
Base vectors, 6
  unit, 7
    scalar products of, 62
    vector products of, 70
Basis, 6

Beam, 182–97
  cantilever, 186, 190–91, 193
  differential relationships, 187
  distributed load on, 110–13, 117–18, 130–31
  loading of, 110, 187, 193
  neutral plane of, 249
  simply-supported, 186, 189–90
  stress in, elastic, 249
  stress resultants in, 182–83
  supports, 186
  types, 186
Bearings:
  journal, 138
  thrust, 95, 137
Bending of beams, 184–97
Bending moment, 184
  diagram, 186–95
    discontinuity in, 188
    slope of, 187
  sign convention, 185
Bending stress in elastic beam, 249
Bernoulli, John, 119, 230
Body (*see* Deformable body, Rigid body)

Bridge, suspension, 197, 205
British system of units, 22, 24, i, ii

Cable:
  catenary, 198, 200–205
  massless, 31
  parabolic, 198–99
Calculator, 27
Cantilever beam, 183, 186, 190–91,
  193
Catenary cable, 198, 200–205
Center of gravity, 19, 116
Center of mass:
  at centroid, for uniform density,
    115, 246, 248
  of collection of particles, 113, 246
  of composite body, 246–47
  of continuous mass distribution,
    113, 115, 246
  of finite body, 246
  in plane of symmetry, 115
  Table A4, v–vii
Center of pressure, 281–82
Centroid:
  of arc, 248, 258–59
    Table A3, iv
  of area, 113, 118, 131, 248, 250–
    52, 255–68, 263–65
    application to elastic beams,
      249
    Table A3, iii–iv
  of load diagram, 113, 118, 131
  by multiple integrals, 262–68
  in plane of symmetry, 115
  by single integral, 255–61
  of triangle, 255–56
  of volume, 115, 248, 259–61,
    265–68
    Table A4, v–vii
Centroidal axes, for moment of inertia, 291
Circle, Mohr's, 277–78
Coefficient of friction, 42–43 (see
  also Friction)
Commutative law:
  of scalar product, 62
  of vector addition, 5
Complete constraint, 126
Components:
  of force, 11, 13, 65–66
  of inertia (rotational), 286, 288,
    294–95
    by single integrals, 294–95
  of moment, 76, 99
  negative, 63
  of vector product, 71
  vector (see Vector components)
  working, 212
Composite arc, 250
Composite area, 250, 270
Composite-body method:
  center of mass, 246–47
  centroid, 246–47, 250
  moments of inertia, 270, 289
  second moment of area, 270
Composite volume, 250, 289

Composition of forces or vectors (see
  Addition of vectors)
Compression, 76
Computer, 27
Concurrent forces, 29
  resultant of, 104
Configuration, 159
  equilibrium, 161, 162, 230
    by potential energy, 218, 223
    stability of (see Stability)
    by virtual displacements, 232–
      36
Connected bodies:
  equilibrium of, 34–35, 153–66
  interaction forces between, 155
Conservative force field (see Force,
  conservative)
Conservative system, 217
  equilibrium of, 218, 223
  stability of, 219–27
Constraint:
  complete and partial, 126–27, 139
  improper, 127–28, 139
  of mechanism, 230
    fixed-guide, 231
  workless, 231, 238
Continuous medium, 2
  fluid, 279–85
Continuum, 2
Conventional symbols for supports,
  120
Conversion factor, with nonconsistent units, 24
Conversion ratios for change of
  units, 24
Conversion of units, 24, ii
Coordinates, generalized, 159, 218,
  222, 231
Coplanar equilibrium equations,
  124–25
Coplanar forces, resultant of, 104,
  106
Cosines:
  direction, 7, 8
  law of, 10
Coulomb friction, 40–43
Couple, 82
  addition of, 88, 90
  combined with perpendicular
    force, 87–88, 89, 92
  equivalent couples, 84
  moment of, 83
  in plane, 84, 85, 89
  reaction often omitted, 121–23
  stress resultants in beams, 182–83
  transmitted by shaft, 86–87
  twisting, 184
  vector, free, 84
    symbol for, 84
  work of, 237–38
Cross product (see Vector product)

Dam gate, force on, 281–82
Datum for potential energy:
  of gravitational field, 214, 216
  of spring, 215

Deformable body, 2
  branch of mechanics, 3
Deformable continuous medium, 2
  fluid, 279–85
Deformations, internal effect of
  force, 15
Degrees of freedom, 159, 217, 231
Density:
  mass, 2, 113
  modeled as continuous, 2
  uniform, 115
Determinant:
  for scalar triple product, 97
  for vector product, 71
Determinate (see Statically determinate)
Diagram (see Active-force, Bend-
  ing-moment, Free-body,
  and Shear diagrams)
Differential relationships in beam,
  187
Dimensions:
  absolute MLT systems, 22
  gravitational FLT systems, 22
  of primitive elements of mech-
    anics, 3
    three independent, 3
Direction:
  force characteristic, 15
  of vector, 4, 7, 8, 12, 51
Direction cosines, 7, 8
Discontinuities in shear and moment
  diagrams, 188
Discriminant, 224
Displacements:
  kinematically admissible, 230
  small, from equilibrium, 220, 230
  virtual (see Virtual displacements)
Distributed force, 15, 103
Distributed load on beam, 110–13,
  117–18, 130–31, 187, 193
Distributive law:
  of scalar product, 61
  of vector product, 70
Dot product (see Scalar product)
Dynamics:
  of machines, 161
  branch of mechanics, 3
Dyne, i, ii

Earth:
  acceleration of gravity on, 20–21,
    26
  mass of, 26
  radius of, 21, 26
Effective-couple vector, 123
Effective-force system, 123
Effective-force vector, 123
Effects of force (see Force)
Elastic beam:
  application of centroid, 249
  bending stress in, 249
Elastic modulus of steel, 26
Elastic spring (see Spring)
Energy, units of, ii, (see also Poten-
  tial energy)

Equilibrium, 3, (see also Stability of equilibrium)
of concurrent forces, 29
configuration of, 161, 162, 218, 223, 230
by virtual displacements, 232–36
of connected bodies, 34–35, 153–66
of a machine, 161, 162
neutral (see Stability)
of a particle, 29–30
coplanar forces, 31
3-dimensional, 50–57
stable (see Stability)
unstable (see Stability)
Equilibrium, rigid body:
component equations of, 124, 139, 140
six independent, 124
three independent in coplanar case, 124–25
coplanar, 124–36
defined, 123
vector equations of, 124, 139, 141
Equilibrium differential relations in beams, 187
Equilibrium equations, 30, 31, 124–25, 139, 140–41
Equipollent sets of forces and couples, 101–3
Equivalent:
dynamically, 102
for rigid-body analysis, 102
statically, 102
Equivalent couples, 84
Equivalent force systems, 103
Euler, L., 115, 119
External effects of force, 15
External forces, 119, 154

Field (see Conservative force field, Force field, and Gravitational field)
First moment, 113, 115
Flexible cable (see Cable)
Fluid, 279
Fluid friction, 40
Fluid statics, 279–85
Force, 3, 14
action-at-a-distance, 15
active, 231, 232, 234
characteristics of, 15
combined with a perpendicular couple, 87–88, 89, 92
concentrated, 15
conservative, 212–16
definition, 213
friction, not a, 213
potential energy of, 212–16
constraint, 231, 232
distributed, 15, 103
effects of, 15
external, 119, 154
external effects of, 15
field, 15

of friction, 42
gravitational, 15, 19–20
intensity (per unit length), 110, 111
interaction, connected bodies, 155
internal, in beam, 182
internal, in truss member, 172, 173
internal effects of, 15, 20
line of action, 15
for given moment, 79, 80
moving to parallel, 87
magnitude of, 15
measurement of, 16
moment of, 73–77
parallelogram law, 16
on pin, 156
point of application, 15
polygon of, 23
shear, in beam, 184–95
on submerged surface, 281–85
triangle of, 35
units of, 23, 24, i, ii
work of, 209 (see also Work)
Forces, concurrent, 29
Four-bar linkages, 162–63
Frame:
equilibrium, analysis of:
coplanar, pin-connected, 153–58
3-dimensional, 164–66
riveted or welded, 160
structure, 159, 160
statically determinate, 160
Free-body diagram, 3, 30, 32, 52, 119
for hanging cable, 198, 204
pin location in, 156
Free-body diagrams for connected bodies, 154, 156, 158, 165
Free-fall acceleration, 20–21 (see also Gravity, acceleration of)
Free vector, 84
Freedom, degrees of, 159, 217, 231
Friction:
angle of, static and kinematic, 42
coefficients of, 42–43
Coulomb, 40–43
dry, static and kinetic, 41–43
fluid, 40
force of, 42
laws of, 40–42
not conservative, 213
problem types, with, 42
static, maximum, 42
torque, 95
viscous, 40

Gas, 279
equation, 285
pressure in, 285
Generalized coordinates, 218, 222, 231
Generalized stresses, 184
Gram, i, ii
Graphical solution example, 33, 35

Gravitation:
force of, 15
Newton's law of, 19
universal constant of, 19, 20
Gravitational field:
inverse-square, 19
potential energy of, 216
work of, 216
uniform, 116
potential energy of, 214
work of, 211, 214
Gravity, 19–21
acceleration of, absolute and apparent, 20–21
latitude dependence, 21
relation to universal gravitation constant, 20
center of, 19–20, 116
formula, international, 21
Gyration, radius of, 271, 272, 292

Hinge:
couple often negligible in, 121–23
friction couple in, 238
Horsepower (unit), ii
Hydrostatics, 280

Idealizations, 30–31
ball-and-socket joint, 50–51
massless cable, 31
physical models, 2
pulley, 31, 37
of support reactions, 119–23, 136–38
two-force member, 31, 34–35, 50, 52
Impending motion, with static friction, 42
Improper constraint, 127–28, 139
Indeterminate (see Statically indeterminate)
Inertia, 16
components (rotational), 286, 288, 294–95
moment of, area (see Second moments)
moment of, mass (see Moment of inertia)
Input force or work, 160
Integral:
arc length, 258–59
area, 255–58, 263–65
double, 263–65, 266
iterated, 114
line, for work, 210
single, 250, 255–62, 266
for moment of inertia, 294–95
triple, 114, 265
volume, 113–15, 259–61, 265–66
work, 210, 212
Internal effects of force, 15, 21
Internal force:
in beam, 182
in truss member, 172, 177–78
International gravity formula, 21

International system of units (SI), 23, i, ii
Inverse-square field (*see* Gravitational field)

Joint:
ball-and-socket, ideal, 50, 51
universal, 137
Joints, method of, for trusses, 173, 174, 176–77, 180
Joule (energy unit J), ii
Journal bearing, 138

Kilogram:
international prototype, 23
mass unit, 23, i, ii
relation to pound-mass, 23
Kilonewton, 25
Kinematically admissible displacement, 230
Kinematics, branch of dynamics, 3
Kinetic friction, 42 (*see also* Friction)
Kinetics, branch of dynamics, 3
Kip, 175

Laws:
of cosines, 10
distributive, 61, 70
of friction, 40–42
of gravitation, Newton's, 19
Kepler's, 18
of lever, 73
of motion, Newton's 18–19
of particle mechanics, 17–19
of sines, 10, 34
Length, units of, ii
Lever arm, 73
Line:
of action (*see* Force)
equations of, 80
Line integral for works, 210
Linear spring (*see* Spring)
Link support, reaction of, 121
Linkage, four-bar, 162
Liquid, 279
density of, 280
hydrostatics of, 280, 285
pressure in, 280, 285
specific weight of, 279
Load, distributed, on beam, 110, 187, 193
general, 112–13, 117–18, 193
triangular, 112, 130–31
uniform, 111

MKS units, i
Machine dynamics, 161
Machine, equilibrium of, 160, 162
potential energy methods, 218–20, 236–37

virtual displacement methods, 230–41
Mass, 3, 16
continuous distribution of, 113, 246
volume integrals for, 113–15
of earth and moon, 26
first moment of, 113, 115
gravitational, 17
inertial, 17
measure of inertia, 16
point, 2, 17 (*see also* Particle)
total, of two parts, 17
units of, 23, 24, i, ii
weight and, 17
Mass center (*see* Center of mass)
Mass moments of inertia (*see* Moment of inertia
Measure numbers of rectangular components, 7
Mechanics:
engineering, branches of, 3
particle, laws, of, 17
physical science of, 1
theory of, 1
variational, 232
Mechanism:
constraints of, 230
equilibrium of, 160, 162
potential energy methods, 218–20, 236–37
virtual displacement method, 230–41
Meganewton, 25
Meter (length unit), ii
Metric units, 23, i, ii
Models:
idealized, in mechanics, 2
physical, 1–2
Modulus, elastic, of steel, 26
Mohr's circle, 277
Moment:
arm, for first moment, 113, 115
bending, 184–95
sign convention, 185
center, change of, 102–3
of a couple, 83
first, 113, 115
of a force, 73–77
about an axis, 74, 77, 97
about a point, 74, 75–76
components of, 74, 76
vector, 75–76
Moment, turning, 77, 83
Moment of inertia, area (*see* Second moments of area)
Moment of inertia, mass, 286–302
centroidal, 291
components, 286, 288, 294–95
of composite body, 289
parallel-axis theorem, 289–90
principal, 294
single integral for, 294–95
Table A4, v–vii
thin-plane approximations, 292–93, 299
units of, 294
Momentum, linear, 123

Moon:
acceleration of gravity on, 26
mass of, 26
radius of, 26
Motion:
accelerated, 3
impending, with static friction, 42
laws of, Newton's 18–19
Multiforce members, 156, 160

Negative of a vector, 6
Neutral equilibrium (*see* Stability of equilibrium)
Neutral plane of beam, 249
Newton, Sir Isaac, 18
laws of motion, 18–19
law of universal gravitation, 18, 19
Newton (unit of force N), 23, i, ii
Normal stress, 182
Numerical accuracy, 27

Orthogonal component, 63
Orthogonal projection, 7
by scalar product, 60–61, 62–63
Orthogonality condition, 64
Orthonormal basis, 6
Output force or work, 121

Parabolic cable, 198–99
Parallel-axis theorems, 271–72, 289–90
Parallel forces, resultant of, 104, 107, 111–13, 117–18
Parallelepiped, volume by scalar triple product, 96
Parallelogram law:
of forces, 16
of vector addition, 5–6
Parameter, solution in terms of, 36–37
Partial constraint, 126–27
Particle, 2, 17
equilibrium of, 29–30
coplanar, 31
3-dimensional, 50–57
finite body treated as, 29
mechanics of a, laws of, 17–19
Perpendicular, common, of two vectors, 72 (*see also* Orthogonality condition)
Pin:
forces, 120, 156
location in free body, 156
Plane of symmetry:
contains centroid, 115
and product of inertia, 287
Point of application of force, 15
Point mass (*see* Particle)
Polar second moment of area, 270, 272
Polygon of forces, 33
Polygon rule of vector addition, 5
Position vector, 75, 76
Positive-definite, 224

Potential energy:
datum, 213, 214, 215, 216
of conservative force, 212–16
and equilibrium, 218, 223
gravitational, 214, 216
maximum, 219, 223–24
minimum, 219, 223–24
quadratic approximation of, 223
of spring, 215
and stability, 219–27
stationary value of, 218, 223
Taylor's series for, 219, 223
variation, 236
virtual change of, 231
Pound:
of force, 22, 23, 24, i, ii
of mass, 22, 23, 24, i, ii
Poundal, i, ii
Power, units of, ii
Pressure in fluid statics, 279
absolute, 280–81
center of, 281–82
gage, 280–81
Primitive elements of mechanics, 3
Principal axes:
of area second moments, 277–78
of inertia, 288
Principal moments of inertia, 294
Principia, Newton's, 18
Principle:
of transmissibility, 25, 102
of virtual displacements, 231
Problem solution procedure, 26
Product:
cross (*see* Vector product)
dot (*see* Scalar product)
of inertia, area, 270, 272, 276, 278
of inertia, mass, 286, 288
scalar (*see* Scalar product)
triple, 96–98
vector (*see* Vector product)
Projection:
orthogonal, 7
by scalar product, 60–61, 62–63
vector component, 7
Pulley, 3, 31, 37

Quadratic function:
approximation of potential
energy, 224
discriminant of, 224

Radius of gyration, 271, 272, 292
Ratio, conversion, for change of
units, 224
Reaction, action and, 19, 154–55
Reactions (*see also* Statically determinate and Statically indeterminate):
of cable supports, 121
couple, often omitted, 121–23
at fixed end, 120
at pin, 120
at roller, 120–21

support, external effect of force,
15
support, idealized, 119–23, 136–
38
support, workless, 238
Rectangular components of vector,
7–8 (*see also* Vector components)
Reduction of force system to resultant, 103 (*see also* Resultant)
examples, 105–8
to a single force, 103, 107–8
special cases, 104–8
to a wrench, 104
Relative position vector, 75, 76
Resolution of force:
into components, 6, 7, 13
into a force and a couple, 87
Resultant:
of distributed load on beam, 110–
13, 117–18, 130–31
force and couple at assigned point,
103, 105
of force system, 103, 104
force on submerged surface, 281–
85
of parallel force system, 104, 107,
110–13, 116–18, 130–31
stress, in beams, 182–83
wrench, 104
Revolution, volume of, 259–61
Right-hand rule, 69, 84
Rigid body, 2
Rigid-body equilibrium:
component equations of, 124,
139, 140
six independent, 124
three independent in coplanar
case, 124–25
coplanar, 124–36
definition of, 123
free-body diagram for, 119
vector equations of, 124, 139, 141
Rotation:
of coordinate axes, 64–65
for area second moments, 276–
78
finite, not a vector, 8–9

SI system of units, 23, i, ii
Sag of a cable, 198, 199
Scalar product, 59
applications of, 60, 62, 63, 64, 209
definition, 60
formula in terms of components,
62
projection by, 62–63
of unit base vectors, 62
Scalar quantity, 4
multiplication of vector by, 6
Scalar triple product, 96
dot and cross interchangeable in,
97
Second moments of area, 268–79,
283, 292
of composite area, 270
maximum and minimum, 277, 278

parallel-axis theorems, 271–72
principal, 277, 278
principal axes for, 277, 278
product of inertia, 270, 272, 276,
277, 292
of rectangle, 269
rotation of axes for, 276–79
single integral for, 273–74
Table A3, iii–iv
Sections, method of:
for beams, 188–95
for trusses, 173, 174–75, 177–78,
180
Series, Taylor's, for potential
energy, 219, 223
Shapes, geometric, properties of,
Tables A3, A4, iii–vii
Shear diagram of beam, 186–95
discontinuity in, 188
Shear force in beam, 184–95
sign convention, 185, 187
Shear stress in beam, 182
Sign conventions:
bending moment, 184, 185
shear force in beam, 185, 187
Significant figures, 27
Simply-supported beam, 186, 189–
90
Sines, law of, 10, 34
Slide rule:
accuracy of, 27
law of sines with, 10
Sliding vector, 102
Slug, 23, i, ii
Smooth surface, 121
Space truss, 180–81
Specific gravity, 285
Specific weight, 279
Spring, linear:
potential energy of, 214–15, 231
virtual work of, 231
work of, 214–15, 231
Spring constant (k), 214
Stability of equilibrium, 161, 217
discriminant for, 224
potential energy and, 219–27
Stable equilibrium (*see* Stability)
Standard atmosphere, 281, 285
Standard form for vector, 7, 51
Static friction (*see* Friction)
Statically determinate reactions,
125–26, 138
Statically indeterminate problems
and method of virtual displacements, 237
Statically indeterminate reactions,
125–26, 138, 143
Statics:
branch of mechanics, 3
fluid, 279–85
Stationary value of potential energy,
218, 223
Stress:
bending in elastic beam, 249
generalized, 184
normal, in beam, 182
Stress analysis:
in beams, 184

in truss, 171, 172
Stress resultants in beams, 182–83
  vector, 182
Stress, shear, in beam, 182
Structures, classified as frames and trusses, 159
Submerged surface, force on, 281–85
Subtraction of vectors, 6
Support reactions (*see* Reactions)
Surface:
  rough, 120
  smooth, 121
  submerged, force on, 281–85
Suspension bridge, 197, 205
Symmetry plane:
  contains centroid, 115
  and product of inertia, 287
System:
  British, of units, 23–24, i
  of connected bodies, 218
  conservative, 217
    equilibrium of, 218, 223
    stability of, 219–27
  constraints of, 230
  International, of units (SI), 23, i
  one-degree-of-freedom, 218, 219–27
  two-degree-of-freedom, 222–26
Système International d'Unités (SI system), 23, i, ii
Systems:
  equipollent, of forces, 101–3
  of units, 22–24, i

Taylor's series for potential energy, 219, 223
Tension, 76
Three-force body equilibrium, 35, 129
Thrust bearing, 95, 137
Time, unit of, i
Torque:
  couple is pure, 83
  frictional, 95
  transmitted by shaft, 86
Torsional shear stress, 87, 95
Traction vector, 182
Transfer formula (*see* Parallel-axis theorem)
Transmissibility, principle of, 25, 102
Triangle, centroid of, 255–56
Triangle rule:
  for particle equilibrium under three forces, 35
  of vector addition, 5
Triple products, 96–98
  mixed, 96
Truss:
  defined, 159, 171

plane, 171–79
simple, 172, 180
space, 180–81
statically determinate, 174, 180
statically indeterminate, 174, 180
symmetrically loaded, 176, 180
zero-force members in, 175–76
Truss analysis:
  method of joints, 173, 174, 176–77, 180
  method of sections, 173, 174–75, 177–78, 180
Two-force body, 31, 34–35, 128
  as member in frame or machine, 156
  in truss, 159

Uniform gravitational field:
  potential energy of, 214
  work of, 211, 214
Unit base vectors, 7
  scalar products of, 62
  vector products of, 70
Unit vector, 6, 7, 11, 12, 51, 63
Units:
  British, 22, i, ii
  change of, 24, ii
  consistent, 22, 23
  conversion factors for, ii
  conversion ratios for change of, 24
  International system (SI), 24, i, ii
  metric, 22, i, ii
  nonconsistent, 24
  SI, 22, 188, i, ii
  systems of, i, ii
  tables of, i, ii
Universal joint, 137
Unstable equilibrium (*see* Stability)

Variation, potential energy, 236
Variational mechanics, 232
Varignon's theorem, 77
Vector:
  addition of, 4, 8
  bound, 4
  components of, 6–8
  couple, 84
  direction of, 4, 7, 8
    through two points, 12, 51
  effective-couple, 123
  effective-force, 123
  equality of, 4
  finite rotation not a, 8–9
  fixed, 4
  free, 84
  magnitude of, 4, 8
  moment of force, 75–76
  multiplication by a scalar, 6
  negative of, 6
  sliding, 4, 102

standard form for, 7, 51
  stress resultant, in beam, 182
  subtraction of, 6
  traction, in beam, 182
  unit, 6, 7, 11, 12, 51, 63
Vector product, 59
  applications of, 60, 72, 75
  defined, 69
  determinant for, 71
  direction of, 69
  formula in terms of components, 71
  magnitude of, 69
  not associative, 70
  not commutative, 70
  triple, 70, 96–98
  of unit base vectors, 70
Vector quantity, 4
Virtual displacements, 230
  method of, 232–36
  principle of, 231
  variation of potential energy, and, 236
Virtual work, 230–31
Viscous friction, 40
Volume:
  centroid of, 248, 259–61, 265–68
  composite, 250
Volume element:
  cylindrical coordinate, 265
  double integral, 266
  single integral, disc, 259–61
  shell, 260–61
  triple integral, 265
Volume of revolution, 259–61

Watt (power unit W), ii
Wedge, 46–48, 49, 50
Weight:
  and mass, 7, 23
  specific, 279
  work done by, 211, 214
Work:
  of conservative force field, 213–16
  of couple, 237–38
  dimensions of, 210
  of force, 209
  input and output, 120
  line integral for, 210, 212
  negative, 211
  of spring, 215
  units of, ii
  of weight, 211, 214
Working component, 212
Workless constraints, 231, 238
Wrench resultant, 104, 110
  negative and positive, 104

Zero-force members in truss, 175–76